Denitrification in Soil and Sediment

FEDERATION OF EUROPEAN MICROBIOLOGICAL SOCIETIES SYMPOSIUM SERIES

Recent FEMS Symposium volumes published by Plenum Press

1990 • MOLECULAR BIOLOGY OF MEMBRANE-BOUND COMPLEXES IN PHOTOTROPHIC BACTERIA
Edited by Gerhart Drews and Edwin A. Dawes
(FEMS Symposium No. 53)

1990 • MICROBIOLOGY AND BIOCHEMISTRY OF STRICT ANAEROBES INVOLVED IN INTERSPECIES HYDROGEN TRANSFER
Edited by Jean-Pierre Bélaich, Mireille Bruschi, and Jean-Louis Garcia
(FEMS Symposium No. 54)

1990 • DENITRIFICATION IN SOIL AND SEDIMENT
Edited by Niels Peter Revsbech and Jan Sørensen
(FEMS Symposium No. 56)

1991 • *CANDIDA* AND CANDIDAMYCOSIS
Edited by Emel Tümbay, Heinz P. R. Seeliger, and Özdem Anğ
(FEMS Symposium No. 50)

Denitrification in Soil and Sediment

Edited by

Niels Peter Revsbech

University of Aarhus
Aarhus, Denmark

and

Jan Sørensen

Royal Veterinary and Agricultural University
Copenhagen, Denmark

PLENUM PRESS • NEW YORK AND LONDON

Library of Congress Cataloging in Publication Data

Denitrification in soil and sediment / edited by Niels Peter Revsbech and Jan Sørensen.

p. cm.—(Federation of European Microbiological Societies symposium series: 56)

"Proceedings of a symposium held under the auspices of the Federation of European Microbiological Societies and the Danish Ministry of Environment, held June 6–9, 1990"—

Includes bibliographical references and index.

ISBN 0-306-43721-X

1. Denitrification—Congresses. 2. Bacteria, Denitrifying—Congresses. I. Revsbech, Niels Peter. II. Sørensen, Jan. III. Federation of European Microbiological Societies. IV. Denmark. Miljoministeriet. V. Series: FEMS symposium: no. 56.

QR111.D46 1991 91-7400

589.9′0526404—dc20 CIP

Proceedings of a symposium held under the auspices of the Federation of European Microbiological Societies and the Danish Ministry of Environment, held June 6–9, 1989, in Aarhus, Denmark

ISBN 0-306-43721-X

A Division of Plenum Publishing Corporation
233 Spring Street, New York, N.Y. 10013

Printed in the United States of America

PREFACE

The formation of atmospheric nitrogen gas by denitrifying bacteria may represent a significant nutrient sink in natural ecosystems. The rate of denitrification has often been difficult to measure in situ, however, and new methodologies should stimulate research on distribution of activity in space and time.

The load of fertilizer nitrogen in modern agriculture has led to increasing nutrient reservoirs in recipient subsoils, aquifers, inland waters and coastal seas. By its conversion of nitrate to atmospheric nitrogen, bacterial denitrification is the only biological process to potentially reduce the impact of increasing nutrient loadings by fertilizer nitrogen in the environment.

As part of a scientific program set up by the Danish Ministry of Environment to study environment cycling of nitrogen, phosphorous and organic matter (NPO program) in the light of agricultural, domestic and industrial activities, a symposium on DENITRIFICATION IN SOIL AND SEDIMENT was held at the University of Århus, Denmark from 6-9 June 1989.

On the basis of lectures given at the symposium, this book contains a number of invited contributions on the regulation of denitrification activity (control of enzyme synthesis and activity) and measurement of in situ rates of denitrification in terrestrial and aquatic environments (control factors, diel and seasonal variations, etc). Emphasis has been placed on including the recent improvements in methodologies and current understanding of process regulation, however the book also contains examples of integrated research on the significance of denitrification in environmental nutrient cycling.

From contributions in Section I on the BIOCHEMISTRY AND GENETICS OF DENITRIFICATION, new information on reactivity and conversion of denitrification intermediates (e.g., nitrite and nitrous oxide) is presented. Chapter 1 is a comprehensive review of important chemical characteristics of nitrogen oxides and relates both to atmospheric chemistry and to chemical and biological denitrification. Chapter 2 addresses one of the key reaction steps in bacterial denitrification - the reduction of nitrite to nitrous oxide (denitrification pathway) - and can be compared with Chapter 4 which deals with the alternative reduction of nitrite to ammonia (nitrate ammonification pathway). Different bacterial species are involved and substantial information is now available on both the synthesis and activity control of several nitrite reductase enzymes. Chapter 3 contains much of the latest information on the terminal reaction steps of denitrification: reduction of nitric oxide to nitrous oxide and further on to atmospheric nitrogen. Studies on the nitric and nitrous oxide reductases have involved identification of active centers, genetic control mechanisms, etc.

Section II on the ECOPHYSIOLOGY OF DENITRIFICATION has two contributions on the role of classical control factors of bacterial denitrification: temperature and the availability of oxygen and nitrate. Chapter 5 demonstrates that selection based on temperature preference and affinity for nitrate plays an important role in natural population dynamics of denitrifying bacteria. The fascinating occurrence in some bacteria of denitrification activity under oxic conditions and of nitrification and denitrification in the same bacterium is addressed in Chapter 6.

An environmental issue related to denitrification is the production of gaseous nitrogen oxides such as nitric and nitrous oxide and their emission into the atmosphere. In Section III, the EMISSION OF N_xO_y COMPOUNDS is covered together with their possible source reactions during microbial metabolism. Soils may be important sources of nitric oxide emission to the atmosphere as described in Chapter 7 (and of nitrous oxide as demonstrated in Chapter 10 of Section IV). A new and interesting stable isotope technology which can be used to determine the source reaction of nitrous oxide emitted from aquatic environments is described in Chapter 8.

Section IV contains six chapters on DENITRIFICATION IN SOIL. The heterogenous microstructure of soils has long been a major obstacle in determining denitrification rates in this environment. Chapters 9, 10 and 11 summarize the recent progress made by acetylene inhibition and stable isotope techniques. The former technique was chosen by the authors of Chapters 12 and 13 to demonstrate the occurrence of denitrification in two different microenvironments: the rhizosphere of the living plant and small organic particles in soil. Both contributions emphasize the importance of small microenvironments with elevated microbial metabolism (hot spots) in developing anoxia and initiating denitrification in agricultural soils. Chapters 14 and 15 present a modelling approach to the study of soil denitrification, in particular its distribution in space and time as influenced by particle size, availability of oxygen and nitrate, etc.

The final section of the book covers DENITRIFICATION IN BIOFILM AND SEDIMENT. As with soil studies, the development of new acetylene inhibition and stable isotope techniques has contributed significantly to recent advances in denitrification studies of aquatic environments. By development of a microsensor of nitrous oxide, Chapters 16 and 17 demonstrate a new application of the acetylene inhibition technique in the study of denitrification on the microscale. Chapter 18 demonstrates that a stable isotope technique can be used to simultaneously determine activities of several processes of nitrogen cycling in sediments. Finally, Chapters 19 and 20 represent examples of integrated studies of denitrification and other nitrogen transformations in the sedimentary environment. This approach, coupled to the modelling of denitrification rates from natural control factors, is important in understanding the significance and variability of the process in situ.

The book is published on behalf of the Federation of European Microbiological Society (FEMS).

We are indebted to the participants who contributed to the symposium on DENITRIFICATION IN SOIL AND SEDIMENT with lectures, posters and manuscripts. We gratefully thank our colleagues and students who assisted us during the meeting.

Niels Peter Revsbech
Jan Sørensen

CONTENTS

BIO-INORGANIC ASPECTS OF DENITRIFICATION: STRUCTURES AND REACTIONS OF N_xO_y COMPOUNDS AND THEIR INTERACTION WITH IRON AND COPPER PROTEINS

Peter M. H. Kroneck and Walter G. Zumft*

Universität Konstanz, Fakultät für Biologie
D-7750 Konstanz, and *Universität Karlsruhe
Lehrstuhl für Mikrobiologie, D-7500 Karlsruhe
Federal Republic of Germany

INTRODUCTION

The nitrogen cycle has received considerable attention in recent years because of its ecological importance (Knowles, 1982; Sprent, 1987; Ullrich et al., 1987; Cole and Ferguson, 1988; Zehnder, 1988). Nitrogen is an essential constituent of many biomolecules which are important for life, e.g. amino acids, proteins, or DNA. In the process of nitrogen fixation dinitrogen is reduced to ammonia. This reaction is catalyzed by nitrogenase, a complex multi-metal enzyme (Smith, 1986). On the other hand, dinitrogen is released into the atmosphere by the activities of a large variety of bacteria (Cole and Ferguson, 1988; Zehnder, 1988). Oxidized nitrogen compounds are used as electron acceptors and are finally reduced to nitrous oxide, nitric oxide, and dinitrogen. Nitrogen oxides are ubiquitous in the environment, and there is a considerable concern about their effect on health (Lee, 1980). Oxidation of reduced (-3) valency nitrogen species may play an important role in the biological formation of carcinogenic nitrosamines (Zehnder, 1988). Model studies suggest the hypothesis that the formation of these carcinogenic compounds may involve metal centres capable of catalysing both oxidation of reduced nitrogen species and subsequent formation of an N-N bond leading to the nitrosamine coordinated to the metal centre (Stershic et al., 1988).

The large number of molecular and ionic species that might be produced from just the atoms of nitrogen and oxygen is remarkable (Stedman, 1979). In addition to the gases N_2O, NO and NO_2 (and its dimeric form N_2O_4), and in addition to the important NO_3^- and NO_2^- anions, more than a dozen other species have been characterized (Laane and Ohlsen, 1980). Many of the latter are short-lived intermediates but might be stabilized at the active site of a metal-dependent enzyme, or can be characterized at low temperature using matrix isolation procedures. Physical techniques such as nuclear magnetic resonance (NMR) (Witanowski and Webb, 1973), electron paramagnetic resonance (EPR) (Palmer, 1982), infrared (IR) and Raman spectroscopy (Laane and Ohlsen, 1980), and isotope labelling (Garber and Hollocher, 1982) appear best suited for characterizing the various species that may be formed.

Denitrification in Soil and Sediment, Edited by N. P. Revsbech and J. Sørensen, Plenum Press, New York, 1990

In this article we wish to address several aspects of the bio-inorganic chemistry of N_xO_y compounds as they are of interest for denitrification and dissimilatory nitrate reduction. The structures and reactions of N_xO_y compounds will be discussed including their coordinative properties towards the transition metals copper and iron. Furthermore, attention will be dedicated to mechanistic aspects of the formation of the N-N bond (conversion of nitrite to nitrous oxide as catalyzed by dissimilatory Fe- and Cu-dependent nitrite reductases) and of the cleavage of the N-O bond (conversion of nitrous oxide to dinitrogen as catalyzed by Cu-dependent nitrous oxide reductase). A first structural model of the active site of nitrous oxide reductase from Pseudomonas stutzeri will be presented.

N_xO_y COMPOUNDS: STRUCTURES, FORMATION AND REACTIONS

Denitrification is the process by which nitrogen atoms of nitrate and nitrite are returned to the atmosphere as nitrous oxide and dinitrogen. The biological process comprises several steps (Scheme 1) including the formation of the N-N bond in N_2O (bond order between 2 and 3) and the cleavage of the N-O bond in N_2O (bond order between 1 and 2) (Colburn, 1966; Zumft and Cardenas, 1979):

$$NO_3^- \rightarrow NO_2^- \searrow \underset{NO}{\rightarrow} \nearrow N_2O \rightarrow N_2$$

Scheme 1. Denitrification pathway, with identified or postulated intermediates.

The first step is catalyzed by a Mo-dependent enzyme (Ferguson, 1987), while in the last step a Cu-containing enzyme with unusual properties is involved (Coyle et al., 1985). The second step has remained controversial, with disagreement as to whether it is carried out by a single enzyme or by two enzymes, with NO as a free obligatory intermediate (Zumft and Cardenas, 1979). Practically nothing is known on the chemical mechanism of the conversion of nitrous oxide to dinitrogen. On the other hand, the reduction of nitrite has been studied in great detail (Zumft and Cardenas, 1979; Kim and Hollocher, 1984; Weeg-Aerssens et al., 1988). The formation of an enzyme-bound nitrosyl complex, $E\text{-}NO^+$, as the first step of the reduction of NO_2^- seems to be generally accepted (Weeg-Aerssens et al., 1988).

Nitrogen can effectively form both σ and π bonds to other nitrogen and oxygen atoms giving a remarkable variety of thermodynamically stable N_xO_y compounds (Laane and Ohlsen, 1982). One should keep in mind the increase in bond energy when raising the bond order from one (N-N single bond) to three (N-N triple bond). This effect becomes even more evident by the comparison with carbon. Whereas the bond energy of the C-C single bond is much higher compared to N-N it is just the opposite for the triple bond (Colburn, 1966).

The reaction shown in Scheme 2 demonstrates how readily the nitrogen-oxygen species are interconverted, and why it is not rare to have many different types of N_xO_y compounds in the same system (Laane and Ohlsen, 1982). This scheme also illustrates some of the difficulties which may arise in the course of mechanistic investigations of denitrification. Thus, for a better understanding of the biochemistry of denitrification it is reasonable to discuss some of the physical

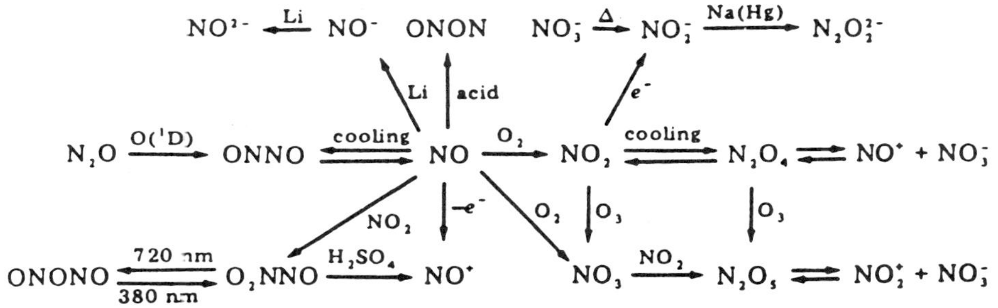

Scheme 2. Reactions of N_xO_y species (Laane and Ohlsen, 1982).

and chemical properties of various nitrogen oxides and nitrogen oxyacids which might be formed during these processes (Scheme 1).

Nitric oxide

Nitric oxide, NO, plays a central role in the formation and interconversion of N_xO_y compounds as documented in Scheme 2. The molecular orbital description of the bonding in NO is similar to that in N_2 or CO (Greenwood and Earnshaw, 1985). It is an odd-electron molecule with the unpaired electron in one of the π^* antibonding orbitals. This accounts for the fact that the interatomic N-O distance (115 pm) is intermediate between the triple-bonded NO^+ (106 pm) and the double-bonded NO species (120 pm). It can donate an electron to form the nitrosonium cation, NO^+ which is isoelectronic to N_2 and CO, or accept an electron to give the nitroxyl anion, NO^-. When NO condenses to a liquid, partial dimerization occurs with formation of $(NO)_2$. Pure NO is colourless both in the liquid and in the crystalline state. NO reacts with molecular oxygen to nitrogen dioxide, NO_2, another odd-electron molecule (Eq. 1). This reaction is fast but slows down with increasing temperature because of the dimerization equilibrium (Eq. 2) (Greenwood and Earnshaw, 1985):

$$2\,NO + O_2 \rightarrow 2\,NO_2 \quad (1)$$
$$NO + NO \rightleftarrows (NO)_2 \quad (2)$$
$$(NO)_2 + O_2 \rightarrow 2\,NO_2 \quad (3)$$

NO also undergoes complex disproportionation reactions leading to the formation of NO_2^-, $N_2O_2^{2-}$, and N_2O (Greenwood and Earnshaw, 1985). Reactions of NO with sulphides, polysulphides, and oxyacids of sulphur are rather complex, and depend very strongly on the reaction conditions (Greenwood and Earnshaw, 1985). Recently, it was proposed that the so-called endothelium-derived relaxing factor (EDRF) is NO (Palmer et al., 1987). Clearly, NO is a versatile complexing agent which can interact with a large variety of metalloproteins and metalloenzymes, e.g. haemoproteins, iron-sulfur proteins or copper proteins as discussed below.

Nitroxyl

The nitroxyl anions, NO^- and NO^{2-}, are formed by reduction of NO with alkali metals, such as lithium (Scheme 2). Products of the type LiNO and Li_2NO were characterized by IR spectroscopy using the matrix isolation technique (Laane and

Ohlsen, 1982). As described before, NO readily loses an electron to form NO^+ which is of considerable interest in atmosphere chemistry. Nitrosonium salts, NO^+X^-, may be prepared via the reaction of N_2O_3, or N_2O_4, with acid or by the action of nitrosyl halides with metal halides (Eq. 4) (Greenwood and Earnshaw, 1985):

$$N_2O_3 + 3\,H_2SO_4 \rightarrow NO^+ + 3\,HSO_4^- + H_3O^+ \qquad (4)$$

Nitroxyl, HNO, has often been postulated as a short-lived intermediate in reactions which give N_2O (Eisenberg and Meyer, 1975; Greenwood and Earnshaw, 1985). A stable osmium complex of HNO, which has the overall formula of $[Os(CO)Cl_2(HNO)(P(C_6H_5)_3]$, was prepared and structurally characterized. HNO is N-bonded with an N-O interatomic distance of 119 pm. HNO is bent both in the free (angle 109°) and in the liganded state (angle 99°) (Greenwood and Earnshaw, 1985). Nitroxyl is formed by decomposition of trioxonitrate monoanion, $HN_2O_3^-$ (oxyhyponitrite) in the reaction with haemoproteins, such as myoglobin, haemoglobin, and cytochrome c (Eq. 5) (Doyle et al., 1988):

$$HN_2O_3^- \rightarrow HNO + NO_2^- \qquad (5)$$

Methaemoglobin, Hb-Fe(III), undergoes reductive nitrosylation with nitroxyl as one of the decay products of trioxonitrate monoanion. HNO causes oxidation of the haemoglobin ß-93 cysteine sulphydryl group (R-SH) and hydroxylamine is liberated (Eqs. 6,7):

$$HNO + R\text{-}SH \rightarrow R\text{-}S\text{-}NHOH \qquad (6)$$
$$R\text{-}SH + R\text{-}S\text{-}NHOH \rightarrow R\text{-}SS\text{-}R + NH_2OH \qquad (7)$$

From the data obtained in these experiments it is concluded that HNO coordinates to Hb-Fe(II). Hb-Fe(II)-HNO then catalyzes the reduction of nitrite and Hb-Fe(II)-NO is formed (Doyle et al., 1988).

Dinitrogen trioxide

Dinitrogen trioxide, N_2O_3, a pale blue solid, can only be obtained at very low temperature; above its melting point (-101°C) it decomposes into NO and NO_2 (Eq. 8):

$$N_2O_3 \rightleftarrows NO + NO_2 \qquad (8)$$

At 25°C, in the gas phase, the dissociation is slightly exergonic with $G^o = -1.59$ kJ mol^{-1} (Greenwood and Earnshaw, 1985). N_2O_3 is a planar molecule with a long N-N bond (186.4 pm). Because of the presence of the very reactive species NO and NO_2 (Eq. 8) studies on the reactivity of N_2O_3 are complicated.

Hyponitrous acid and hyponitrites

The chemistry and properties of hyponitrous acid, $H_2N_2O_2$, and the hyponitrite ion, $N_2O_2^{2-}$, have been extensively studied and reviewed (Hughes, 1968). Consistent with the formula HO-N=N-OH it yields N_2O with sulphuric acid. Aqueous solutions are unstable in the pH range 4-14 due to base-catalyzed decomposition via $HO\text{-}N{=}N\text{-}O^-$ to form N_2O again. Hyponitrites are best made by the reduction of nitrates or nitrites using Na(Hg) or Mg(Hg). This implies that the salts of hyponitrous acid are moderately stable towards reductants but can act as reducing agents themselves. Vibrational spectroscopy data indicate that the hyponitrite ion has the trans-configuration in salts such as $Na_2N_2O_2$ whereas when liganded to transition metals the cis-configuration seems to be favoured (see discussion below). The electrophilic catalysis of the breakdown of trans-hyponitrite monoanion into N_2O by aldehydes and ketones in aqueous solution, and that of trans-hyponitrite dianion by CO_2 was recently reported (Loechler et al., 1987). On the basis of kinetic arguments the isomerization to cis-hyponitrite is proposed followed by the rapid conversion to nitrous oxide.

Nitrous oxide

Nitrous oxide, N_2O, is made by the careful thermal decomposition of ammonium nitrate at about 250°C. The reaction mechanism is rather complex and involves a redox reaction between (-3) and (+5) nitrogen species with NO_2^+ as the key intermediate (Greenwood and Earnshaw, 1985). N_2O is not regarded as the anhydride of hyponitrous acid since $H_2N_2O_2$ is not formed when N_2O is dissolved in water. A similar situation is found for the reactivity of carbon monoxide in water. N_2O is a linear asymmetrical molecule and rather unreactive as expected for 16-electron triatomic species. It is isoelectronic to CO_2, and the N-N and the N-O distances are rather short in agreement with the calculated bond orders of 2.73 (N-N) and 1.61 (N-O) (Jug, 1978). N_2O is thermodynamically unstable; above 600°C it decomposes to N_2 and O_2 by fission of the weaker N-O bond. The activation energy of this process is around 250 $kJmol^{-1}$ (Jones, 1975). With hydrogen, N_2O reacts to yield N_2 and H_2O in a process which is more exothermic than the H_2/O_2 reaction. Copper catalyzes the reaction between N_2O and CO which produces CO_2 and N_2 between 200-250°C (Schwab and Drikos, 1940). Copper and cuprous oxide seem to be effective catalysts of the cleavage of the N-O bond of N_2O (Dell et al., 1953a; Dell et al., 1953b; Scholten and Konvalinka, 1969). N_2O reacts rapidly with a number of transition-metal complexes under very mild conditions; Cobalt(I) complexes catalyze the reduction of nitrous oxide to dinitrogen by borohydride (Banks et al., 1969; Pu et al., 1969). In the case of vitamin B_{12} derivatives the reaction proceeds as outlined in Eq. 9 (Pu et al., 1969):

$$2\ \mathrm{Co(I)vitB_{12}} + \mathrm{N_2O} \rightarrow 2\ \mathrm{Co(II)vitB_{12}} + \mathrm{N_2} \qquad (9)$$

A series of Co(I), Ni(I), and Cu(I) complexes with unsaturated nitrogen macrocycles which were produced by the pulse radiolysis technique reacted very rapidly with N_2O to yield N_2 (Buxton et al., 1976; Tait et al., 1976a; Tait et al., 1976b; Jubran et al., 1985). In the case of the Ni(I) complex of 1,4,8,11-tetra-azacyclotetradecane, $Ni(I)L^+$, the following stoichiometry was observed (Eq. 10) (Jubran et al., 1985):

$$\mathrm{Ni(I)L^+} + \mathrm{N_2O} + 2\ \mathrm{H_3O^+} \rightarrow \mathrm{Ni(III)L^{3+}} + \mathrm{N_2} + 3\ \mathrm{H_2O} \qquad (10)$$

No evidence was obtained in these studies for the addition of N_2O into the coordination sphere of the metal. Most recently, oxygen atom transfer from nitrous oxide in the coordination sphere of hafnium and zirconium compounds was reported (Vaughan et al., 1987; Vaughan et al., 1988). Thus, with a very few exceptions when N_2O reacts with transition-metal complexes nitrogen is extruded and oxo species are formed (Eq. 11):

$$M + N_2O \rightarrow O{=}M^{2+} + N_2 \quad (11)$$

These O-atom transfer reactions should proceed in a very clean fashion since the sole byproduct is molecular nitrogen.

N_2O is also reduced to N_2 by nitrogenase (Liang and Burris, 1988). Actually, it was the first reported example of reduction of a substrate other than N_2 by the Mo,Fe enzyme (Mozen and Burris, 1954). N_2 is competitive with N_2O, and N_2O is competitive with acetylene. On the other hand, C_2H_2 is noncompetitive with N_2O (Liang and Burris, 1988).

Nitrous acid and nitrites

Although "nitrous acid", HNO_2, has not been isolated as a pure compound it can be readily obtained by acidification of cooled solutions of nitrites. Even at room temperature the disproportionation reaction (Eq. 12) has to be taken into account:

$$3\ HNO_2\ (aq.) \rightarrow H_3O^+ + NO_3^- + 2\ NO \quad (12)$$

It is a weak acid, $pK_a = 3.3$ at 25°C, comparable to formic acid. Concentrated solutions of HNO_2 (ca. 0.05-0.1 M) in mineral acid (ca. 4 M) will dehydrate to N_2O_3. At an even higher concentration of mineral acid, ionization to nitrosonium occurs (Eq. 13) (Deno et al., 1959):

$$H^+ + HNO_2 \rightarrow NO^+ + H_2O \quad (13)$$

HNO_2 and NO_2^- are bent molecules with interatomic N-O distances of 117.7 and 143.3 pm (HNO_2) and 114-124 pm in nitrites (Greenwood and Earnshaw, 1985). The aqueous solution chemistry of nitrous acid and nitrites has been extensively studied and reviewed. Reduction by SO_2 leads to NO and N_2O whereas the reaction with hydrogen sulphide yields ammonia. Hydrazinium salts give azides which then react with further HNO_2 to yield N_2O and N_2 (Greenwood and Earnshaw, 1985).

Nitric acid and nitrates

Pure nitric acid, HNO_3, can only be obtained in the solid state. Both in the gas and liquid phases the molecule decomposes spontaneously to NO_2 (Eq. 14):

$$2\ HNO_3 \rightarrow 2\ NO_2 + H_2O + 1/2\ O_2 \quad (14)$$

In addition, the pure liquid undergoes autoproteolysis followed by rapid loss of water. These equilibria cause a very fast exchange of N atoms between the various species in solution as documented by the presence of one single ^{15}N NMR signal at the weighted average position of HNO_3, NO_2^+, and NO_3^- (Greenwood and Earnshaw, 1985).

COORDINATION CHEMISTRY OF N_xO_y COMPOUNDS

Some of the molecules shown in scheme 2 are good ligands with high affinities towards biologically relevant transition-metals, such as Fe, Cu, and Mo. Nitrous oxide, on the other hand, behaves as a rather poor ligand. There is spectral evidence for $[Ru(II)(NH_3)_5N_2O]^{2+}$ in equilibrium with N_2O in aqueous solution (Diamantis and Sparrow, 1970). At high pressures of N_2O a microcrystalline solid with the formula $[Ru(II)(NH_3)_5N_2O](BF_4)_2$ x H_2O was isolated which decomposed in high vacuum to give N_2, N_2O and H_2O as gaseous products and a solid, $[Ru(II)(NH_3)_5N_2](BF_4)_2$ (Diamantis and Sparrow, 1970). To our knowledge, this is the "only" known isolated complex containing coordinated N_2O. The linkage of N_2O in this complex is still a controversial issue. Molecular orbital calculations on N_2O, and the isoelectronic cyanate anion, NCO^-, bound to a pentacoordinate metal fragment indicate that (a) N-linkage complexes are more stable than O-linkage complexes, and (b) that other transition-metal complexes of N_2O should be stable (Tuan and Hoffmann, 1985).

Hyponitrite, $N_2O_2^{2-}$, can act as a bidentate ligand in either a bridging or a chelating mode. The red dinuclear cation, $[(Co(NH_3)_5NO)_2]^{4+}$ contains a bridging cis-hyponitrite (Hoskins et al., 1969). The chelate type of binding was found for the yellow platinum complex $[Pt(O_2N_2)(P(C_6H_5)_3)_2]$ (Bhaduri et al., 1969). Note again the cis-configuration which is not found in ordinary salts of hyponitrous acid (Greenwood and Earnshaw, 1985), and which was proposed as the activated configuration for the electrophilic breakdown of trans-hyponitrite to N_2O (Loechler et al., 1987).

Both the nitrite and the nitrate group are versatile ligands with numerous modes of coordination (Hathaway and Billing, 1970; Addison et al., 1971). Nitrite can coordinate in at least five different ways. Nitro-nitrito isomerism was intensively studied in the early days of coordination chemistry (Hathaway and Billing, 1970). The bidentate chelating mode is observed in the cupric complexes $[Cu(II)(Bpy]_2(O_2N)]NO_3$ and $[Cu(II)(Bpy)](O_2N)_2$ (Greenwood and Earnshaw, 1985; Simmons et al., 1987). From model studies on the reduction of nitrite to ammonia catalyzed by osmium, ruthenium and iron complexes it is deduced that nitrite binds in the nitro mode followed by the conversion to the nitrosyl complex in the first step (Barley et al., 1987) (Eq. 15):

$$[Fe(II)(H_2O)(TMPyP)]^{4+} + 2\ H^+ + NO_2^- \rightarrow$$

$$[Fe(II)(NO^+)(TMPyP)]^{5+} + H_2O \qquad (15)$$

The most common mode of coordination of the nitrate group, which is isoelectronic with both borate, BO_3^{3-}, and carbonate, CO_3^{2-}, is the symmetric bidentate mode. It is observed in many transition metal compounds (Addison et al., 1971). In addition to the unidendate binding mode different modes of bridging in polynuclear complexes were reported. In the dimer $[Cu(II)(NO_3)_4(py)_4]$py two Cu centres are bridged by two nitrato ligands, and each Cu(II) is coordinated to a second nitrate in the bidendate mode (Greenwood and Earnshaw, 1985).

By far the most versatile and interesting ligand among the N_xO_y compounds discussed here is nitric oxide. NO readily reacts with numerous transition metal compounds to give nitrosyl complexes. These complexes are frequently formed in reactions involving other N_xO_y species (Enemark and Feltham, 1974; Eisenberg and Meyer, 1975; McCleverty, 1979; Butler et al., 1988). The "classical" examples are $[Fe_2S_2(NO)_4]^{2-}$ and $[Fe_4S_3(NO_7)]^-$, the nitroprusside anion $[Fe(CN)_5NO]^{2+}$, and the "brown ring" complex $[Fe(H_2O)_5NO]^{2+}$ which is formed during the qualitative analysis for nitrate (Greenwood and Earnshaw, 1985). NO exhibits a variety of

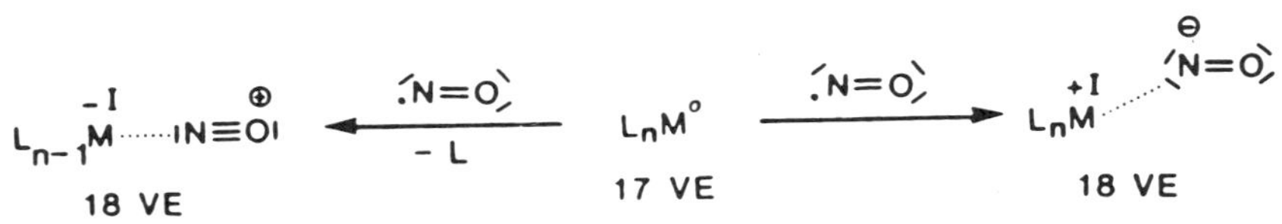

Scheme 3. Linear ("NO^+", isoelectronic with CO) and bent ("NO^-", isoelectronic with O_2) binding mode of nitric oxide (Elschenbroich and Salzer, 1988).

coordination geometries, e.g. linear, bent, doubly bridging and triply bridging, and sometimes more than one mode occurs within the same complex. NO has one electron more than CO and can act as a three-electron donor (scheme 3, "reductive" binding, linear). A preliminary one-electron transfer from NO to the metal centre gives NO^+ which behaves as a two-electron σ-donor, two-electron π-acceptor analogous to CO. This type of linear binding tends to occur with the earlier transition metals that more readily accomodate the larger number of electrons supplied by the ligand. In the bent mode (scheme 3, "oxidative" binding, bent) the ligand acts as a one-electron donor and is considered as the NO^- anion which is isoelectronic with O_2 (Elschenbroich and Salzer, 1988). Non-linear coordination is generally observed with the later transition elements in which the low-lying orbitals on the metal are already filled. Note that bending withdraws an electron pair from the metal centre to the nitrogen atom thus creating a vacant coordination site. Such a vacancy might be an important feature for the binding of a second metal centre, or the attack of a substrate during catalysis. In $[Fe(CN)_5(NO)]^{2+}$ the bond angle Fe-N-O is 178° in agreement with a rather short N-O interatomic distance (IR frequency v (N-O) 1935 cm^{-1}) whereas in $[Co(NH_3)_5(NO)]^{2+}$ this angle is 119° (v (N-O) 1610 cm^{-1}) (Greenwood and Earnshaw, 1985).

INTERACTION OF N_xO_y WITH IRON AND COPPER PROTEINS

In addition to the fact that N_xO_y compounds are effective substrates of metal-dependent enzymes of the denitrification pathway (scheme 1) nitric oxide and nitrite have found their place as valuable exogenous ligands and probes for the active site of numerous metalloproteins and metalloenzymes. Nitric oxide is the most versatile coordinating agent; it is a strong field ligand and it reacts with high-spin ferrous haem (Fe(II)(h.s.)) compounds rapidly and with high affinity to yield low-spin ferrous (Fe(II)(l.s.)) nitrosyl complexes (Eq. 16):

$$\text{Fe(II)(h.s.)} + \text{NO } (S=1/2) \rightarrow \text{Fe(II)(l.s.)-NO } (S=1/2) \quad (16)$$

The electronic picture of Fe(II)(l.s.)-NO can be represented as low-spin 3 d^7 state with one electron transferred from NO to the metal (see previous section). An accurate description of the nitrosyl Fe(II) centre which also accounts for the stereochemistry of the Fe(II)-NO interaction is more complex as discussed elsewhere in greater detail (Palmer, 1982). NO binds to the ferrous haem with a bent geometry, as does O_2. The haem-NO adduct gives a characteristic EPR spectrum which makes it such an important tool for probing the metal environment. The spectrum is centered around g=2 and is anisotropic, with both axial and rhombic cases. This g-anisotropy is small, the extreme values of g_x and g_y being around 2.08 and 1.95 respectively, whereas g_z is relatively constant at g=2.005 (Palmer, 1982). A special feature comes from the so-called hyperfine structure which is frequently observed in the g_z region. This splitting of the line at g_z is due to a strong interaction between the unpaired electron and the nitrogen of the nitrosyl function (both ^{14}N and ^{15}N isotopes have nuclear magnetic moments with I=1 and 1/2 respectively). EPR spectroscopy of NO derivatives of iron proteins has become a widely used technique to clarify the stereochemistry of the haem pocket and the electronic structure of the iron centre in many haemoproteins, such as cytochrome c oxidase (Stevens et al., 1979; Ohnishi et al., 1982), cytochrome c' from photosynthetic, denitrifying and nitrogen-fixing bacteria (Yoshimura et al., 1986; Yoshimura et al., 1987; Suzuki et al., 1987), horse heart cytchrome c (Yoshimura and Suzuki, 1988), microsomal prostaglandin H synthase (Karthein et al., 1987), bovine cytochrome P-450 (Tsubaki et al., 1987), and several myoglobins and model haem compounds (Sharma et al., 1987). Nitric oxide was also used to probe the non-haem Fe(II) centres in bleomycin (Suguira et al., 1982) and soybean lipoxygenase-1 (Nelson, 1987). With lipoxygenase a species is formed displaying an EPR spectrum characteristic of a nearly axial S=3/2 electronic spin system arising from the Fe(II)-NO complex (Nelson, 1987).

In addition to the investigation of structural aspects, rates for the reaction of NO with ferric haemoproteins and model compounds were measured (Sharma et al., 1987). The NO combination rates are markedly affected by the nature of the distal ligand. Thus, replacement of the distal E7 histidine in elephant myoglobin by glutamine leads to a 500 to 1000-fold increase of this rate. In contrast to ferrohaemoproteins that are five-coordinated, the ferric forms can either be five- or six-coordinated with a water molecule bound to Fe(III). In the first case NO would just add to the Fe(III) site whereas with water bound in the sixth position its displacement by NO would dominate the kinetics (Sharma et al., 1987). These results are in agreement with the idea that the orientation of the axial ligand is important for the "fine-tuning" of the physical properties of iron porphyrinate derivatives, e.g. structure, spin state, EPR properties and redox potentials (Scheidt and Lee, 1987).

Furthermore, synthetic models for the [2Fe-2S] and [4Fe-4S] clusters, respectively, of natural redox proteins can be nitrosylated under mild conditions with nitric oxide or nitrite (Butler et al., 1988). Model complexes of the type $[Fe_2(SR)_2(NO)_4]$ were shown to undergo rapid exchange of thiolate ligands with R*-SH and $R^*\text{-}S^-$. In addition, the mononuclear complexes $[Fe(NO)_2(SR)_2]^-$ also undergo rapid exchange of the nitrosyl ligands in the presence of isotopically labeled nitrite. A coordinated N_2O_3 intermediate was postulated for this ligand exchange process (Butler et al., 1988). The nitrosylation of the [4Fe-4S] centre in cells of Clostridium botulinum by nitrite was achieved (Reddy et al., 1983). The EPR signal observed at g=2.035 is characteristic of a $[Fe(NO)_2X_2]$ species, most likely of the $[Fe(NO)_2(SR)_2]^-$ type (Butler et al., 1988). The transfer of nitrosyl groups from Fe-S-NO to atoms other than iron (e.g. formation of nitrosamines) is problematical and remains a controversial issue (Butler et al., 1988). The antimicrobial activity of nitrosyl complexes of iron-sulfur clusters of the type $[Fe_4S_3(NO)_7]$ against both anaerobic and aerobic microorganisms was shown almost four decades ago (Dobry and Boyer, 1945). The central role of the complexes $[Fe(NO)_2(SR)_2]^-$ in the reaction chemistry of Fe-S-NO clusters and their formation in vitro and in vivo suggests that the antimicrobial activity of NO_2^- is not only because of its interaction with iron-sulfur proteins of the electron transport chain but also specifically because of the formation of $[Fe(NO)_2(SR)_2]^-$ species (Butler et al., 1988).

The interaction of NO with copper proteins can lead to complex reaction pathways as described for laccase (4 Cu/M_r) (Martin et al., 1981), cytochrome c oxidase (3 Cu,2Fe/monomeric catalytic unit) (Boelens et al., 1984), or ascorbate oxidase (8 Cu/M_r) (Van Leeuwen et al., 1975). Note the distinct reactivities of the three Cu types (Malkin and Malmstroem, 1970) towards NO. Type-2 Cu ("non-blue") seems rather inert against NO, Type-1 Cu ("blue") forms EPR-silent complexes with NO in the Cu(II) state which can be photodissociated at 77 K in the case of azurin, ceruloplasmin, and ascorbate oxidase (Gorren et al., 1987). We obtained similar results for the reaction of nitrous oxide reductase (8 Cu/M_r) from Pseudomonas stutzeri with NO using the appropriate length of time and partial pressure of NO (Riester et al., 1989).

Upon longer exposure to NO, the type-1 Cu (mononuclear) becomes reduced whereas the type-3 Cu pair shows rapid redox reactions with NO. This centre also binds NO. Nitrite is the reaction product in the case of fungal laccase (Martin et al., 1981) and deoxyhaemocyanin (Verplaetse et al., 1979). NO also reacts as an oxidizing agent towards reduced tree laccase and nitrous oxide was identified as the main reaction product (Gorren et al., 1987). This observation could account for the finding that dithionite-reduced nitrous oxide reductase becomes reoxidized in the presence of NO (Riester et al., 1989). At low concentrations of NO and short reaction times the Cu site responsible for the purple chromophore might bind one NO molecule as reported for type-1 Cu sites (Boelens et al., 1984). Antiferromagnetic coupling of the electron spins of Cu(II) and NO would explain the disappearance of the characteristic multiline EPR signal of nitrous oxide reductase (Riester et al., 1989). Surprisingly, form V of nitrous oxide reductase (isolated from a transposon Tn5-induced mutant with defective chromophore biosynthesis, 2 Cu/M_r) did not react with NO, which was rather unexpected in view of the fact that in this form of the enzyme we observed the mixed-valence [Cu(1.5)...Cu(1.5)] centre with the narrow hyperfine splitting of approximately 3.5 mT, typical for type-1 Cu sites (Malkin and Malmstroem, 1970). Nitrite was formed during the reaction between nitrous oxide reductase and nitric oxide despite the exclusion of dioxygen. We investigated therefore, the effect of nitrite and nitrate on the optical and EPR properties of nitrous oxide reductase (Riester et al., 1989). Nitrate proved to be rather inert; nitrite, on the other hand, caused marked spectral changes as found for other exogenous ligands such as thiocyanate and cyanate which are known to coordinate to binuclear type-3 Cu centres (Solomon et al., 1983). Nitrous oxide, the natural substrate of nitrous oxide reductase, and acetylene, the "classical" inhibitor of N_2O reduction (Hyman and Arp, 1988), did not give any significant spectral effects with the purified enzyme both in the oxidized and in the reduced state (Riester, 1989). Interestingly, nitrous oxide was recently reported to inhibit reversibly oxygen utilization by bovine heart and bean mitochondria (Sowa et al., 1987). Consequently, interactions of the anaesthetic N_2O with bovine heart cytochrome c oxidase were investigated (Einarsdottir and Caughey, 1988). Oxidase activity of the isolated enzyme was partially and reversibly reduced by N_2O, but it had no effects on the visible Soret spectra or on the reaction site. N_2O molecules were shown by IR spectroscopy to occupy sites within the oxidase; there was no evidence found for N_2O serving as a ligand to the metal centres of cytochrome c oxidase. It does not serve as a source for O-transfer as discussed earlier in this article for a series of Co(I), Ni(I) and Cu(I) model complexes.

Nitric oxide and nitrite were also employed to probe the electronic and stereochemical environment of the type-3 Cu pairs in tyrosinase, haemocyanin and laccase (Verplaetse et al., 1979; Martin et al., 1981; Solomon et al., 1983). NO_2^- acts as a bridging ligand with the half-met form, [Cu(I)...Cu(II)], of tyrosinase and haemocyanin but not laccase. In the case of methaemocyanin NO_2^- gave the magnetic dipole-coupled [Cu(II)...Cu(II)] pair as demonstrated by the appearance of a 7-line hyperfine pattern in the EPR signal at half field around g=4 (Verplaetse et al., 1979).

Nitrite was an effective oxidant of Cu(I)-plantacyanin (type-1 Cu) (Nerrissian et al., 1985) and Cu(I)-deoxyhaemocyanin (type-3 Cu pair) (Van der Deen and Hoving, 1977). NO may have been the real oxidizing agent in these experiments produced by the well-known disproportionation reaction of nitrous acid (Eq. 12).

MECHANISTIC ASPECTS OF DENITRIFICATION: N-N BOND FORMATION AND N-O BOND FISSION

Molybdenum (nitrate reductase), iron (nitrite reductase, nitric oxide reductase) and copper (nitrite reductase, nitrous oxide reductase) are the key transiton metals of the denitrification pathway (Zumft and Cardenas, 1979, see also article by Zumft and Kroneck, chapter 3 in this book). The different enzymic steps (Scheme 1) of the denitrification pathway in various bacteria including the chemistry of potential N_xO_y intermediates and properties of the purified enzymes have been recently reviewed (Henry and Bessières, 1984). However, the main effort has been directed towards the mechanism of nitrite reduction by the haem cd_1-containing nitrite reductases from Pseudomonas aeruginosa (Kim and Hollocher, 1984) and from Pseudomonas stutzeri (Aerssens et al., 1986; Weeg-Aerssens et al., 1987; Weeg-Aerssens et al., 1988). Recently, first mechanistic studies on the copper-containing nitrite reductase from "Achromobacter cycloclastes" were reported (Hulse and Averill, 1989).

Molybdenum plays the central role in the reduction of nitrate to nitrite. Nitrate reductase belongs in the family of molybdoenzymes, often referred to as the molybdenum hydroxylases (Bray, 1988). It catalyses the $2e^-/2H^+$ reduction of NO_3^- to NO_2^- (E_o' + 420 mV) with NAD(P)H or ferredoxins as electron donors. The cofactor is a complex of molybdenum and molybdopterin (Johnson and Rajagopalan, 1982). Recent EXAFS experiments gave the minimal coordination sites $Mo(VI)O_2(SR)_{2-3}$ and $Mo(IV)O(SR)_{3-4}$ in the oxidized and reduced state, respectively (Cramer et al., 1984). Two types of mechanisms were proposed: (a) the coupled electron-proton transfer (Stiefel, 1973) that involves coordination of NO_3^- at the Mo(IV) centre and concomitant transfer of electrons and protons to the substrate (Eq. 17), and (b) the oxo transferase mechanism (Eq. 18) (Craig and Holm, 1989).

$$NO_3^- + 2\,H^+ + 2\,e^- \longrightarrow NO_2^- + H_2O \tag{17}$$

Nitrate Reductase

$$Mo(IV)O\text{-}E + NO_3^- \longrightarrow Mo(VI)O_2\text{-}E + NO_2^- \tag{18}$$

Results from model studies demonstrate that reduction of nitrate by atom transfer is a plausible (but unproven) pathway in the mechanism of action of molybdenum-containing nitrate reductases (Craig and Holm, 1989).

Two types of mechanisms for the reduction of NO_2^- to N_2O have been proposed, mainly on the basis of nitrogen isotope labelling and ^{18}O exchange experiments (Kim and Hollocher, 1984; Aerssens et al., 1986; Weeg-Aerssens et al., 1987; Weeg-Aerssens et al., 1988; Shearer and Kohl, 1988). There is conclusive evidence, however, that nitrite reduction by whole cells (Aerssens et al., 1986; Weeg-Aerssens et al., 1988), and the purified cd_1-haem containing nitrite reductase from Pseudomonas aeruginosa (Kim and Hollocher, 1984) proceeds via an enzyme-bound ferrous nitrosyl intermediate , E-NO^+. Such an intermediate was also proposed for the Cu-containing nitrite reductase from "Achromobacter cycloclastes" (Hulse and Averill, 1989). The nitrosyl intermediate, which is common to both mechanisms, arises from the dehydration of coordinated nitrite (Eq. 19):

$$\text{E-Fe(II)} + NO_2^- \rightarrow \text{E-Fe(II)-}NO_2^- \xrightarrow{2\,H^+ \searrow \;\; \nearrow H_2O} \text{E-Fe(II)-}NO^+ \qquad (19)$$

According to one mechanism ("parallel mechanism", Hollocher and coworkers) (Kim and Hollocher, 1984) E-NO^+ is reduced to E-NO^- and nitroxyl is then released from the iron centre followed by a dimerization of two HNO molecules (Eq. 20):

$$\text{E-Fe(II)-}NO^+ \rightarrow \text{E-Fe(II)-}NO^- \xrightarrow[4\,H^+ \nearrow \;\; \searrow 2\,H_2O]{NO^- \searrow} \text{E-Fe(II)} + N_2O \qquad (20)$$

In another mechanism ("sequential mechanism", Averill and coworkers) (Weeg-Aerssens et al., 1988) NO_2^- makes a nucleophilic attack on E-NO^+ forming E-(N_2O_3) (Eq. 21). The coordinated dinitrogen trioxide becomes reduced and N_2O is finally released via oxyhyponitrite, $N_2O_3^{2-}$, and cis-hyponitrite, cis-$N_2O_2^{2-}$, as enzyme-bound intermediates.

$$\text{E-Fe(II)-}(N_2O_3) \xrightarrow{\text{Red.}} \text{E-Fe(II)-}(N_2O_3)^{2-} \xrightarrow{\text{Red./2 }H^+ \searrow \;\; \nearrow H_2O} \text{E-Fe(II)-}(N_2O_2)^{2-}$$

$$\xrightarrow{2H^+ \searrow \;\; \nearrow H_2O} \text{E-Fe(II)-}(N_2O) \rightarrow \text{E-Fe(II)} + N_2O \qquad (21)$$

Thus, in the "parallel mechanism" of nitrite reduction two NO_2^- molecules are reduced in parallel; the N-N bond is then formed from reduced intermediates. In the "sequential mechanism" one of the nitrogen atoms is still in the oxidized +3 valence state when the N-N bond is formed. Recent results from nitrogen isotopic fractionation and ^{18}O exchange experiments with whole cells from Pseudomonas stutzeri tend to rule out the "parallel mechanism" (Shearer and Kohl, 1988). However, depending on the organism, E-NO^+ might have several fates: (a) reduction and release of NO which is taken up by a membrane-bound nitric oxide reductase (see Zumft and Kroneck, this book), (b) nucleophilic attack by NO_2^- to produce enzyme-bound N_2O_3, or (c) further reduction followed by the release and dimerization of nitroxyl. The major argument brought forth in favour of nitroxyl as an intermediate has been the positional isotopic equivalence of nitrogen in $^{14,15}N_2O$ produced by the reduction of $^{15}NO_2^-$ and ^{14}NO (Garber and Hollocher, 1982; Kim and Hollocher, 1984). This implies the production of a symmetric intermediate, e.g. the mononitrogen species HNO, or an effectively symmetric dinitrogen species consistent with the "symmetrical mechanism", e.g. cis-hyponitrite coordinated to the ferrous haem (Weeg-Aerssens et al., 1988). Cyclic voltammetry of Fe(II) porphyrins complexed to NO (Por-Fe(II)-NO) revealed a reversible one-electron oxidation and a reversible one-electron reduction (Olson et al., 1982). The potentials for oxidation of a series of representative Porphyrin-Fe(II)-NO compounds in organic solvents are extremely positive. The stability of the ferrous form is reflected in the relative inertness of

Porphyrin-Fe(II)-NO to air oxidation and displacement by other ligands. The oxidation of Porphyrin-Fe(II)-NO, on the other hand, leads to an increase in lability of the nitrosyl ligand (Wayland and Olson, 1974; Olson et al., 1982). In contrast, the dissociation of NO from Porphyrin-Fe(II)-NO is very slow (Chang, 1985) whereas the rate of complexation of NO with haemproteins is very fast, in the case of haemoglobin it was found to be at the diffusion controlled limit for protein molecules (Goretski and Hollocher, 1988). These findings would account for the production and consumption of NO by denitrifiers, and other observations reported in the literature (Weeg-Aerssens et al., 1988).

In contrast to rather detailed investigations of the reaction mechanism of $\underline{cd_1}$ heme-containing enzymes, practically nothing is known on the mechanism of action of nitrous oxide reductase (N_2OR) and the fission of the N-O bond. The dinitrogen extrusion mechanism and formation of oxometal species (Eq. 11) seems an attractive working hypothesis to be tested. As discussed most recently (see also Zumft and Kroneck, this book) there is spectroscopic and biochemical evidence accumulating on N_2OR (8 Cu/monomeric catalytic unit) from <u>Pseudomonas stutzeri</u> that the cupric site is similar to the "Cu_A" site of cytochrome <u>c</u> oxidase (3 Cu, 2 Fe/monomeric catalytic unit) (Kroneck et al., 1988). It is intriguing that both enzymes couple the reduction of a small molecule to ATP synthesis (eqs 22,23). Both enzymes appear to be linked to the formation of a membrane potential, and to proton translocation (Allen and Van Niel, 1952; Koike and Hattori, 1975; McCarthy and Ferguson,1981).

$$N_2O + 2\,e^- + 2\,H^+ \xrightarrow{N_2OR} N_2 + H_2O \qquad (22)$$

$$O_2 + 4\,e^- + 4\,H^+ \xrightarrow{COX} H_2O + H_2O \qquad (23)$$

Several spectroscopically distinct intermediates of the reaction between reduced cytochrome <u>c</u> oxidase and dioxygen were characterized at lower temperature by the so-called triple-trapping technique (Chance et al., 1975). The first intermediate is probably a species in which dioxygen is bound to ferrous haem a_3 and Cu_B is reduced (Clore and Chance, 1978). In the next step $(Fe_{a3}\text{-O-O-}Cu_B)^{2+}$ may be formed after rapid electron transfer from the "Cu_A" site or Fe_A (Blair et al., 1985). N_2O, via N-coordination (Tuan and Hoffmann, 1985) might bind to a Cu centre of N_2OR yet to be identified.

On the basis of biochemical and physical properties a first structural model for the active site of the multicopper enzyme N_2OR from <u>Pseudomonas stutzeri</u> has been proposed (Riester, 1989). This model is based on the following premises: (a) N_2OR consists of two identical subunits each containing 4 Cu atoms similar to the situation of the plant copper enzyme ascorbate oxidase (Messerschmidt et al., 1989), (b) a cysteine(CySH)/cystine(CySSCy) redox couple is involved in the structural organization of the active site, and (c) each Cu centre is coordinated on the average to at least one sulphur ligand (Scott et al., 1989). For native N_2OR (form I, oxidized) two distinct Cu sites have been characterized (centres A and C). The mononuclear centre A is best described by $[Cu(II)N(His)_2S(Cys)_2]$ as proposed for the "Cu_A" site of cytochrome <u>c</u> oxidase (Kroneck et al., 1989). In view of its reduced EPR activity, magnetic interaction with another paramagnet is likely. Centre A is relatively labile towards dioxygen and changes in the chemical environment. One of the cysteine ligands is irreversibly oxidized (centre A*) which would account for the low activity of N_2OR purified in the presence of oxygen (Riester et al., 1989).

By contrast, centre C is dinuclear, and is represented by the mixed-valence $S=1/2$ species [Cu(1.5)...Cu(1.5)] with the unpaired electron delocalized between two equivalent Cu nuclei. Again it is assumed that each Cu of centre C is coordinated by at least one N(His) and one S(Cys) residue. Centre B is formed by the reduction of native N_2OR yielding the blue form N_2OR III (Riester et al., 1989). This centre shows rather unusual properties for a "reduced" Cu species, i.e. it exhibits a blue chromophore and a broad, featureless EPR signal in the frequency range 2.4-35 GHz (Kroneck et al., 1988). Its resistance towards further reduction and its spectral properties together with the results from quantitative determinations of sulphydryl and disulphide groups in both N_2OR I and N_2OR III led to the conclusion that reduction of a proximal disulphide residue might contribute to the stabilization of centre B (Riester, 1989).

Both centres A and C react very rapidly with a large variety of reducing agents, e.g. dithionite, Cr(II) or Ru(II) complexes, and photochemically generated reductants (Riester et al., 1989). On the other hand, the formation of centre B proceeds rather slowly and therefore seems to be of no physiological importance. Centre A may be the entrance point for reducing equivalents after reaction of N_2OR with the (so far unknown) physiological electron donor. This would leave the dinuclear mixed-valence species [Cu(1.5)...Cu(1.5)] as the binding site of the substrate N_2O, or other small exogenous ligands (Riester, 1989).

SUMMARY

Nitrogen is an essential constituent of many biomolecules which are important for life, e.g. amino acids, proteins, or DNA. In the process of nitrogen fixation dinitrogen is reduced to ammonia. This reaction is catalyzed by nitrogenase, a complex multi-metal enzyme. Dinitrogen is released into the environment by the activities of a large variety of bacteria. Oxidized nitrogen compounds are used as electron acceptors and are finally reduced to dinitrogen, nitrous oxide and ammonia.

The large number of molecular and ionic species that may be formed from just the atoms of nitrogen and oxygen is remarkable. Because nitrogen can effectively form both σ-and π-bonds to other nitrogen or oxygen atoms, many molecules of the general formula N_xO_y are thermodynamically stable. In addition to the gases N_2O, NO and NO_2 (and its dimeric form N_2O_4), and in addition to the important NO_3^- and NO_2^- anions more than a dozen other species have been characterized. Many of the latter are short-lived intermediates but might be stabilized at the active site of a metal-dependent enzyme, or can be characterized at low temperatures.

In this article we have addressed several aspects of the bio-inorganic chemistry of N_xO_y-compounds as these might be of interest for important processes such as denitrification, or dissimilatory nitrate reduction. The structures and reactions of these N_xO_y compounds were discussed, also including their coordinative properties towards the transition metals copper and iron. Furthermore, special attention was dedicated to mechanistic aspects of the formation of the N-N bond (conversion of nitrite to nitrous oxide as catalyzed by dissimilatory Fe- and Cu-dependent nitrite reductases) and to the cleavage of the N-O bond (conversion of nitrous oxide to dinitrogen as catalyzed by Cu- dependent nitrous oxide reductases). A first structural model of the active site of the multi-copper enzyme nitrous oxide reductase from Pseudomonas stutzeri was presented.

ACKNOWLEDGMENT

This work was supported by Deutsche Forschungsgemeinschaft and Fonds der Chemischen Industrie.

REFERENCES

Addison, C.C., Logan, N., Wallwork, S.C., and Garner, C.D., 1971, Structural aspects of coordinated nitrate groups, Quart. Rev., 25: 289-322.

Aerssens, E., Tiedje, J.M., and Averill, B.A., 1986, Isotope labeling studies on the mechanism of N-N bond formation in denitrification, J. Biol. Chem., 261: 9552-9556.

Allen, M.B., and Van Niel, C.B., 1952, Experiments on bacterial denitrification, J. Bacteriol., 64: 397-412.

Banks, R.G.S., Henderson, R.J., and Pratt, J.M., 1969, Reactions of gases in solution. Part III. Some reactions of nitrous oxide with transition metal complexes, J. Chem. Soc.(A): 2886-2889.

Barley, M.H., Rhodes, M.R., and Meyer, T.J., 1987, Electrocatalytic reduction of nitrite to nitrous oxide and ammonia based on the N-methylated, cationic iron porphyrin complex $[Fe(III)(H_2O)(TMPyP)]^{5+}$, Inorg. Chem., 26: 1746-1750.

Bhaduri, S., Johnson, B.F.G., Pickard, A., Raithby, P.R., Sheldrick, G.M., and Zuccaro, C.I., 1977, Nitrosyl coupling in transition-metal complexes: The molecular structure of $[(PPH_3)_2Pt(N_2O_2)]$, J. Chem. Soc. Chem. Commun., 354-355.

Blair, D.F., Witt, S.N., and Chan, S.I., 1985, Mechanism of cytochrome c oxidase-catalyzed dioxygen reduction at low temperatures. Evidence for two intermediates at the three-electron level and entropic promotion of the bond-breaking step, J. Am. Chem. Soc., 107: 7389-7399.

Boelens, R., Rademaker, H., Wever, R., and Van Gelder, B.F., 1984, The cytochrome c oxidase-azide-nitric oxide complex as a model for the oxygen-binding site, Biochim. Biophys. Acta., 765: 196-209.

Bray, R.C., 1988, The inorganic biochemistry of molybdoenzymes, Quart. Rev. Biophys., 21: 299 - 329.

Butler, A.R., Glidewell, C., and Min-Hsin, L., 1988, Nitrosyl complexes of iron-sulfur clusters, Adv. Inorg. Chemistry, 32: 335-393.

Buxton, G.V., Sellers, R.M., and Cracken, D.R., 1976, Pulse radiolysis study of the monovalent cadmium, cobalt, nickel and zinc in aqueous solution. Part 2. Reactions of the monovalent ions, J. Chem. Soc. Faraday Trans., 72: 1464-1476.

Chance, B., Saronio, C., and Leigh, Jr. J.S., 1975, Functional intermediates in the reaction of cytochrome oxidase with oxygen, Proc. Natl. Acad. Sci. (USA) 72: 1635-164

Chang, C.K.J., 1985, On the structure of heme d_1, J. Biol. Chem., 260: 9520-9522.

Clore, G.M. and Chance, E.M.,1978, The mechanism of reaction of ferricyanide-pretreated mixed-valence state membrane-bound cytochrome oxidase with oxygen at 173 K, Biochem. J., 173: 811-820.

Colburn, C.B., 1966, Developments in inorganic nitrogen chemistry. Vol. 1 and 2. Elsevier Publishing Company, Amsterdam, London, New York

Cole, J.A., and Ferguson, S.J., (ed.), 1988, The nitrogen and sulphur cycles. Cambridge University Press, Cambridge, New York, New Rochelle, Melbourne, Sidney

Coyle, C.L., Zumft, W.G., Kroneck, P.M.H., Koerner, H., and Jakob, W., 1985, Nitrous oxide reductase from denitrifying Pseudomonas perfectomarina. Purification and properties of a novel multicopper enzyme, Eur. J. Biochem., 153: 459-467.

Craig, J.A. and Holm, R.H., 1989, Reduction of nitrate to nitrite by molybdenum-mediated atom transfer. A nitrate reductase analogue reaction system, J. Am. Chem. Soc., 111: 2111-2115.

Cramer, S.P., Solomonson, L.P., Adams, M.W.W., and Mortenson, L.E.J., 1984, Molybdenum sites of Escherichia coli and Chlorella vulgaris nitrate reductase, J. Am. Chem. Soc., 106: 1467-1471.

Dell, R.M., Stone, F.S., and Tiley, P.F., 1953a, The adsorption of oxygen and other gases on copper, Trans. Farad. Soc., 49: 195-201.

Dell, R.M., Stone, F.S., and Tiley, P.F., 1953b, The decomposition of nitrous oxide on cuprous oxide and other oxide catalysts, Trans. Farad. Soc., 49: 201-209.

Delwiche, C.C., 1981, Denitrification, nitrification, and atmospheric nitrous oxide, John Wiley & Sons, New York, Chichester, Brisbane, Toronto

Deno, N.C., Berkheimer, H.E., Evans, W.L., and Peterson, H.J., 1959, Carbonium ions.VII. An evaluation of the H_R acidity function in aqueous perchloric and nitric acids, J. Am. Chem. Soc., 81: 2344-2347.

Diamantis, A.A. and Sparrow, G.J., 1970, Nitrous oxide complexes: The isolation of pentammine(dinitrogenoxide) ruthenium(II) tetrafluoroborate, J. Chem. Soc. Chem. Commun., 819-820

Dobry, A. and Boyer, F., 1945, Sur le nitrosulfure de fer ou sel de Roussin - action antiseptique, Ann. Inst. Pasteur, 71: 455 - 462.

Doyle, M.P., Mahapatro, S.N., Broene, R., and Guy, J.K., 1988, Oxidation and reduction of hemoproteins by trioxodinitrate(II). The role of nitrosyl hydride and nitrite, J. Am. Chem. Soc., 110: 593-599.

Einarsdottir, O. and Caughey, W.X., 1988, Interactions of the anesthetic nitrous oxide with bovine heart cytochrome c oxidase, J. Biol. Chem., 263: 9199-9205.

Eisenberg, R. and Meyer, C.D., 1975, The coordination chemistry of nitric oxide, Acc. Chem. Res., 8: 26-34.

Elschenbroich, Ch., and Salzer, A., 1988, Organometallchemie, 2nd ed, B.G. Teubner, Stuttgart.

Enemark, J.H. and Feltham, R.D., 1974, Principles of structure, bonding, and reactivity for metal nitrosyl complexes, Coord. Chem. Rev., 13: 339-406.

Ferguson, S.J., 1987, Denitrification: a question of the control and organization of electron and ion transport, Trends Biochem. Sci., 12: 354-357.

Fu-Tai Tuan, D. and Hoffmann, R., 1985, N vs. O linkage and σ vs π bonding in transition-metal complexes of N_2O and NCO^-, Inorg. Chem., 24: 871-876.

Garber, E.A. and T.C. Hollocher, 1982, Positional isotope equivalence of nitrogen in N_2O produced by the denitrifying bacterium Pseudomonas stutzeri, J. Biol. Chem., 257: 4705-4708.

Goretski, J. and Hollocher, T.C., 1988, Trapping of nitric oxide produced during denitrification by extracellular hemoglobin, J.Biol.Chem., 263: 2316-2323.

Gorren, A.C.F., De Boer, E., and Wever, R., 1987, The reaction of nitric oxide with copper proteins and the photodissociation of copper-NO complexes, Biochim. Biophys. Acta., 916: 38-47.

Greenwood, N.N., and Earnshaw, A., 1985, Chemistry of the elements. Pergamon Press, Oxford, New York, Toronto, Sidney, Paris, Frankfurt.

Hathaway, B.J. and Billing, D.E., 1970, The electronic properties and stereochemistry of mono-nuclear complexes of the copper(II) ion, Coord. Chem. Rev., 5: 143-207.
Henry, Y., and Bessières, P., 1984, Denitrification and nitrite reduction: Pseudomonas aeruginosa nitrite-reductase, Biochimie, 66: 259-289.
Hoskins, B.F., Whillans, F.D., Dale, P.H., and Hodgkin, D.C., 1969, The structure of the red nitrosylpentaammine cobalt(III) cation, J. Chem. Soc. Chem. Commun., 69-70.
Hughes, M.N., 1968, Hyponitrites, Quart. Rev., 22: 1-13.
Hulse, C.L. and Averill, B.A., 1989, Evidence for a copper-nitrosyl intermediate in denitrification by the copper-containing nitrite reductase of "Achromobacter cycloclastes", J. Am. Chem. Soc., 111: 2322-2323.
Hyman, M.R., and Arp, D.J., 1988, Acetylene inhibition of metalloenzymes, Anal. Biochem., 173: 207-220.
Johnson, J.L. and Rajagopolan, K.V., 1982, Structural and metabolic relationship between molybdenum cofactor and urothione, Proc. Natl. Acad. Sci. (USA) 79: 6856-6860.
Jones, K., 1975, The chemistry of nitrogen. Pergamon Press, Oxford, New York, Toronto, Sidney, Paris, Braunschweig.
Jubran, N., Ginzburg, G., Cohen, H., Koresh, Y., and Meyerstein, D., 1985, Stabilization of the monovalent nickel complex with 1,4,8,11-tetraazacyclotetradecane in aqueous solution by N- and C-methylation. An electrochemical and pulse radiolysis study, Inorg. Chem., 24: 251-258.
Jug, K., 1978, Bond order orbitals and eigenvalues, J. Am. Chem. Soc., 100: 6581-6586.
Karthein, R., Nastainczyk, W., and Ruf, H., 1987, EPR study of ferric native prostaglandin H synthase and its ferrous NO derivative, Eur. J. Biochem., 166: 173-180.
Kim, C-H. and Hollocher, T.C., 1984, Catalysis of nitrosyl transfer reactions by dissimilatory nitrite reductase (cytochrome c,d_1), J. Biol. Chem., 259: 2092-2099.
Knowles, R., 1982, Denitrification, Microbiol .Rev., 46: 43-70.
Koike, I. and Hattori, A., 1975, Energy yield of denitrification: An estimate from growth yield in continuous cultures of Pseudomonas denitrificans under nitrate-, nitrite-, and nitrous oxide limited conditions, J. Gen. Microbiol., 88: 11-19.
Kroneck, P.M.H., Antholine, W.E., Riester, J., and Zumft, W.G., 1988, The cupric site in nitrous oxide reductase contains a mixed-valence [Cu(II),Cu(I)] binuclear center: a multifrequency electron paramagnetic resonance investigation, FEBS Lett., 242: 751-762.
Kroneck, P.M.H., Antholine, W.E., Riester, J., and Zumft, W.G., 1989, The nature of the cupric site in nitrous oxide reductase and of Cu_A in cytochrome c oxidase, FEBS Lett., 242: 212-213.
Laane, J. and Ohlsen, J.R., 1980, Characterization of nitrogen oxides by vibrational spectroscopy, Prog. Inorg. Chemistry, 27: 465-513.
Lee, S.D., 1980, Nitrogen oxides and their effects on health. Ann Arbor Science, Ann Arbor.
Liang, J. and Burris, R., 1988, Interactions among N_2, N_2O, and C_2H_2 as substrates and inhibitors of nitrogenase from Azotobacter vinelandii, Biochemistry, 27: 6726-6732.
Loechler, E.L., Schneider, A.M., Schwartz, D.B., and T.G. Hollocher, 1987, The covalent electrophilic catalysis of the breakdown of hyponitrite to nitrous oxide by aldehydes, ketones, and carbon dioxide, J.Am.Chem.Soc., 109: 3076-3087.

Malkin, R. and Malmstroem, B.G., 1970, The state and function of copper in biological systems, Adv.Enzymol., 33: 177-244.
Martin, C.T., Morse, R.H., Kanne, R.M., Gray, H.B., Malmstroem, B.G., and Chan, S.I., 1981, Reactions of nitric oxide with tree and fungal laccase, Biochemistry, 20: 5147-5155.
McCarthy, J.E.G. and Ferguson, S.J., 1981, Estimation with an ion-selective electrode of the membrane potential in cells of Paracoccus denitrificans from the uptake of the butyltriphenylphosphonium cation during aerobic and anaerobic respiration, Biochem. J., 196: 311-321.
McCleverty, J.A., 1979, The reactions of nitric oxide coordinated to transition metals, Chem.Rev., 79: 53-76.
Messerschmidt, A., Rossi, A., Ladenstein, R., Huber, R., Bolognesi, M., Gatti, M., Marchesini, A., Petruzzelli, R., and Finazzi-Agro,A., 1989, X-Ray crystal structure of the blue oxidase ascorbate oxidase from zucchini. Analysis of the polypeptide fold and a model of the copper sites and ligands, J.Mol.Biol., 206: 513-529.
Mozen, M.M. and Burris, R.H., 1954, The incorporation of ^{15}N-labelled nitrous oxide by nitrogen fixing agents, Biochim.Biophys.Acta., 14: 577-578.
Nelson, M.J., 1987, The nitric oxide complex of ferrous soybean lipoxygenase-1. Substrate, pH, and ethanol effects on the active-site iron, J.Biol.Chem., 262: 12137-12142.
Nerrissian, A.M., Babayan, M.A., Sarkassian, L.K., Samkhanian, E.G., and Nalbandyan, R.M., 1985, Studies on plantacyanin.II. NMR data, redox properties reaction with nitrite and the formation of complex with plastocyanin, Biochim.Biophys.Acta., 830: 195-205.
Ohnishi,T., LoBrutto, R., Salerno, J.C., Bruckner, R.C., and Frey, R.F., 1982, Spatial relationship between cytochrome a and a_3, J.Biol.Chem., 257: 14821-14825.
Olson, L.W., Schaeper, D., Lancon, D., and Kadish, K.M., 1982, Characterization of several novel iron nitrosyl porphyrins, J.Am.Chem.Soc., 104: 2042-2044.
Palmer, G., 1982, Electron paramagnetic resonance of hemoproteins, In:"Iron porphyrins" (Lever, A.B.P. and Gray, H.B., ed.) Vol. II, pp. 43-88, Addison-Wesley Publishing Company, London, Amsterdam, Don Mills,Ontario, Sidney, Tokyo.
Palmer, R.M.J., Ferrige, A.G., and Moncada, S., 1987, Nitric oxide release accounts for the biological activity of endothelium-derived relaxing factor, Nature, 327: 524-526.
Pu, L.S., Yamamoto, A., and Ikeda, S., 1969, Catalytic reduction of nitrous oxide with nitrogentris (triphenylphosphine) cobalt hydride, J.Chem.Soc.Chem. Commun., 189-190.
Reddy, D., Lancaster, J.R., and Cornforth, D.P., 1983, Nitrite inhibition of Clostridium botulinum: Electron spin resonance detection of iron-nitric oxide complexes, Science, 221: 769-770.
Riester, J., 1989, Distickstoffoxid-Reduktase aus Pseudomonas stutzeri, ein Multikupferenzym in denitrifizierenden Bakterien: Physikochemische Untersuchungen zur Struktur und katalytischen Funktion. Ph.D. Thesis, Universitaet Konstanz.
Riester, J., Zumft, W.G., and Kroneck, P.M.H., 1989, Nitrous oxide reductase from Pseudomonas stutzeri. Redox properties and spectroscopic characterization of different forms of the multicopper enzyme, Eur.J.Biochem., 178: 751-762.
Scheidt, W.R. and Lee, Y.J., 1987, Recent advances in the stereochemistry of metallo tetrapyrroles, Struct. & Bonding, 64: 1-72.
Scholten, J.J.F. and Konvalinka, J.A., 1969, Reaction of nitrous oxide with copper surfaces, Trans.Farad.Soc., 65: 2465-2473.

Schwab, G.-M. and Drikos, G., 1940, Katalytische Reaktion von Stickoxydul mit Kohlenoxyd an Kupfer und Kupferoxyd, Z.Phys.Chem. A, 186: 348-358.
Scott, R.A., Zumft, W.G., Coyle, C.L., and Dooley, D.M., 1989, Pesudomonas stutzeri N_2O reductase contains Cu_A-type sites, Proc.Natl.Acad. Sci.(USA), 86: 4082 - 4086.
Sharma, V., Traylor, T.G., Gardiner, R., and Mizukami, H., 1987, Reaction of nitric oxide with heme proteins and model compounds of hemoglobin, Biochemistry, 26: 3837-3843.
Shearer, G. and Kohl, D.H., 1988, Nitrogen isotopic fractionation and ^{18}O exchange in relation to the mechanism of denitrification of nitrite by Pseudomonas stutzeri, J.Biol.Chem., 263: 13231-13245.
Simmons, C.J., Hathaway, B.J., Amornjarusiri, K., Santarsiero, B.D., and Clearfield, A., 1987, The first determination of the energy difference between solid-state conformers by x-ray diffraction: 1. The crystal structure of the Pseudo-Jahn-Teller complex (nitrito)bis(2,2'-bipyridyl)copper(II)-nitrate at 20, 100, 165 and 296 K and of its isostructural zinc(II) analogue at 295 K. 2. The possibility of using x-ray diffraction to characterize adiabatic potential energy surfaces and relative ligand strengths, J.Am.Chem.Soc., 109: 1947-1958.
Smith, B.E., 1986, Biologische Stickstoff-Fixierung, Nachr.Chem.Tech.Lab., 34: 308-317.
Solomon, E.I., Penfield, K.W., and Wilcox, D.E., 1983, Active sites in copper proteins. An electronic structure overview, Struct. & Bonding, 53: 1-57.
Sowa, S., Dong, A., Roos, E.E., and Caughey, W.S., 1987, The anesthetic nitrous oxide affects dioxygen activity by bovine heart and bean seed mitochondrial particles, Biochem.Biophys.Res.Commun., 144: 643-648.
Sprent, J.I., 1987, The ecology of the nitrogen cycle. Cambridge University Press., Cambridge, New York, New Rochelle, Melbourne, Sidney.
Stedman, G., 1979, Reactions of inorganic nitrogen compounds, Adv.Inorg.Radiochem., 22: 113 - 170.
Stershic, M.T., Keefer, L.K., Sullivan, B.P., and Meyer, T.J., 1988, Formation of complexed nitrosamines by oxidation of coordinated ammonia in the presence of secondary amines, J.Am.Chem.Soc., 110: 6884-6885.
Stevens, T.H., Brudvig, G.W., Bocian, D.F., and Chan, S.I., 1979, Structure of cytochrome $\underline{a}_3$-Cu_{a3} couple in cytochrome $\underline{c}$ oxidase as revealed by nitri oxide binding studies, Proc.Natl.Acad.Sci.(USA), 76: 3320-3324.
Stiefel, E.I., 1973, Proposed molecular mechanism for the action of molybdenum in enzymes: Coupled proton and electron transfer, Proc.Natl.Acad.Sci.(USA), 70: 988-992.
Suguira, Y., T. Suzuki, H. Kawabe, H. Tanaka, and K. Watanabe,1982, Spectroscopic studies on bleomycin-iron complexes with carbon monoxide, nitric oxide, azide, cyanide and comparison with iron-porphyrin complexes. Biochim.Biophys.Acta., 716: 38-44.
Suzuki, S., Yoshimura, T., Nakahara, A., Iwasaki, H., Shidara, S., and Matsubara, T., 1987, Electronic and magnetic circular dichroism spectra of pentacoordinate nitrosylhemes in cytochromes c' from non-photosynthetic bacteria and their model complexes, Inorg.Chem., 26: 1006-1008.
Tait, M., Hoffman, M.Z., and Hayon, E., 1976a, The reactivity of cobalt(I) complexes containing unsaturated macrocylic ligands in aqueous solution, J.Am.Chem.Soc., 98: 86-93.
Tait, A.M., Hoffman, M.Z., and Hayon, E., 1976b, Reactivity of nickel(I) complexes containing 14-membered macrocyclic ligands in aqueous solution, Inorg.Chem., 15: 934-939.
Tsubaki, M., Hiwatashi, A., Ichikawa, Y., and Hori, H., 1987, Electron paramagnetic resonance study of ferrous cytochrome $P\text{-}450_{scc}$-nitric oxide

complexes: Effects of cholesterol and its analogues, Biochemistry, 26: 4527-4534.
Ullrich, W.R., Aparicio, P.J., Syrett, P.J., and Castillo, F., (ed.), 1987, Inorganic nitrogen metabolism. Springer Verlag, Berlin, Heidelberg, New York, London, Paris, Tokyo.
Van der Deen, H., and Hoving, H., 1977, Nitrite and nitric oxide treatment of Helix pomatia hemocyanin: Single and double oxidation of the active site, Biochemistry, 16: 3519-3525.
Van Leeuwen, F.X.R., Wever, R., Van Gelder, B.F., Avigliano, V., and Mondovi, B., 1975, The interaction of nitric oxide with ascorbate oxidase, Biochim.Biophys.Acta., 403: 285-291.
Vaughan, G.A., Rupert, P.B., and Hillhouse, G.L., 1987, Selective O-atom transfer from nitrous oxide to hydride and aryl ligands of bis(pentamethylcyclopentadienyl)hafnium derivatives, J.Am.Chem.Soc., 109: 5538-5539.
Vaughan, G.A., Hillhouse, G.L., Lum, R.T., Buchwald, S.L., and Rheingold, A.L., 1988, Oxygen-atom transfer from nitrous oxide. Synthesis and structure of a zirconocene oxametallacyclobutene complex, J.Am.Chem.Soc., 110: 7215-7217.
Verplaetse, J., Van Tornont, P., Defreyn, G., Witters, R., and Lontie, R., 1979, The reaction of nitrogen monoxide and of nitrite with deoxyhemocyanin and methemocyanin of Helix pomatia, Eur.J.Biochem., 95: 327-331.
Wayland, B.B. and Olson, L.W., 1974, Spectroscopic studies and bonding model for nitric oxide complexes of iron porphyrins, J.Am.Chem.Soc., 96: 6037-6041.
Weeg-Aerssens, E., Tiedje, J.M., and Averill, B.A., 1987, The mechanism of microbial denitrification, J.Am.Chem.Soc., 109: 7214-7215.
Weeg-Aerssens, E., Tiedje, J.M., and Averill, B.A., 1988, Evidence from isotope labeling studies for a sequential mechanism for dissimilatory nitrite reduction, J.Am.Chem.Soc., 110: 6851-6856.
Witanowski, M., and Webb, G.A., 1973, Nitrogen NMR. Plenum Press, London, New York.
Yoshimura, T., Suzuki, S., Nakahara, A., Iwasaki, H., Masuko, M., and Matsubara, T., 1986, Spectral properties of nitric oxide complexes of cytochrome c' from Alcaligenes sp. NCIB 11015, Biochemistry, 25: 2436-2442.
Yoshimura, T., Suzuki, S., Nakahara, A., Iwasaki, H., Matsubara, T., and Iwaizumi, M., 1987, Effect of sodium dodecyl sulfate on spectral properties of the nitric oxide complex of ferrous cytochrome c' from Alcaligenes sp. NCIB 11015, Inorg. Chim.Acta., 136: L17-L19.
Yoshimura, T. and Suzuki, S., 1988, The pH dependence of the stereochemistry around the heme group in NO-cytochrome c (horse heart), Inorg.Chim.Acta., 152: 241-249.
Zehnder, A.J.B. (ed.), 1988, Biology of anaerobic microorganisms. John Wiley & Sons, New York, Chichester, Brisbane, Toronto, Singapore.
Zumft, W.G., and Cardenas, J., 1979, The inorganic biochemistry of nitrogen. Bioenergetic processes, Naturwissenschaften, 66: 81-88.

DISTRIBUTION AND DIVERSITY OF DISSIMILATORY NO_2^- REDUCTASES IN DENITRIFYING BACTERIA

Mark S. Coyne and James A. Tiedje

Department of Crop and Soil Sciences
Michigan State University
East Lansing, MI 48824, U.S.A

INTRODUCTION

Denitrification is the biological reduction of nitrogenous oxides to gaseous products during anaerobic bacterial growth. Some organisms, Thiosphaera pantotropha in particular, appear to denitrify in aerobic conditions (Robertson and Kuenen, 1984). However, denitrification can usually be delimited by these criterion (Tiedje, 1988):

(1) It is sensitive to oxygen inhibition;
(2) Growth yield will be proportional to the amount of nitrogenous oxide present;
(3) Eighty percent or more of the available NO_3^- or NO_2^- will accumulate as N_2O or N_2;
(4) NO_3^- or NO_2^- will be rapidly dissimilated without NH_4^+ accumulation;
(5) A dissimilatory NO_2^- reductase (dNir) will be present.

The key step which characterizes denitrification is the irreversible reduction of NO_2^- to NO and N_2O by dissimilatory NO_2^- reductase (dNir). This paper will focus primarily on the diversity of dNirs in terms of their characteristics, distribution, immunological comparisons, and evolutionary divergence.

CHARACTERISTICS OF NO_2^- REDUCTASES

Denitrifiers and dNir types

Dissimilatory NO_2^- reductase is found in organisms which cover the gamut of physiological groups: general aerobic organotrophs, oligocarbophiles, fermenters, halophiles, thermophiles, sporeformers, magnetotactic bacteria, dinitrogen-fixers (free and symbiotic), plant and animal pathogens, phototrophs, hydrogen oxdizers, sulfur oxidizers, and NH_4^+ oxidizers (Tiedje, 1988). It is easier to note those groups in which dNir is absent than those in which it is present:

Denitrification in Soil and Sediment, Edited by N. P. Revsbech and J. Sørensen, Plenum Press, New York, 1990

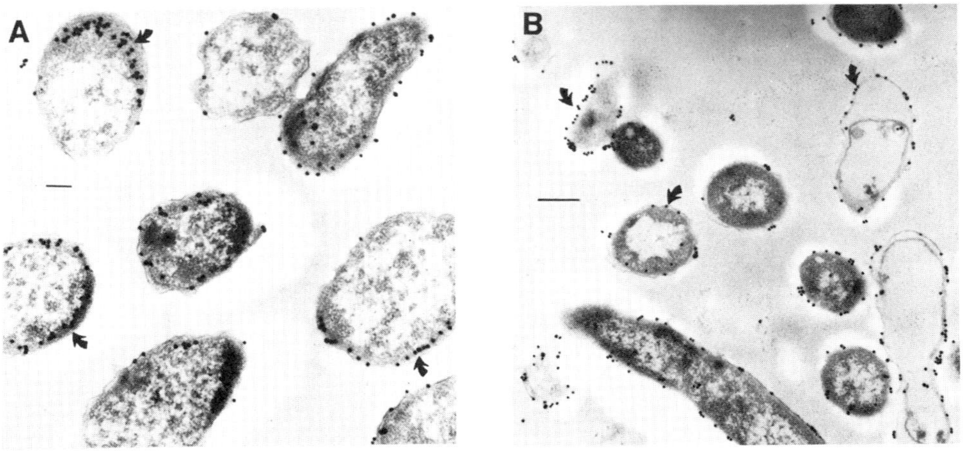

Figure 1. Immunogold labeling of anaerobically-grown denitrifiers with 15 nm colloidal gold probes. A, Achromobacter cycloclastes, Bar = 0.15 μm; B, Pseudomonas aeruginosa, Bar = 0.3 μm. Arrows point to localization of the gold particles in the periplasmic space or at cell membranes.

(1) Obligate anaerobes;
(2) Gram-positive organisms other than Bacillus spp.;
(3) The Enterobacteriaceae.

Nitrite dissimilation in these groups, if it occurs, is mediated by enzymes distinct from the respiratory NO_2^- reduction associated with denitrification (Tiedje, 1988).

Two basic dNir types are known: one contains copper centers (Cu-dNir); the other hemes c and d_1 (cd_1-dNir). The bacteria from which they have been purified are few considering the many different species in which they are found. Heme cd_1-dNir has been purified from Alcaligenes faecalis (Iwasaki and Matsubara, 1971), Paracoccus (Micrococcus) denitrificans (Newton, 1969), P. halodenitrificans (Grant and Hochstein, 1984), Pseudomonas aeruginosa (Gudat et al., 1973; Yamanaka and Okunuki, 1963), P. stutzeri (formerly P. perfectomarina) (Liu et al., 1983), and Thiobacillus denitrificans (LeGall et al., 1979; Sawhney and Nicholas, 1978). It has also been detected in Azospirillum brasilense (Lalande and Knowles, 1987), and an Erythrobacter strain (Shioi et al., 1988).

Nonheme Cu-dNir has been purified from Achromobacter cycloclastes (Hulse and Averill, 1988; Iwasaki et al., 1975; Kashem et al., 1987), A. xylosoxidans (synonymous with Alcaligenes sp. NCIB 11015 and A. denitrificans ss xylosoxidans) (Masuko et al., 1984), A. faecalis strain S6 (Kakutani et al., 1981), Nitrosomonas europaea (Dispirito et al., 1985; Miller and Wood, 1983; Ritchie and Nicholas, 1974), P. aureofaciens (Zumft et al., 1987), and Rhodobacter (Rhodopseudomonas) sphaeroides f. sp. denitrificans (Michalski and Nicholas, 1985). It has been identified in P. denitrificans (Radcliffe and Nicholas, 1968; Michalski and Nicholas, 1988) and Thiosphaera pantotropha (Robertson, 1988).

Table 1. Selected physical characteristics of Cu-dNirs

Enzyme Source	Mass (kDa)	Subunit (Structure)	Cu types	pI	References
A. cycloclastes	69	36.8 x 2	I&II	nd	Liu et al., 1986
A. xylosoxidans	70	37.0 x 2	I	8.4	Iwasaki et al., 1963 Masuko et al., 1984
A. faecalis S6	110	30.0 x 4	I&II	4.5	Kakutani et al., 1981
N. europaea	120	40.0 x 4	I&II	4.63	Miller & Wood., 1983
P. aureofaciens	85	40.0 x 2	I	6.05	Zumft et al., 1987
R. sphaeroides	80	37.5 x 1 39.5 x 1	I II	nd	Michalski & Nicholas, 1985

Localization

Both dNir types have been associated with the cytoplasmic membrane as would be anticipated from their presumed role in energy generation (Koike and Hattori, 1975). However, their exact orientation is not yet clear. Cu-dNir in R. sphaeroides is localized in the periplasm (Sawada and Satoh, 1980), but no other Cu-dNirs have been similarly examined. Heme cd_1-dNirs have been reported as loosely bound cytoplasmic membrane proteins (Sawhney and Nicholas, 1978); as soluble proteins (Mancinelli et al., 1986); or both (Zumft et al., 1979). If membrane bound, they have been noted on the cytoplasmic (Saraste and Kuronen, 1978) or periplasmic (Wood, 1978) aspect of the cytoplasmic membrane. Immunogold labeling of denitrifying cells, suggests that Cu-dNir in A. cycloclastes ATCC 21921 (Figure 1A) and cd_1-dNir from P. aeruginosa ATCC 10145 (Figure 1B) are both found in the periplasm, either loosely associated with cell membranes or as soluble proteins (M.S. Coyne, A. Arunakumari, H.S. Pankratz, and J.M. Tiedje, unpublished data).

Enzyme characteristics

The essential differences between Cu- and cd_1-dNirs are in terms of molecular weight and metal content. The molecular weights of purified heme cd_1-dNirs have been fairly homologous, generally around 120,000 Dalton, although, other molecular weights ranging from 90,000 to 140,000 have been reported (Hochstein and Tomlinson, 1988). The enzymes are dimers composed of identical subunits, each of which contains one cytochrome c and a cytochrome unique to NO_2^- reductase, heme d_1. The cd_1-dNir isolated from P. stutzeri JM300 has a pI of 5.4 (E. Weeg-Aerssens, R. Ye, J.M. Tiedje, R. Hausinger, and B. Averill, unpublished data). The pI of P. aeruginosa cd_1-dNir was reported as 6.9 (Barber et al., 1976) and from P. denitrificans as 3.85 (Newton, 1969). All cd_1-dNirs have cytochrome oxidase activity, although this is presumably not their main physiological function since the K_m for O_2 is high relative to other cytochrome oxidases, and NO_2^- is a much better substrate for the enzyme than O_2 (Sawhney and Nicholas, 1978).

There are two forms of Cu-dNir distinguished by characteristic EPR signals from their Cu centers. One form has both type I and type II copper centers, the

Table 2. Amino acid composition of <u>Achromobacter</u> <u>cycloclastes</u> Cu-dNir

Amino Acid	Iwasaki et al. 1975	Liu et al., 1986	Hulse 1989 (PhD Thesis)
Lys	30	30	27
His	25	25	24
Arg	22	23	26
Asp	66	71	61
Thr	53	63	48
Ser	13	10	12
Glu	54	61	45
Pro	44	47	37
Gly	63	69	55
Ala	64	63	57
Cys	2	2	2
Val	63	57	52
Met	71	12	4
Ile	22	19	19
Leu	44	50	40
Tyr	24	22	nd
Phe	26	26	24
Trp	6	5	nd

[nd] Not determined

other form has type I copper centers only. Type I Cu is presumably the catalytic site. Type II Cu may be an artifact, since it tends to disappear with enzyme purification. <u>Achromobacter</u> <u>cycloclastes</u> Cu-dNir was recently crystallized by Turley et al. (1988). In contrast to earlier reports from their lab and others, they found only two Cu atoms/protein, and only one type of Cu (type I), which existed in identical environments. There is no obvious immunological difference between enzymes with different Cu centers (Michalski and Nicholas, 1988).

Cu-dNirs are heterogeneous with respect to several parameters: molecular weight, subunit composition, and pI (Table 1). So far, the only Cu-dNir with a basic pI occurs in <u>A.</u> <u>xylosoxidans</u>. Like the <u>cd_1</u>-dNirs, Cu-dNirs have cytochrome oxidase activity (with the exception of <u>P.</u> <u>aureofaciens</u> Cu-dNir). If <u>A.</u> <u>cycloclastes</u> Cu-dNir serves as a model, the subunits are bound via a cysteine-thiolate-copper bond (C. Hulse, 1989, Ph.D. Thesis, University of Virginia). Total amino acid analyses of <u>A.</u> <u>cycloclastes</u> Cu-dNir are shown in Table 2. Though other residues differ, the presence of just two cysteine residues per enzyme suggests the functional importance of these groups.

DISTRIBUTION OF NO_2^- REDUCTASES IN DENITRIFYING BACTERIA

<u>Immunoscreening</u>

Polyclonal antibodies raised against specific dNir types have facilitated the identification and comparison of dNirs between different strains and genera. Previous studies were limited in scope to a few organisms, or only one dNir type (Körner et al., 1987; Michalski and Nicholas, 1988). As a first step to investigate the microbial ecology of dNirs, we instigated a general survey to study the

Table 3. Effect of diethyldithiocarbamate (DDC) on N_2O evolution by heme cd$_1$- and nonheme Cu-dNirs

Strain/dNir type	N2O Evolution (nKat mg-1)a		
	- DDC	+ DDC	% Inhibition
cd1-type			
P. aeruginosa	132	114	14
P. stutzeri	120	114	5
Cu-type			
A. cycloclastes	198	0	100
A. xylosoxidans	300	24	92
A. faecalis	156	0	100

[a] Enzyme activity is expressed in units of Katals (Kat), the conversion of one mol of NO_2^- per sec, N_2O, 1 nKat = 10-9 Kat

distribution of cd$_1$- and Cu-dNirs in denitrifiers, and determine which was the predominant type in nature. Polyclonal antibodies were raised to both dNir types: Cu-dNir from A. cycloclastes and cd$_1$-dNir from, P. aeruginosa and P. stutzeri. The survey included commonly studied denitrifiers from culture collections, numerically dominant denitrifiers from soils, sediments, and sludges of six countries (Gamble et al., 1977), and isolates from four sites in North America which were not originally selected for their ability to denitrify, but which were identified as denitrifiers using simple physiological tests (Tiedje, 1982).

Several fundamental trends were apparent when over 170 denitrifying strains were screened in Western immunoblots. The molecular weights of Cu-dNirs were heterogeneous (Fig. 2), that of cd$_1$-dNirs more homogeneous (data not shown). Antibodies raised against specific dNirs showed differential binding to dNirs of other strains. Antibodies to A. cycloclastes Cu-dNir did not recognize Cu-dNir from numerous strains. Antibodies to P. aeruginosa and P. stutzeri cd$_1$-dNir recognized many of the same cd1-dNirs,but not all. No strain had more than one type of dNir.

Chemical assay for NO2- reductase type

The chelating agent, diethyldithiocarbamate (DDC), will specifically inhibit Cu-dNirs during in vitro assays of dNir activity (Shapleigh and Payne, 1985). Consequently, we used DDC inhibition assays to confirm the immunological results, and identify the dNir type in immunologically unreactive strains. Our results with strains of known dNir type showed that Cu-dNir activity is virtually halted by 10 mM DDC while cd$_1$-dNir activity is relatively unaffected (Table 3).

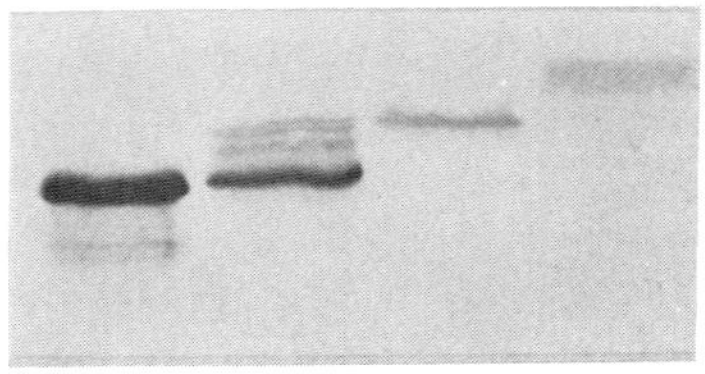

Figure 2. Western immunoblot of denitrifiers containing Cu-dNirs. Lane 1, A. cycloclastes purified Cu-dNir; 2, A. cycloclastes crude extract; 3, A. faecalis #43; 4, A. faecalis #40. Molecular mass indicators are in kDa.

Distribution of NO_2^- reductase types

By using a combination of polyclonal antibodies to the two dNir types, and DDC inhibition assays in cases where there was ambiguity in the results, we could identify the dNir type in virtually all the strains examined (Table 4). The dNir type of several species for which this was not previously known were determined in these immunoscreening/chemical assays. Heme cd_1-dNir was found in Aquaspirillum itersonii, Flavobacterium sp., and Pseudomonas fluorescens. P. fluorescens is ecologically significant because it is the numerically dominant type of culturable denitrifier from soils, sediments, and sludges (Gamble, 1977). Cu-dNir was identified in A. eutrophus, Bacillus sp., Bradyrhizobium japonicum, Corynebacterium nephridii, and Rhizobium spp. Quantitatively, heme cd_1-dNir was the dominant type in both of the culture collections obtained from environmental samples, rather than from laboratory strains. Alcaligenes faecalis was the dominant species with a Cu-dNir. Tables 4 and 5 summarize what we know about dNir distribution among denitrifiers.

The dNir type is not restricted within a species, while it is restricted within a strain. That is, no organism had both heme cd_1- and Cu-dNir. That Alcaligenes faecalis has members with cd_1- or Cu-dNir may simply reflect ambiguities in classification. Most Pseudomonas spp. contain cd_1-dNir, so it is interesting to find some species (P. aureofaciens and P. denitrificans) in which Cu-dNir appears. This may only reflect the use of the Pseudomonas genus as a dumping ground for strains which are difficult to characterize by classical methods, but are otherwise unrelated. Relationships based on bacterial phylogenies from RNA homologies (De Vos and De Ley, 1983; Fox et al., 1980; Ohkubo et al., 1986) or DNA homologies (Palleroni et al., 1973) do not help predict whether an organism will have Cu- or cd_1-dNir. Perhaps this is an indication that mechanisms exists in nature for denitrifiers to acquire dNir if they have lost this function, and there is a strong selection for denitrification.

Numerically, cd_1-dNirs were the most common type. This may reflect a bias in the isolation procedure, since the numerically dominant denitrifiers in readily cultured environmental samples are typically pseudomonads, and 75% of the pseudomonads we examined contained cd_1-dNir. In one other study of numerically dominant denitrifiers from soil samples, Bacillus sp. were the dominant organism obtained (Abd-el-Malek et al., 1974). Our evidence suggests that Bacillus sp. contain Cu-dNir. This implies that Cu-dNirs may be numerically

Table 4. Occurrence of cd_1- and Cu-dNirs in culture collections

Culture group	Number Examined	dNir type	
		% cd_1[a]	% Cu[a]
Laboratory cultures	23	40	56
Numerically[b] dominant denitrifiers	100	64	32
Aerobic soil[c] enrichments	51	92	8

[a] dNir type based on immunological reactivity and/or response to DDC inhibition
[b] Four strains grew too poorly to determine dNir type
[c] Isolates not originally selected for denitrification ability

dominant in some environments. However, we found no evidence for correlation between dNir type and geographic location or soil type, cd_1-dNirs predominated in all.

To reiterate an observation which Gamble et al. (1977) made in their study, it is unfortunate that most physiological and genetic studies on denitrification (particularly dNir) have been done with only a few strains (P. aeruginosa and P. denitrificans, to name two) which don't accurately reflect the true microbial populations. Nitrite reductase has yet to be purified from P. fluorescens. Another reason to suspect that Cu-dNirs are significant in the environment, at least certain environments, is suggested by the diversity of species and physiological groups in which it is found (Table 1 and 5). Cu-dNirs are found in a greater diversity of bacteria and a greater diversity of physiological groups than are cd_1-dNirs. Halobacterium denitrificans could be significant in this respect, since it represents the only Archaebacterial denitrifier. It contains Cu-dNir which is sufficiently different from A. cycloclastes Cu-dNir to preclude immunological recognition. Whether this reflects the environments in which it is found, or the evolutionary distance between Archaebacteria and Eubacteria awaits resolution.

IMMUNOBLOT ANALYSIS

Comparisons of antibody affinity

Qualitative immunoblotting procedures, although not a replacement for protein purification, do provide rapid information about the relatedness of dNirs. Körner et al. (1987) used polyclonal antibodies to look at the relatedness of N_2O reductases and dNirs of P. perfectomarina isolates (now P. stutzeri strain Zobell). Antibodies raised against P. perfectomarina cd_1-dNir reductase crossreacted poorly with the other cd_1-dNirs. Michalski and Nicholas (1988) immunologically compared NO_3^-, NO_2^-, and N_2O reductases in NO_3^- respirers, and denitrifiers

Table 5. Distribution of heme $\underline{cd}_1$- and Cu-dNir in bacteria

Species/Strain	dNir type	Reference
A. cycloclastes	Cu	Hulse et al., 1988
A. xylosoxidans	Cu	Masuko et al., 1984
A. eutrophus	Cu	Coyne et al., 1989
A. faecalis strains	$\underline{cd}1$/Cu	Coyne et al., 1989
A. odorans	Cu	Coyne et al., 1989
A. itersonii	$\underline{cd}1$	Coyne et al., 1989
A. brasilense	$\underline{cd}1$	Lalande & Knowles, 1987
Bacillus sp.	Cu	Coyne et al., 1989
B. azotoformans	Cu	Coyne et al., 1989
B. japonicum	Cu	Coyne et al., 1989
C. nephridii	Cu	Coyne et al., 1989
Erythrobacter sp.	$\underline{cd}1$	Shioi et al., 1988
Flavobacterium sp.	$\underline{cd}1$	Coyne et al., 1989
Halobacterium denitrificans	Cu	L.I. Hochstein (pers. comm)
N. europaea	Cu	Miller & Wood, 1983
P. denitrificans	$\underline{cd}1$	Newton, 1969
P. halodenitrificans	$\underline{cd}1$	Grant & Hochstein, 1984
P. aeruginosa	$\underline{cd}1$	Yamanaka, 1963
P. aureofaciens	Cu	Zumft et al., 1987
P. denitrificans	Cu	Michalski & Nicholas, 1988
P. fluorescens (II & IV)	$\underline{cd}1$	Coyne et al., 1989
P. stutzeri	$\underline{cd}1$	Körner et al., 1987
Rhizobium sp.	Cu	Coyne et al., 1989
R. sphaeroides	Cu	Michalski & Nicholas, 1985
T. denitrificans	$\underline{cd}1$	Sawhney & Nicholas, 1978
Thiosphaera pantotropha	Cu	Robertson, 1988

which were predominantly of the Cu-dNir type. Cu-dNirs in the organisms they tested were very similar in terms of molecular weight and antibody affinity. There were no immunological similarities between $\underline{cd}_1$- and Cu-dNirs. We noted, during the Western immnoscreening survey, that numerous Cu-dNirs showed no affinity for antibodies raised against A. cycloclastes Cu-dNir and that the antibody affinity $\underline{cd}_1$-dNirs exibited, seemed related to the $\underline{cd}_1$-dNir of the strain against which those antibodies were raised. In consequence, we tried to discern whether there was greater diversity in $\underline{cd}_1$-dNirs than was apparent from published reports.

Immunoblot analysis is a quantitative measure of immunological similarity between proteins based on antibody binding (Howe and Hershey, 1981). Its great advantage over microcomplement fixation assays is the ability to use crude cell extracts for the analysis of specific proteins. When equal amounts of a given protein are electrophoretically bound to nitrocellulose filters, and incubated with saturating concentrations of antibodies, the degree of antibody binding (as measured by densitometry) is a quantitative measure of similarity based on common epitopes. Immunoblot distance is a reflection of that similarity. It is the ratio of antibody binding to a type strain vs. antibody binding to a test strain, and is given by the equation (Howe and Hershey, 1984):

$$\text{Immunoblot Distance} = 100 \times \log(\text{Value of Type Strain})/(\text{Value of Test Strain})$$

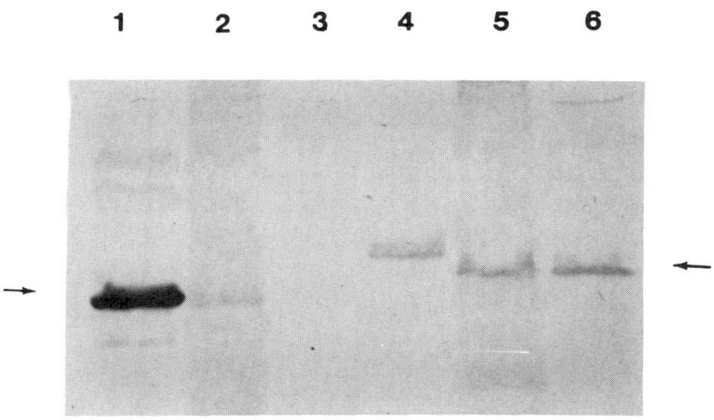

Figure 3. Western immunoblot of denitrifier crude extracts containing Cu-dNirs incubated with A. cycloclastes Cu-dNir antiserum. Lane 1, A. cycloclastes ATCC 21921; 2, A. xylosoxidans NCIB 11015; 3, P. aureofaciens ATCC 13985; 4, A. faecalis #40; 5, A. faecalis #43; 6, A. eutrophus #154. Arrows point to Cu-NO_2^- reductases.

The divergence among Cu- and cd_1-dNirs were estimated by this method. We also wanted to know whether there were differences in cd_1-dNir antibody binding based on which cd_1-dNir the antibodies were raised against, and whether divergence among the different dNir types was at the same rate.

There appeared to be real divergence in both Cu- and cd_1-dNirs quite apart from physical differences in subunit size and structure. Strains with Cu-dNirs were selected to reflect differences in molecular mass and type of Cu atom present (either I or II). Four of the five strains tested were recognized by A. cycloclastes Cu-dNir antiserum (Fig. 3). The exception was P. aureofaciens. The immunoblot distances revealed one cluster of Cu-dNirs which shared equal numbers of common epitopes with A. cycloclastes Cu-dNir (Fig. 4). The Cu-dNir in A. xylosoxidans, like P. aureofaciens, has type I Cu only, so Cu type is not a significant factor in recognition (Michalski and Nicholas 1988). This would be expected if Cu atoms are sequestered in the active site of the enzyme which is not available as an epitope. The Cu type of the two other strains in this cluster, both A. faecalis strains, is not known. They differ in terms of molecular mass (see Figs. 2 and 3).

Alcaligenes eutrophus Cu-dNir shared some common epitopes with A. cycloclastes Cu-dNir, but based on this analysis, fewer than the Cu-dNirs in other denitrifiers. The Cu type of this strain is also not known. The immunological relatedness of Cu-dNirs in the few strains which were examined can be represented as follows: A. cycloclastes >> A. faecalis #43 = A. xylosoxidans > A. faecalis #40>> A. eutrophus >> P. aureofaciens.

We chose strains with cd_1-dNirs to reflect the two types of cd_1-dNir antisera which were available (to P. aeruginosa and P. stutzeri cd_1-dNir), and P. fluorescens, potentially the dominant denitrifier in soils. A different picture of divergence emerged with cd_1-dNirs. We noted a doublet of immunologically recognized proteins in P. fluorescens and isolate #2.4.E. The significance of this is not known, but the doublet is characteristic of P. fluorescens strains examined by Western immunoblots (Figure 5). Immunoblot distances calculated after

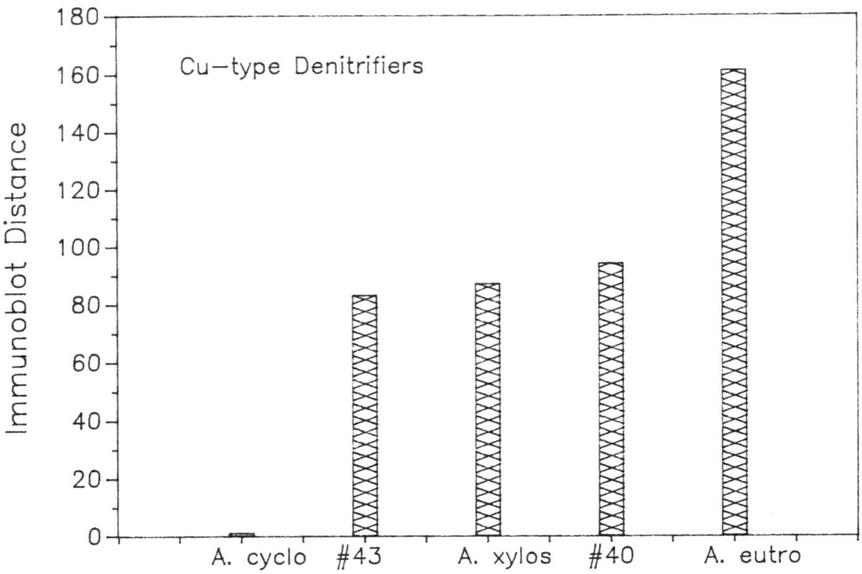

Figure 4. Immunoblot distance of bacteria which crossreact with antiserum to A. cycloclastes Cu-dNir.

binding with P. aeruginosa cd_1-dNir antiserum suggested that dNirs in P. aeruginosa, P. fluorescens, and a fluorescent-like isolate, #2.4.E., shared many common epitopes (Figure 6). In contrast, dNirs in P. stutzeri and another isolate #1.6.L., shared fewer common epitopes with the dNirs in this first group. The same analysis could not be done with antisera to P. stutzeri cd_1-dNir. While the antibodies strongly recognized P. stutzeri cd_1-dNir, there was very little recognition of cd_1-dNirs in the other strains. This suggests that there is a suite of epitopes in P. stutzeri cd_1-dNir which is distinct from the epitopes present in P. aeruginosa cd_1-dNir.

Divergence of NO_2^- reductases

Proteins which differ by more than 40% in amino acid sequence generally lack cross-reactivity in complement fixation tests and immunodiffusion assays (Cocks and Wilson, 1972). The immunoblot distance between cd_1-dNirs was less than the immunoblot distance separating clusters of Cu-dNirs. Protein sequence differences derived from immunoblot distances suggest that the other Cu-dNirs differ from A. cycloclastes Cu-dNir by some 20-30% of the protein sequence (Cocks and Wilson, 1972). The difference between cd_1-dNirs is less than 10%. The extent to which these differences reflect the time for divergence, the conservation of structure, or the diversity of organisms is unclear given this small sample size.

In the same functional class of proteins, sequence evolution goes on at approximately the same rate (Wilson et al., 1977). In a given protein, evolution will depend on the compatibility of sequence change with biochemical function and the liklihood that the organism will survive without that biochemical function. That Cu-dNirs appear to diverge more than cd_1-dNirs may not reflect a longer evolutionary history, only a greater capacity of nonheme Cu-dNirs to accept

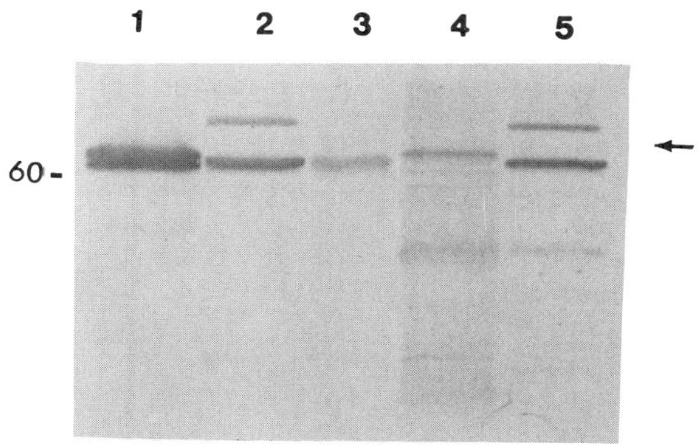

Figure 5. Western immunoblot of denitrifier crude extracts containing cd_1-dNirs incubated with P. aeruginosa cd_1-dNir antiserum. Lane 1, P. aeruginosa ATCC 10145; 2, P. fluorescens ATCC 33512; 3, P. stutzeri JM300; 4, Isolate 1.6.L.; 5, Isolate 2.4.E. Molecular mass indicator is in kDa. Arrow indicates one of the characteristic proteins in P. fluorescens-like cd_1-dNir.

sequence changes without losing function compared to cd_1-dNirs. There are no obligate denitrifiers, and in the presence of O_2, dNir synthesis and activity is strongly inhibited (Tiedje, 1988). Consequently, loss of dNir activity due to sequence substitution is not a lethal change in the aerobic environments to which most denitrifiers have adapted.

EVOLUTIONARY DEVELOPMENT

Two views have emerged regarding the geochemical arrival of denitrification in microbial evolution. One view holds that denitrifying enzymes arose prior to oxygenic respiration (Mancinelli and McKay, 1988). Another, that it arose afterwards (Broda, 1975; Jones, 1985). The key issue is essentially whether primordial environments had respirable nitrogenous oxides.

The place of dNirs is central to this dispute. The post-oxygenic view suggests that since dNirs have oxidase activity, the enzyme arose after O_2 appeared, and as a functional component of an electron transfer system originating in photosynthetic organisms (Broda, 1975; Jones, 1985). The pre-oxygenic view, in turn, argues that dNir oxidase activity does not result in bacterial energy generation, besides having a physiologically high K_m (Mancinelli and McKay, 1988). Rather, dNir could be envisioned as a preliminary version of oxidase which evolved in anaerobic environments. The direct ancestor of denitrifiers may have been some primordial purple nonsulfur photodenitrifier (Betlach, 1982) since, with the exception of Bacillus and Halobacterium, denitrification is confined to bacteria in the purple nonsulfur group (Woese, 1987). In closely related species, denitrification might have been lost with time, as has photosynthesis in some species of closely related photosynthetic and nonphotosynthetic bacteria.

There is no clear answer to why both enzymes evolved. Nor is there a clear answer for the appearance of both Cu- and cd_1-dNir in the same genus or in closely related genera (Ohkubo et al., 1986). One possibility is horizontal gene transfer via a plasmid. Nitrite reductase is located on a plasmid in one

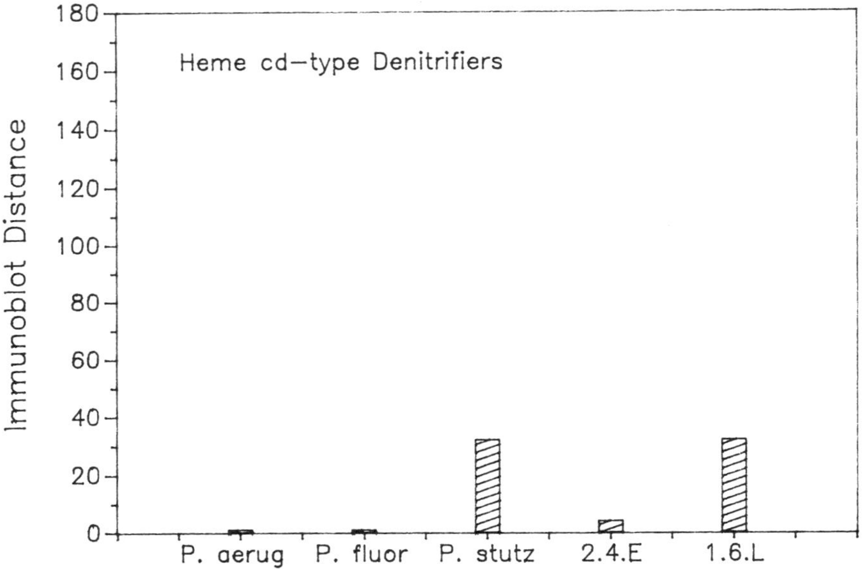

Figure 6. Immunoblot distance of bacteria which crossreact with antiserum to P. aeruginosa cd_1-dNir.

Alcaligenes eutrophus strain (Römermann and Friedrich, 1985). Truncated denitrifiers are common (Jeter and Ingraham, 1981). Pseudomonas spp. which once lost dNir enzyme may have regained it by plasmid acquisition, albeit, the dNir they regained was not the same type of dNir they originally possessed. Perhaps there is still a positive selection for dNir despite the current O_2-rich environment. Certainly, organisms in transiently anaerobic environments could benefit from either the additional energy or the removal of by dNir of toxic NO_2^- levels.

SUMMARY

The latest weapon in the arsenal of experimental techniques to study denitrification are polyclonal antibodies raised to specific denitrification enzymes. Soon, enzyme specific DNA probes will join that arsenal. They are tools that make us just a little bit dangerous in the sense that we can now raise far more questions than our techniques permit us to answer. We can identify the dNir type, using the new methods, in virtually any culturable denitrifier, but are unable to explain why they should possess the dNir-type which they have. Nor can we say why cd_1-dNir predominates in natural systems. Does cd_1-dNir confer a competitive advantage? We do not know. We can detect subtle changes in dNirs between denitrifiers, but as yet do not know if they represent significant evolutionary changes.

From the viewpoint of microbial ecology, the most interesting aspects of dNirs are their distribution, variability (or lack thereof), and the mechanisms which favored development of two unique enzymes to exploit the same environmental niche. We have just begun to examine relationships among denitrifying bacteria at the molecular level. Characterization of the denitrifying enzymes they possess is the first of many steps to understand the cellular mechanisms underlying denitrifier diversity, activity, and ecology.

REFERENCES

Abd-el-Malek, Y., Hosny, I., and Emam, N.F., 1974, Evaluation of media used for enumeration of denitrifying bacteria. Zentralbl. Bakteriol. Parasitenkd. Infektionskr. Hyg. Abt. 2, 129:415.

Barber, D., Parr, S.R., and Greenwood, C., 1976, Some spectral and steady-state kinetic properties of Pseudomonas cytochrome oxidase. Biochem. J., 157:431.

Betlach, M.R., 1982, Evolution of bacterial denitrification and denitrifier diversity. Antonie van Leeuwenhoek. J. Microbiol, 48:585.

Broda, E., 1975, The history of inorganic nitrogen in the biosphere, J. Mol. Evol, 7:87.

Cocks, G.T., and Wilson, A.C., 1972, Enzyme evolution in Enterobacteriaceae, J. Bacteriol., 110:793.

Coyne, M.S., Arunakumari, A., Tiedje, J.M., and Averill, B.A., 1989, Immunological identification and distribution of dissimilatory heme cd1 and nonheme copper nitrite reductases in denitrifying bacteria, Appl. Environ. Microbiol., 55:2924.

De Vos, P., and De Ley, J., 1983, Intra- and intergeneric similarities of Pseudomonas and Xanthomonas ribosomal ribonucleic acid cistrons, Int. J. Syst. Bact., 33:487.

Dispirito, A.A., Taaffe, L.R., Lipscomb, J.D., and Hooper, A.B., 1985, A 'blue' copper oxidase from Nitrosomonas europaea, Biochem. Biophys. Acta., 827:320.

Fox, G.E., et al., 1980, The phylogeny of prokaryotes, Science, 209:457.

Gamble, T.N., Betlach, M.R., and Tiedje, J.M., 1977, Numerically dominant denitrifying bacteria from world soils, Appl. Environ. Microbiol., 33:926.

Grant, M.A., and Hochstein, L.I., 1984, A dissimilatory nitrite reductase in Paracoccus halodenitrificans, Arch. Microbiol., 137:79.

Gudat, J.C., Singh, J., and Wharton, D.C., 1973, Cytochrome oxidase from Pseudomonas aeruginosa. I. Purification and some properties, Biochim. Biophys. Acta., 292:376.

Hochstein, L.I., and Tomlinson, G.A., 1988, The enzymes associated with denitrification, Ann. Rev. Microbiol., 42:231.

Howe, J.G., and Hershey, J.W.B., 1981, A sensitive immunoblotting method for measuring protein synthesis initiation factor levels in lysates of Escherichia coli, J. Biol. Chem., 256:12836.

Howe, J.G., and Hershey, J.W.B., 1984, The rates of evolutionary divergence of initiation factors IF2 and IF3 in various bacterial species determined quantitatively by immunoblotting, Arch. Microbiol., 140:187.

Hulse, C.L., Tiedje, J.M., and Averill, B.A., 1988, A spectrophotometric assay for dissimilatory nitrite reductases, Anal. Biochem., 172:420.

Iwasaki, H., Shidara, S., Suzuki, H., and Mori, T., 1963, Studies on denitrification. VII. Further purification and properties of denitrifying enzyme. J. Biochem., 53:299.

Iwasaki, H., and Matsubara, T., 1971, Cytochrome c-557 (551) and cytochrome cd of Alcaligenes faecalis, J. Biochem., 69:847.

Iwasaki, H., and Matsubara, T., 1972, A nitrite reductase from Achromobacter cycloclastes, J. Biochem., 71:645.

Iwasaki, H., Noji, S., and Shidara, S., 1975, Achromobacter cycloclastes nitrite reductase: The function of copper, amino acid composition, and ESR spectra, J. Biochem., 78:355.

Jeter, R.M., and Ingraham, J.L., 1981, The denitrifying procaryotes, in: "The Procaryotes," M.P. Starr, ed., Springer-Verlag, New York.

Jones, C.W., 1985, The evolution of bacterial respiration, in: "Evolution of Prokaryotes," K.H. Schleifer, and Stackerbrandt, E., eds., Academic Press, London.

Kakutani, T., Watanabe, H., Arima, K., and Beppu, T., 1981, Purification and properties of a copper-containing nitrite reductase from a denitrifying bacterium, Alcaligenes faecalis strain S6, J. Biochem., 89:453.

Kashem, M.A., Dunford, H.B., Liu, M-Y., Payne, W.J., and LeGall, J., 1987, Kinetic studies of the copper nitrite reductase from Achromobacter cycloclastes and its interaction with a blue copper protein, Biochem. Biophys. Res. Comm., 145:563.

Koike, I., and Hattori, A., 1975, Energy yield of denitrification: An estimate from growth yield in continuous cultures of Pseudomonas denitrificans under nitrate-, nitrite, and nitrous oxide-limited conditions, J. Gen. Microbiol., 88:11.

Körner, H., Frunzke, K., Döhler, K., and Zumft, W.G., 1987, Immunochemical patterns of distribution of nitrous oxide reductase and nitrite reductase (cytochrome cd_1) among denitrifying pseudomonads, Arch. Microbiol., 148:20.

LaLande, R., and Knowles, R., 1987, Cytoplasmic content of Azospirillum brasilense sp7 grown under aerobic and denitrifying conditions, Can. J. Microbiol., 33:151.

LeGall, J., Payne, W.J., Morgan, T.V., and DerVartanian, D., 1979, On the purification of nitrite reductase from Thiobacillus denitrificans and its reaction with nitrite under reducing conditions, Biochem. Biophys. Res. Comm., 87:355.

Liu, M-C., Payne, W.J., Peck, H.D., and LeGall, J., 1983, Comparison of cytochromes from anaerobically and aerobically grown cells of Pseudomonas perfectomarinus. J. Bacteriol., 154:278.

Liu, M-Y., Liu, M-C., Payne, W.J., and LeGall, J., 1986, Properties and electron transfer specificity of copper proteins from the denitrifier Achromobacter cycloclastes, J. Bacteriol., 166:604.

Mancinelli, R.L., Cronin, S., and Hochstein, L.I., 1986, The purification and properties of a cd-cytochrome nitrite reductase from Paracoccus halodenitrificans, Arch. Microbiol., 145:202.

Mancinelli, R.L., and McKay, C.P., 1988, The evolution of nitrogen cycling, Origins of Life and Evolution of the Biosphere, 18:311.

Masuko, M., Iwasaki, H., Sakurai, T., Suzuki, S., and Nakahara, A., 1984, Characterization of nitrite reductase from a denitrifier, Alcaligenes sp. NCIB 11015. A novel copper protein, J. Biochem., 96:447.

Michalski, W.P., and Nicholas, D.J.D., 1985, Molecular characterization of copper-containing nitrite reductase from Rhodopseudomonas sphaeroides forma sp. denitrificans, Biochim. Biophys. Acta., 828:130.

Michalski, W.P., and Nicholas, D.J.D., 1988, Immunological patterns of distribution of bacterial denitrifying enzymes, Phytochemistry, 27:2451.

Miller, D.J., and Wood, P.M., 1983, The soluble cytochrome oxidase of Nitrosomonas europaea, J. Gen. Microbiol., 129:1645.

Newton, N., 1969, The two-haem nitrite reductase of Micrococcus denitrificans, Biochim. Biophys. Acta., 185:316.

Ohkubo, S., Iwasaki, H., Hori, H., and Osawa, S., 1986, Evolutionary relationship of denitrifying bacteria as deduced from 5S rRNA sequences, J. Biochem., 100:1261.

Palleroni, N.J., Kunisawa, R., Contopoulou, R., and Doudoroff, M., 1973, Nucleic acid homologies in the genus Pseudomonas, Int. J. Syst. Bact., 23:333.

Radcliffe, B.C., and Nicholas, D.J.D., 1968, Some properties of a nitrite reductase from Pseudomonas denitrificans, Biochim. Biophys. Acta., 153:545.

Ritchie, G.A.F., and Nicholas, D.J.D., 1974, The partial characterization of purified nitrite reductase and hydroxylamine oxidase from Nitrosomonas europaea, Biochem. J., 138:471.

Robertson, L.A., 1988, "Aerobic denitrification and heterotrophic nitrification in Thiosphaera pantotropha and other bacteria," Ph.D. Thesis, University of Delft.

Robertson, L.A., and Kuenen, J.G., 1984, Aerobic denitrification: a controversy revived, Arch. Microbiol., 139:351.

Römermann, D., and Friedrich, B., 1985, Denitrification by Alcaligenes eutrophus is plasmid dependent, J. Bacteriol., 162:852.

Saraste, M., and Kuronen, T., 1978, Interaction of Pseudomonas cytochrome cd_1 with the cytoplasmic membrane, Biochem. Biophys. Acta., 513:117.

Sawada, E., and Satoh, T., 1980, Periplasmic location of dissimilatory nitrate and nitrite reductase in a denitrifying phototrophic bacterium Rhodopseudomonas sphaeroides forma sp. denitrificans, Plant Cell Physiol., 21:205.

Sawhney, V., and Nicholas, D.J.D., 1978, Sulphide-linked nitrite reductase from Thiobacillus denitrificans with cytochrome oxidase activity: Purification and properties, J. Gen. Microbiol., 106:119.

Shapleigh, J.P., and Payne, W.J., 1985, Differentiation of cd_1 cytochrome and copper nitrite reductase production in denitrifiers. FEMS Microbiol. Lett., 26:275.

Shioi, Y., Doi, M., Arata, H., and Takamiya, K., 1988, A denitrifying activity in an aerobic photosynthetic bacterium, Erythrobacter sp. strain OCH.114, Plant Cell Physiol., 29:861.

Tiedje, J.M., 1982, Denitrification, in: "Methods of Soil Analysis: Part 2 - Chemical and Microbiological Properties," A.L. Page, ed., ASA Inc., Madison.

Tiedje, J.M., 1988, Ecology of denitrification and dissimilatory nitrate reduction to ammonia, in: "Biology of Anaerobic Microorganisms," A.J.B. Zehnder, ed., Wiley, New York.

Turley, S., Adman, E.T., Sieker, L.C., Liu, M.-Y., Payne, W.J., and LeGall, J., 1988, Crystallization of nitrite reductase from Achromobacter cycloclastes, J. Mol. Biol, 200:417.

Wilson, A.C., Carlson, S.S., and White, T.J., 1977, Biochemical evolution, Ann. Rev. Biochem., 46:573.

Woese, C.R., 1987, Bacterial evolution, Microbiol. Rev., 51:221.

Wood, P.M., 1978, Periplasmic location of the terminal reductase in nitrite respiration, FEBS Lett., 92:214.

Yamanaka, T., and Okunuki, K., 1963, Crystalline Pseudomonas cytochrome oxidase: I. Enzymic properties with special reference to the biological specificity, Biochim. Biophys. Acta., 67:379.

Zumft, W.G., Sherr, B.F., and Payne, W.J., 1979, A reappraisal of the nitric-oxide-binding protein of denitrifying Pseudomonas, Biochem. Biophys. Res. Comm., 88:1230.

Zumft, W.G., Gotzmann, D.J., and Kroneck, P.M.H., 1987, Type 1, blue copper proteins constitute a respiratory nitrite-reducing system in Pseudomonas aureofaciens, Eur. J. Biochem., 168:301.

METABOLISM OF NITROUS OXIDE

Walter G. Zumft and Peter M. H. Kroneck*

Lehrstuhl für Mikrobiologie der Universität, D-7500 Karlsruhe 1, and *Fakultät für Biologie der Universität D-7750 Konstanz 1, Federal Republic of Germany

INTRODUCTION

Nitrous oxide (N_2O) is a firmly established inorganic metabolite of denitrification. It is either the end product of the process or the obligatory intermediate antecedent to N_2. Ample evidence for its role comes from physiological, biochemical, and genetic work as well as from isotope studies. Involvement of N_2O in denitrification is now documented to such an extent, that the existence of a copper and/or iron-containing metalloflavoprotein (EC 1.7.99.2), described to form N_2 from NO (Chung and Najjar, 1956; Fewson and Nicholas, 1961), must be questioned. N_2O is also a product of denitrification by nitrifiers and of NO_3^- and NO_2^- metabolism of non-denitrifying microorganisms.

Until very recently, investigation of the enzymology of the denitrification pathway had proceeded only to the level of NO_2^-, with the cytochrome $\underline{cd}_1$-type NO_2^- reductase representing the overall best studied enzyme. The enzymes involved in the formation and reduction of N_2O had not been investigated because N_2O-forming activity in most organisms is asociated with the cytoplasmic membrane. This results for isolation and purification in the usual problems with membrane proteins. N_2O reductase, on the other hand, is a labile enzyme and often impossible to demonstrate in cell extracts. The first preparations of purified nitric oxide (NO) reductase (EC 1.7.99._) (Heiss et al., 1989) and N_2O reductase (EC 1.7.99.6) (Coyle et al., 1985; Riester et al., 1989; Zumft and Matsubara, 1982) were obtained from a marine strain of Pseudomonas stutzeri ZoBell (formerly P. perfectomarina; a representative of the ecologically important group of denitrifying pseudomonads). Here we shall discuss the biochemistry and genetics of formation and reduction of N_2O by denitrifying bacteria. Recent reviews of the physiology and biochemistry of the denitrification process are by Hochstein and Tomlinson (1988), Stouthamer (1988a), and Zumft et al. (1988b); bioenergetic aspects are reviewed by Ferguson (1988), and Stouthamer (1988b); and geochemical aspects of N_2O are dealt with by Kaplan and Wofsy (1985).

FORMATION OF NITROUS OXIDE BY NITRIC OXIDE REDUCTASE

N_2O may be formed in denitrifying bacteria by more than one metabolic route. The possibility of N_2O formation by NO_2^- reductase is being discussed

Denitrification in Soil and Sediment, Edited by N. P. Revsbech and J. Sørensen, Plenum Press, New York, 1990

elsewhere in this book. Here we consider the formation of N_2O from NO under the catalysis of NO reductase. Nitric oxide is readily utilized by intact cells or cell extracts of denitrifying bacteria (Baalsrud and Baalsrud, 1954; Cox and Payne, 1973; Firestone et al., 1979; Frunzke and Zumft, 1986; Grant et al., 1984; Matsubara, 1971; Miyata et al., 1969; Payne et al., 1982; Pichinoty et al., 1979; Zumft and Frunzke, 1982). The product depends on the organism and the reaction conditions, and is either N_2O or N_2. Nevertheless, denitrifying bacteria usually cannot be grown with NO as the electron acceptor. There is a single, unconfirmed report of growth of Bacillus sp. on NO (Pichinoty et al., 1979). Bacterial growth exhibited a lag phase of 40 to 60 h and apparently no adaptation of cells took place, since the same lag phase occurred on transfer of the bacteria to fresh medium. NO in higher concentration might be toxic even to denitrifying bacteria, and some of them may have developed protective mechanisms and exhibit tolerance to NO. Cytochrome c' of Alcaligenes xylosoxidans subsp. xylosoxidans (formerly called "Pseudomonas denitrificans" or Alcaligenes sp. NCIB 11015) forms an NO-complex which may function to scavenge free NO from NO_2^- reduction (Yoshimura et al., 1988). Whether NO reductase could have such a function in addition to a role in energy conservation, cannot yet be decided from the available data.

Mixed bacterial populations evolve NO only under anaerobic conditions (Wijler and Delwiche, 1954). Cells grown under limited oxygen supply express little NO reductase activity, leading to NO accumulation from NO_2^- (Frunzke and Zumft, 1986; Matsubara, 1971; Matsubara and Zumft, 1982). NO-dependent proline uptake by Paracoccus denitrificans is only observed with anaerobically grown cells (Garber et al., 1982), as is NO-induced proton translocation by this bacterium and by Rhodobacter sphaeroides forma sp. denitrificans, "Achromobacter cycloclastes", and Bradyrhizobium japonicum (Shapleigh and Payne, 1985). At the same time NO inhibits NO_2^- reductase (Dhesi and Timkovich, 1984) and N_2O reductase (Frunzke and Zumft, 1986; Matsubara et al., 1982).

Most studies have found NO reductase activity in a particulate fraction or bound to the membrane. This has been shown for Alcaligenes faecalis (Matsubara and Iwasaki, 1971), P. stutzeri (Heiss et al., 1989; Zumft and Frunzke, 1982; Zumft and Vega, 1979), Paracoccus halodenitrificans (Grant et al., 1984), and Pa. denitrificans (Carr et al., 1989). Soluble activities were reported for P. stutzeri (Chung and Najjar, 1956) and Pseudomonas aeruginosa (Fewson and Nicholas, 1961) but lack confirmation. A 'soluble' activity from P. stutzeri (Cox et al., 1971) was subsequently shown to be sensitive to Triton X-100. This effect was taken to indicate the membrane-bound nature of the enzyme and was found also in "Achromobacter cycloclastes", P. aeruginosa, and Pa. denitrificans (Shapleigh et al., 1987). Contrasting with the data for Pa. denitrificans by Ferguson and coworkers (Carr and Ferguson, 1988; Carr et al., 1989), which clearly refer to a membrane-bound enzyme, 3/4 of the NO-reducing activity of the same bacterium was reported by Hoglen and Hollocher (1986) to be soluble and associated with a lipoprotein of M_r 38 000 subunit mass. A soluble, minor NO-reducing activity (less than 20% of the cell-free activity) of A. faecalis was attributed to cytochrome cd_1. This protein converted NO to N_2O with a specific activity of about 0.33 μmol per min per mg protein, which is about 80-fold less than its NO_2^--reducing activity (Matsubara and Iwasaki, 1972). Further evidence for the distinct nature of NO_2^--reductase and NO reductase comes from work with mutants. Cytochrome cd_1-less mutants continue to reduce NO both in vivo and in vitro with activities comparable to the wild type (Zumft et al., 1988a).

The first highly purified NO reductase was obtained from P. stutzeri ZoBell

Table 1. Properties of nitric oxide reductase from Pseudomonas stutzeri[a]

Property	Numerical value or observation
Cellular location	Cytoplasmic membrane
Mol. mass (kDa)	180 (Triton X-100 micelle by gelfiltration)
Subunits (kDa)	38 and 17 (SDS polyacrylamide-gel electrophoresis)
Metal atoms/M_r	6 Fe/55 000
S^{2-}	Not detected
Prosthetic groups	Heme c and b plus nonheme Fe
UV/VIS maxima (nm)	
Enzyme as isolated	280, 411.5, 537
Reduced form	420.5, 522.5, 552.5, 560(s)
Spec. act.[b]	40 (in the presence of phospholipids)
pH optimum	4.8
Electron donor	Phenazine methosulfate (PMS)/ascorbate
K_m (NO) μM	60
K_m (PMS) μM	2

[a] Data from Heiss et al. (1989).
[b] μmol NO reduced per min per mg protein.

(Heiss et al., 1989). It was shown to be a membrane-bound cytochrome bc complex (Fig. 1), with subunits of approximately 38 kDa and 17 kDa associated with heme b and heme c, respectively. The basic properties of this NO reductase are shown in Tab. 1. In addition to heme iron, the enzyme appears to have more than an equivalent amount of nonheme iron which may be crucial to enzymatic catalysis. Ferrous heme-nitrosyl complexes might be kinetically incompetent and require a redox system of a too low potential (Weeg-Aerssens et al., 1988). The presence of a second iron center could possibly solve this difficulty. As is frequently observed with membrane-bound enzymes, the activity of NO reductase is markedly stimulated by phospholipids (Heiss et al., 1989). Of a number of compounds tested, L-α-phosphatidylglycerol increased NO reductase about threefold, whereas other phospholipids stimulated as crude preparations only, but not as pure compounds. The effect of phospholipids could be mimicked by certain detergents such as octylglucopyranosides or dodecylmaltoside, but not by Triton X-100 or CHAPS.

Heme groups are associated with NO-reducing preparations from Pa. halodenitrificans (Grant et al., 1984) and R. sphaeroides (Urata and Satoh, 1985). However, activities of these preparations were low and purification of an NO reductase from these two organisms has not yet been reported. Nitric oxide reduction is sensitive to the inhibitors of the cytochrome bc_1 complex, antimycin A and myxothiazol (Carr et al., 1989; Itoh et al., 1989; Shapleigh and Payne, 1985; Urata and Satoh, 1985). The cytochrome bc_1 complex co-purifies with NO reductase activity but was recently separated from chromatophores by a modified detergent procedure with dodecylmaltoside (Itoh et al., 1989). The nature of the enzyme is still unresolved. The work both with Rhodobacter sp. and Paracoccus sp., however, demonstrates that NO reductase must be closely associated with the cytochrome bc_1 complex. It was also shown that cytochrome c_2 acts as the electron donor for NO reduction in Rhodobacter sp. (Itoh et al., 1989).

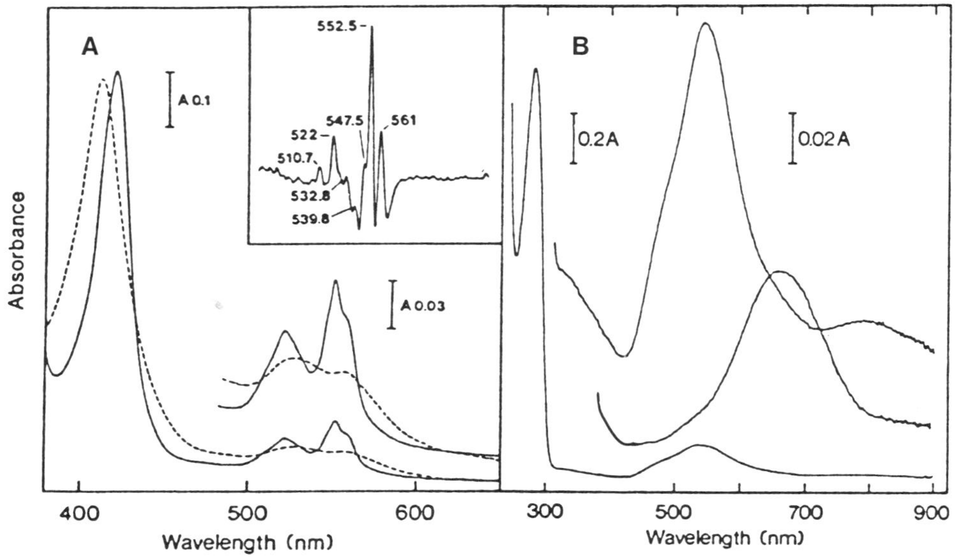

Fig. 1. Electronic absorption spectra of NO reductase (A) and N_2O reductase (B) from P. stutzeri ZoBell. (A) The dotted line shows the protein as isolated; the solid line shows the protein after addition of solid dithionite; solvent is 20 mM Tris-HCl (pH 8.5), 0.08% Triton X-100, 0.08% n-octyl-β-D-thioglucopyranoside. Insert, second derivative spectrum of the dithionite-reduced minus ferricyanide-oxidized form. (B) Oxidized (form I) and reduced, blue species (form III) of N_2O reductase, as isolated under anaerobic conditions. Data from Heiss et al. (1989), and Zumft et al. (1985a) with permission.

Inhibitor and binding studies with the purified NO reductase from P. stutzeri indicated that, although a cytochrome bc complex it must be different from cytochrome bc_1. Fractionation and purification of the cytochrome co-type oxidase of P. stutzeri also differentiated this enzyme from NO reductase (Heiss et al., 1989). Pa. halodenitrificans has a membrane-bound cytochrome cd_1 as NO_2^- reductase (Mancinelli et al., 1986). Treatment of membranes from this organism with CHAPSO and fractionation by ammonium sulfate separates cytochrome cd_1 from NO reductase (Grant et al., 1984) and supports the distinct nature of these two enzymatic activities. Finally, NO is non-enzymatically reduced by iron(II)-ascorbate or $NADH_2$-$FMNH_2$ (Zumft and Frunzke, 1982). Such non-enzymatic N_2O formation may have been involved in earlier work. The reactivity and easiness of transformation of NO to N_2O requires a constant awareness of non-enzymatic processes in all experimental work employing NO.

The reduction of NO by whole cells is associated with proton translocation and hence the generation of an electrochemical gradient. Nitric oxide reduction has recently been shown to be coupled to ATP synthesis in a cell-free system of Pa. denitrificans (Carr et al., 1989).

Denitrification is not the only source for N_2O. Several genera of the enterobacteria, which apparently lack denitrification, form N_2O. The property is also found in Azotobacter vinelandii, a few Gram-positive bacteria, cyanobacteria, green algae, yeasts and fungi (for review see Tiedje, 1988). It is likely that the mechanisms of N_2O formation among these different groups are diverse. Most of the N_2O producers have either respiratory or assimilatory NO_3^--metabolizing capabilities and often N_2O production is stronger from NO_2^- than from NO_3^-. N_2O evolution by the nitrifying bacteria appears to be part of denitrification *sensu stricto* and is covered elsewhere in this book.

REDUCTION OF NITROUS OXIDE TO DINITROGEN

Many denitrifying bacteria utilize N_2O and sustain an independent respiratory process with this gas as the electron acceptor instead of O_2. Relevant observations were made long ago by Beijerinck and Minkman (1910). Representative examples of denitrifying bacteria that grow at the expense of N_2O are *Alcaligenes* sp. (Iwasaki et al., 1980; Matsubara, 1971), *Agrobacterium* sp. (Pichinoty et al., 1977a); *Azospirillum* sp. (Tibelius and Knowles, 1984), *Bacillus* sp. (Pichinoty et al., 1979), *Cytophaga* sp. (Adkins and Knowles, 1986), *Flavobacterium* sp. (Pichinoty et al., 1976), *Pseudomonas* sp. (Bonin et al., 1987; Garcia et al., 1977; Koike and Hatori, 1975; Matsubara, 1971; Matsubara et al., 1982; Stanier et al., 1966), *Thiobacillus versutus* (Wood and Kelly, 1983), and *Thiosphaera pantotropha* (Robertson and Kuenen, 1983). Notable exceptions are *P. aeruginosa* (Bazylinski et al., 1986b; Carlson and Ingraham, 1983) and *Aquaspirillum magnetotacticum* (Bazylinski and Blakemore, 1983), which cannot grow, or grow only to a limited extent with N_2O. This effect may be traced to two phenomena. For one, N_2O reductase is derepressed under N_2O to a lesser degree than in the presence of nitrate, as shown by immunochemical quantitation of the enzyme (Körner and Zumft, 1989; Zumft et al., 1985b; see also below). On the other hand, exogenous N_2O causes, by unknown means, a loss of enzyme activity in *P. aeruginosa* (Snyder et al., 1987).

Even though these effects may be operative to a variable extent in ecologically important denitrifiers, anaerobic enrichment under N_2O has revealed a number of denitrifying bacilli (Pichinoty et al., 1979), a *Flavobacterium* sp. (Pichinoty et al., 1976), and denitrifying pseudomonads (Pichinoty et al., 1977b). This technique has certainly not been fully exploited. Particularly, its application could lead to the isolation of more strains that are N_2O utilizers, but not denitrifiers. Existing examples of such organisms are limited, although early Beijerinck and Minkman (1910) found this property in a *Spirillum* sp. isolate. Presently, *Wolinella succinogenes* (Yoshinari, 1980) and *R. capsulatus* (McEwan et al., 1985) belong in this category, as well as a vibroid-shaped magnetotactic bacterium (Bazylinski et al., 1988).

Several denitrifying bacteria terminate anaerobic respiration at N_2O. Examples are *Aquaspirillum itersonii* (Krieg, 1976), various denitrifying rhizobia (Daniel et al., 1987; Poth and Focht, 1985), nitrifying bacteria (Freitag et al., 1987; Poth and Focht, 1985) with the exception of a *Nitrosomonas* sp. which forms N_2 (Poth, 1986), *Chromobacterium violaceum* (Bazylinski et al., 1986a), *Corynebacterium nephridii* (Hart et al., 1965), and *P. chlororaphis* and *P. aureofaciens* among the pseudomonads (Greenberg and Becker, 1977). In the latter case, immunochemical evidence and DNA-DNA hybridization support the view that the genetic information for N_2O reductase is missing in these bacteria (Körner et al., 1987; S. Lochelt and W. G. Zumft, unpublished data), which likely holds also for other such cases.

Expression of N_2O reductase in *P. stutzeri* is regulated in response to a nitrogenous substrate and oxygen (Fig. 2) with 3.8 mg dissolved O_2 per liter being the critical threshold level for increased synthesis (Körner and Zumft, 1989). However, low levels of the enzyme are constitutively expressed in

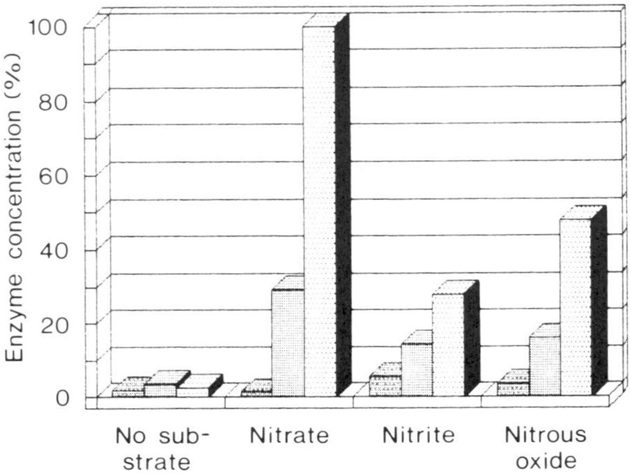

Fig. 2. Expression of N_2O reductase in response to oxygen and the presence of nitrogenous oxides in a magnesium-limited chemostat culture of P. stutzeri. The enzyme concentration was determined by quantitative immunoelectrophoresis. $NaNO_3$ and $NaNO_2$ were supplied each at 0.5 g/liter; N_2O was supplied by sparging. Hatched, stippled, or dotted columns (from left to right in each block) denote values of air saturation of 100, 17, and 0%, respectively (data from Körner and Zumft (1989) with permission).

P. stutzeri, i.e., under full aerobiosis and independent of a nitrogenous respiratory substrate. Anaerobic conditions together with nitrate cause the highest expression of N_2O reductase. Interestingly, the inducing effect of nitrite is weakest of all the nitrogenous oxides, indicating that signals controlling expression of N_2O reductase are preferentially acting at the level of nitrate. The lower amount of N_2O reductase formed under N_2O has already been mentioned. Oxygen immediately inhibits N_2O respiration, but at a partial pressure of O_2 below 10 mbar, N_2O and O_2 are utilized simultaneously as electron acceptors (Fig. 3).

The indirect approach for identifying N_2O reductase, after repeated failures in various laboratories, was by first determining the trace metal requirement for N_2O respiration in metal deficient cultures (Iwasaki et al., 1980; Matsubara et al., 1982). Knowledge of the Cu requirement for N_2O respiration led to the isolation of a high M_r copper protein (Matsubara and Zumft, 1982) which had a low, although significant N_2O-reducing activity (Zumft and Matsubara, 1982). The only other metalloenzyme acting as N_2O reductase is nitrogenase (Jensen and Burris, 1986), which cannot be blocked by C_2H_2 and which, incidentally, also acts as an NO_2^- reductase (Vaughn and Burgess, 1989). The correct identification of N_2O reductase and its Cu protein nature was initially questioned (Hollocher, 1982; Snyder and Hollocher, 1984), a view that, in the light of compelling evidence from P. stutzeri (Coyle et al., 1985; Zumft et al., 1985b), could not be substantiated (Snyder and Hollocher, 1987). An apparent separation of N_2O-reducing activity from the Cu protein (Snyder and Hollocher, 1984) was shown to occur only with crude extracts under aerobic conditions (Zumft et al., 1985a).

The principal properties of N_2O reductases are listed in Tab. 2. All enzymes known so far are multi-copper proteins; however, it is evident from Table 2 that N_2O reductases from different organisms do not follow a uniform molecular pattern. The enzymes from P. stutzeri and Pa. denitrificans are dimeric

Table 2. Molecular properties of nitrous oxide reductase from denitrifying and nitrous oxide-utilizing bacteria

	Pseudomonas stutzeri	*Alcaligenes xylosoxidans*	*Rhodobacter sphaeroides*	*Paracoccus denitrificans*	*Rhodobacter capsulatus*	*Wolinella succinogenes*
Cellular location	periplasm	periplasm	periplasm	? (soluble)	periplasm	? (soluble)
Mol. mass (kDa)	140	120	95	143	-	162
M_r of subunits	74 (α_2)	-	75 (α)	70 (α_2)	76	88 (α_2)
UV/VIS absorbance						
Form as isolated (nm)	350, 480s 540, 780	360s, 480s 545, 650	481, 534 635, 743	350, 480s 550, 620, 780	482, 542 635, 762	410, 528
With dithionite	650	650, 750	635, 743	665	642	416, 520
Prosthetic groups	8 Cu	5 Cu	4 Cu	8 Cu	-	6 Cu, 2 heme *c*
EPR (as isolated)						
g_{II}	2.18	2.24	2.27	2.22	-	-
g_I	2.03	2.07	2.03	2.04	-	-
A_{II} (mT)	3.83 (7)[a]	16 (3?)	12.7 (4)	3.5 (4)	-	-
K_m (N_2O) μM	-	-	-	6.7	-	7.5
K_m (BV) μM	-	-	-	0.9	-	4
References	b	c	d	e	f	g

[a]no. of lines resolved at 9.32 GHz.

References: b, Coyle et al., 1985; Riester et al., 1989; c, Matsubara and Sano, 1985; d, Michalski et al., 1986; e, Snyder and Hollocher, 1987; f, McEwan et al., 1985; g, Teraguchi and Hollocher, 1989.

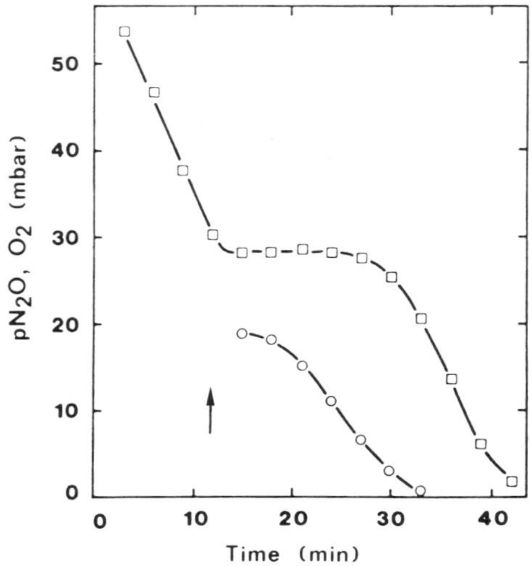

Fig. 3. Inhibition of N_2O respiration by O_2 in P. stutzeri and co-respiration of both gases at low partial pressure of O_2. N_2O-grown cells, equivalent to 8.1 mg protein, were assayed by gas chromatography for N_2O respiration (□). After 12 min 900 μl air was injected into the test vial (arrow) and the reduction of O_2 also followed (○) (K. Frunzke and W. G. Zumft, unpublished).

proteins with about 8 Cu atoms per dimer. The M_r of N_2O reductase from A. xylosoxidans is close to that of P. stutzeri, but its Cu content is only about half. It is not known whether this enzyme is monomeric or dimeric. The monomeric protein of Rhodobacter sp. has approximately half the mass and Cu content of the P. stutzeri enzyme. Despite these molecular differences, one observes very similar electronic absorption properties and electron paramagnetic resonance (EPR) parameters for these proteins, indicating prosthetic Cu in similar, though sometimes slightly perturbed environments.

The nature of Cu-bonding has been investigated in detail with the N_2O reductase from P. stutzeri (see below, and Chapter 1 by Kroneck and Zumft in this book). The protein occurs in different forms discernible by their electronic absorption (Fig. 1B), EPR properties, Cu content, and activity profiles (Coyle et al, 1985; Riester et al., 1989; Zumft et al., 1985a; Zumft and Matsubara, 1982). The purple form, N_2O reductase I, probably comes closest to the protein with an intact chromophore. The Cu content of this form is the highest found. Form I exhibits a remarkable seven-line hyperfine splitting in its EPR spectrum. Most enzymes from other sources are obtained as form II, the pink form, and show a lower resolution in their EPR spectra, a lower absorbance around 540 nm and an additional chromophore at 640 nm. A blue enzyme, form III, which is catalytically inactive, is obtained on addition of reductants. This form shows some features of a type I Cu, but is not reducible by a large variety of reductants (Riester et al., 1989). An inactive enzyme form V, obtained from a mutant with defective chromophore synthesis, does not yield this blue chromophore on reduction, but rather produces a colorless form.

W. succinogenes is a non-denitrifying bacterium that nevertheless respires N_2O. Its N_2O reductase has been purified and shown to contain two heme c prosthetic groups in addition to six Cu atoms per protein dimer (Table 2). EPR data to assess the nature of the Cu chromophore are not yet available (Teraguchi and Hollocher, 1989). Other than the heme-free N_2O reductase, this enzyme is only sluggishly inhibited by C_2H_2. Also, N_2O reduction of Cytophaga johnsonae is not inhibited by C_2H_2 (Adkins and Knowles, 1986). The structural data of the metal centers are presently not detailed enough to allow an explanation for this differential behavior.

Genetic approaches have complemented the biochemical work on N_2O reductase. Thus, generation of transposon Tn5-induced mutants of P. stutzeri that were deficient in the high M_r copper protein, early proved that it is essential for N_2O respiration (Zumft et al., 1985b). At the same time, insertion mutants opened the avenue to recombinant DNA and cloning techniques which recently led to the identification of a gene cluster involved in N_2O respiration (Viebrock and Zumft, 1987) and elucidation of the primary structure of the enzyme (Viebrock and Zumft, 1988).

The dominant denitrifying bacteria in soils belong to biovar B of Pseudomonas fluorescens (Gamble et al., 1977). Prelimary data with this organism indicate that it has a cytochrome cd_1-type NO_2^- reductase (W. G. Zumft, unpublished) and an N_2O reductase similar to that of P. stutzeri. Partial purification yielded a Cu enzyme with the characteristic absorbance properties and high M_r (Matsubara and Zumft, 1982). The N_2O reductase of P. fluorescens also cross-reacts immunochemically with an antiserum to the P. stutzeri enzyme (Körner et al., 1987). The structural gene for N_2O reductase of P. stutzeri ZoBell, nosZ, hybridizes with genomic DNA of the type strain of P. stutzeri and with strain JM300, as well as with P. aeruginosa, P. fluorescens DSM 50415, P. mandocina, Alcaligenes eutrophus, and Pa. denitrificans, (S. Lochelt and W. G. Zumft, unpublished). From these data, one can conclude that enzymes similar to that characterized from P. stutzeri have a wide distribution among denitrifying bacteria. When it comes to developing suitable hene probes for the screening of denitrification enzymes it will be of importance that enzymes share conserved common domains.

N_2O reduction is plasmid-borne in A. eutrophus H16, representing the only example so far of denitrification enzymes being located extrachromosomally (Römermann and Friedrich, 1985). The nosZ gene of P. stutzeri hybridizes with the pGE136 subclone of the pHG1 megaplasmid of A. eutrophus and partial sequencing shows similarities between the two N_2O reductases (S. Löchelt, B. Schneider, B. Friedrich, and W. G. Zumft, unpublished). No evidence for an extrachromosomal location of the genes for denitrification is available with P. stutzeri. Attempts to identify plasmids by standard methods and by those designed to preserve megaplasmids have been negative.

THE COPPER SITES OF NITROUS OXIDE REDUCTASE

From the very beginning of the spectroscopic characterization of N_2O reductase (Coyle et al., 1985), it was evident that the features of this bacterial multi-copper enzyme were not readily described within the classical framework of three types of Cu proteins (Malkin and Malström, 1970). There is now compelling evidence that part of the EPR-active copper site of N_2O reductase has properties of the Cu_A of cytochrome c oxidase, a structure that has been perceived to be unique among the Cu proteins. The Cu_A site has the metal ligated by two cysteine and two histidine residues (Li et al., 1987). The primary structure of N_2O reductase reveals at its carboxy terminus a highly conserved consensus sequence with that of the Cu_A-binding subunit II of cytochrome c oxidase, containing three of the four Cu ligands (Scott et al., 1989; Viebrock and Zumft, 1988). A multiplicity of spectroscopic data can be interpreted best by assuming a Cu_A site in N_2O reductase.

The electron spin echo (ESE) envelope modulation frequencies and the modulation depth of N_2O reductase are strikingly similar to that of cytochrome c oxidase (Jin et al., 1989). All frequencies of cytochrome c oxidase are also observed in the spectrum of N_2O reductase. The ESE data specifically suggest for N_2O reductase a highly covalent Cu(II) site with imidazole ligation. The magnetic circular dichroism spectrum of N_2O reductase is again different from other Cu proteins, but highly similar to cytochrome c oxidase, reflecting similar electronic structures of the respective Cu sites (Scott et al., 1989). Resonance Raman spectroscopy has led to the proposal of a $Cu^{II}S_2(cys)_2N(his)$ site for N_2O reductase (Dooley et al., 1987). Recent extended X-ray absorption fine structure (EXAFS) data of N_2O reductase are fully consistent with such a model (Scott et al., 1989). Cytochrome c oxidase and N_2O reductase show a rather small edge shift upon reduction, which can be taken as indirect evidence for sulfur ligation. Although the interpretation of the EXAFS data is problematic because of multiple Cu sites in N_2O reductase, acceptable fits require a Cu-(N,O) interaction at about 2.0 Å, and a Cu-(S,Cl) interaction at 2.27 Å for the oxidized enzyme compared to the average Cu-(S,Cl) distance of 2.28 Å for cytochrome c oxidase (Li et al., 1987). A long Cu-(S,Cl) interaction, of unknown origin, of about 2.6 Å in the oxidized N_2O reductase and about 2.7 Å in the reduced protein has also been found in cytochrome c oxidase (Scott et al., unpublished results cited in Scott et al., 1989). The present state of analysis certainly does not preclude the presence of significant Cu-Cu interaction in N_2O reductase. Preliminary EXAFS data showing Cu imidazole interactions have been reported for the A. xylosoxidans Cu protein (Matsubara and Sano, 1985).

The seven-line hyperfine splitting of form I of N_2O reductase is very sensitive to ligands, the pH, and oxygen (Jin et al., 1989; Riester et al., 1989). It is temperature dependent and not observable above 100 K. Similar hyperfine splittings have not so far been demonstrated in enzyme samples from other sources (Matsubara and Sano, 1985; Michalski et al., 1986; Snyder and Hollocher, 1987). Our current explanation for the seven-line EPR spectrum is the presence of a mixed valence complex, [Cu(1.5)..Cu(1.5)] (Jin et al., 1989; Kroneck et al., 1988; Riester et al., 1989). Most of the Cu in N_2O reductase is EPR-silent and presumably corresponds to usual type III Cu (Coyle et al., 1985; Jin et al., 1989; Riester et al., 1989; Snyder and Hollocher, 1987). The blue form of the enzyme, which is not reducible by strong reductants, may be a Cu(II) center that is stabilized by a disulfide bridge.

The Cu-Cu interaction seen in the seven line spectrum may be easily perturbed during the isolation procedure. If such an interaction occurs between the two subunits (e.g., of two mononuclear Cu sites) loss of this interaction is easy to conceive. Since all N_2O reductase preparations are being assayed with artificial electron donors of very low redox potential, perturbation of the prosthetic group need not necessarily be reflected in a decrease of enzymatic activity. The hyperfine splitting in N_2O reductase is resolved best at low frequency (S and L-band spectra). Again spectra of cytochrome c oxidase from beef heart show an impressive degree of similarity with N_2O reductase in their low-frequency EPR spectra (Kroneck et al., 1989).

COPPER INCORPORATION AND PROCESSING

Mutational analysis of strain JM300 and the ZoBell strain of P. stutzeri have provided first insights into nos-related functions, as well as identification of the structural gene for N_2O reductase. The frameshift mutagen ICR-191 led either to a loss of the enzyme (and in a few instances to the synthesis of a new protein (NosE) of unknown function) or to the loss of an outer membrane protein (NosA) (Mokhele et al., 1987). NosA has been obtained in homogeneous form; it is a channel-forming protein when inserted into black lipid bilayers (Lee et al., 1989). Although the channel lacks specificity for Cu, NosA binds Cu and is repressed by the presence of this metal in the medium and by aerobic conditions. At 3 μM the level of NosA decreased already substantially, and no protein could be detected at 11 μM Cu; other transition metals (Mn, Co, Ni, Zn, and Fe) had no effect on NosA expression. NosA defective mutants cause the accumulation of catalytically inactive apoenzyme, lacking Cu. These are strong indications that NosA is involved in Cu-processing for N_2O reductase; its precise role is presently undetermined (however, see below). Immunochemical screening for NosA in several other denitrifying bacteria, including different strains of P. stutzeri, has been negative, perhaps because of a high specificity of the antiserum used for screening, thus making P. stutzeri JM300 the only example where such a component has been identified, so far.

Random insertional mutagenesis by transposon Tn5 was successfully employted with P. stutzeri (Zumft et al., 1985b). Tn5 insertion resulted in apoenzyme-synthesizing strains as one class out of three different phenotypes. The kanamycin resistance of Tn5 can be used as selection marker for cloning and thus with this type of mutagenesis the mutation can be physically mapped. Nineteen mutants were analyzed and found to have Tn5 insertions clustered within a single DNA fragment of about 8 kilobase pairs (kbp) (Viebrock and Zumft, 1987). The biochemical phenotypes of these mutants indicated that this region harbored the structural gene for N_2O reductase, a putative regulatory function, and an extended region of approx. 3.5 kbp, whose mutagenesis led to apoenzyme-producing strains.

The entire nos-coding region identified from random Tn5 mutagenesis of P. stutzeri was cloned by plasmid integration and excision (Viebrock and Zumft, 1988). The genetic organization of the nos region was largely deduced from sequencing (Fig. 4). Upstream from the structural gene nosZ is an open reading frame, ORF6, that codes for a protein of minimal M_r 64 000 (the translational start site has not yet been determined). Mutagenesis of this gene leads to strains lacking NosZ, or synthesizing only traces of it. At the present stage of analysis, we presume that this gene is of a regulatory nature, although other hypotheses are also being pursued. Since mutational loss of NosA leads to apoenzyme-synthesizing strains, we do not believe that this gene represents nosA. A presumed single transcriptional unit is found downstream of nosZ, consisting of three different genes, designated nosDFY, and the start of an open reading frame, ORF4 (Fig. 4). Tn5 insertions within the genes nosDFY all yield an identical phenotype of apoenzyme-synthesizing strains. MiniMu-lac fusions of this region show that these genes, like nosZ, are regulated by anaerobiosis (H. Cuypers and W. G. Zumft, unpublished)

A key to the possible function of these genes comes from a sequence analysis of the nosF gene. Its gene product, NosF, shows a clear similarity to the superfamily of ATP/GTP-binding proteins. These proteins form part of bacterial transport systems for small molecules (Higgins et al., 1986) and a limited number

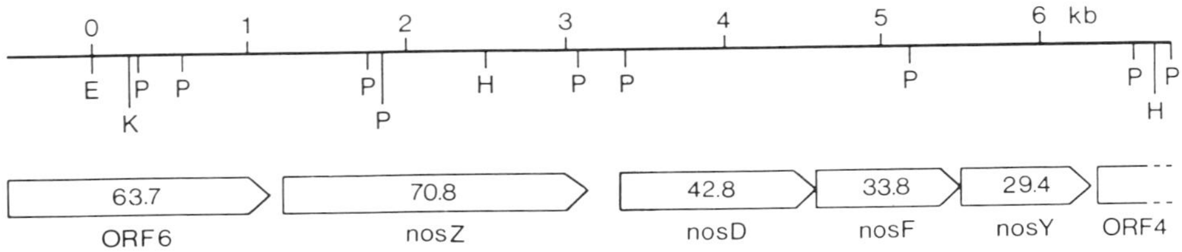

Fig. 4. Molecular organization of the nos-coding region of P. stutzeri. The top bar shows a partial restriction map covering the nos region previously identified by transposon mutagenesis; E, EcoRI; H, HindIII; K, KpnI; and P, PstI (Viebrock and Zumft, 1987). Positions of the structural gene, nosZ, and chromophore genes, nosDFY, and the flanking open reading frames (ORF), together with the masses of identified (NosZ) and presumed gene products (NosDFY and ORF6) are given in the lower half (A. Viebrock, C. Braun, and W. G. Zumft, unpublished).

of functions apparently unrelated to transport in prokaryotic and eukaryotic cells (Higgins et al., 1988). In bacterial transport systems the nucleotide-binding component is associated with an integral membrane component, and this in turn with a periplasmic binding-protein (for review see Ames, 1986). The nosY gene product is predicted from hydropathy analysis to be an extremely hydrophobic protein possibly with six transmembrane helices. The product of the nosD gene could represent the periplasmic component, if we carry the analogy to bacterial transport systems further. In particular, the organization of the system for vitamin B_{12} transport (Friedrich et al., 1986) or the ferrihydroxamate uptake system (Burkhardt and Braun, 1987) have similar organizational features to the Cu-processing part of the nos region. The outer membrane receptor of these transport systems could have its equivalent in the product of the nosA gene. However, despite the striking similarity of the nos chromophore region to bacterial transport systems for small molecules, the pathway of Cu transport as well as the mechanism of Cu insertion into N_2O reductase still remain open. Both will provide challenging subjects for further investigation.

N_2O reductase is a periplasmic enzyme. The protein is most likely synthesized with an export signal peptide of 35 residues (Viebrock and Zumft, 1988). Direct fractionation of the periplasm of R. capsulatus (McEwan et al., 1985), R. sphaeroides (Urata et al., 1982) and P. stutzeri (Minagawa and Zumft, 1988) and indirect evidence from proton stoichiometries (Alefounder et al., 1983; Boogerd et al., 1981) are in agreement and support a periplasmic location. Copper-deficient cells are unable to sustain N_2O respiration (Iwasaki et al., 1980; Iwasaki and Terai, 1982; Matsubara et al., 1982), but synthesize inactive N_2O reductase (Matsubara et al., 1982; Minagawa and Zumft, 1988). The enzyme can be reactivated by exogenous Cu without de novo protein biosynthesis. Mutants affected in the chromophore region still export apoenzyme into the periplasm. This, however, cannot be reactivated because of the mutational defect in the Cu-processing apparatus. Copper insertion is temperature dependent and can be inhibited by several other transition metals (Minagawa and Zumft, 1988). Correct Cu-processing will thus have a direct effect on N_2O respiration. It has been suggested for marine environments that N_2O emissions could possibly be related to a low Cu concentration rather than a lack in denitrification capacity (Iwasaki et al., 1980).

CONCLUSIONS

The study of the biochemistry of N_2O metabolism is relatively recent compared to the century-long history of denitrification research. The genetics of this process is still in its infancy. The already recognizable diversity of systems for the formation and reduction of N_2O will probably lead to further discoveries when extending research to the great variety of denitrifying bacteria. Interesting groups for further investigations are certainly the nitrifying bacteria, Gram-positive representatives, or denitrifying archaebacteria. In as much as more structures of denitrification enzymes will become available, the puzzling question of the distribution and the evolutionary origin of denitrification (Broda, 1975) may become explorable from a comparative set of structural data.

ACKNOWLEDGMENTS

Work in the authors laboratories was supported by the Deutsche Forschungsgemeinschaft and Fonds der Chemischen Industrie.

LITERATURE CITED

Adkins, A. M., and R. Knowles. 1986. Denitrification by Cytophaga johnsonae strains and by a gliding bacterium able to reduce nitrous oxide in the presence of acetylene and sulfide. Can. J. Microbiol. 32:421-424.

Alefounder, P. R., A. J. Greenfield, J. E. G. McCarthy, and S. J. Ferguson. 1983. Selection and organisation of denitrifying electrontransfer pathways in Paracoccus denitrificans. Biochim. Biophys. Acta 724:20-39.

Ames, G. F.-L. 1986. Bacterial periplasmic transport systems: structure, mechanism, and evolution. Annu. Rev. Biochem. 55:397-425.

Baalsrud, K., and K. S. Baalsrud. 1954. Studies on Thiobacillus denitrificans. Arch. Mikrobiol. 20:34-62.

Bazylinski, D. A., and R. P. Blakemore. 1983. Denitrification and assimilatory nitrate reduction in Aquaspirillum magnetotacticum. Appl. Environ. Microbiol. 46:1118-1124.

Bazylinski, D. A., R. B. Frankel, and H. W. Jannasch. 1988. Anaerobic magnetite production by a marine, magnetotactic bacterium. Nature (London) 334:518-519.

Bazylinski, D. A., E. Palome, N. A. Blakemore, and R. P. Blakemore. 1986a. Denitrification by Chromobacterium violaceum. Appl. Environ. Microbiol. 52:696-699.

Bazylinski, D. A., C. K. Soohoo, and T. C. Hollocher. 1986b. Growth of Pseudomonas aeruginosa on nitrous oxide. Appl. Environ. Microbiol. 51:1239-1246.

Beijerinck, M. W., and D. C. J. Minkman. 1910. Bildung und Verbrauch von Stickoxydul durch Bakterien. Zentralbl. Bakteriol. Abt. 2, 25:30-63.

Bonin, P., M. Gilewicz, and J. C. Bertrand. 1987. Denitrification by a marine bacterium Pseudomonas nautica strain 617. Ann. Inst. Pasteur/Microbiol., 138:371-383.

Boogerd, F. C., H. W. van Versefeld, and A. H. Stouthamer. 1981. Respiration-driven proton translocation with nitrite and nitrous oxide in Paracoccus denitrificans. Biochim. Biophys. Acta 638:181-191.

Broda, E. 1975. The evolution of the bioenergetic processes. Pergamon Press, Oxford.

Burkhardt, R., and V. Braun. 1987. Nucleotide sequence of the fhuC and fhuD genes involved in iron(II) hydroxamate transport: domains in FhuC homologous to ATP-binding proteins. Mol. Gen. Genet. 209:49-55.

Carlson, C. A., and J. L. Ingraham. 1983. Comparison of denitrification by Pseudomonas stutzeri, Pseudomonas aeruginosa, and Paracoccus denitrificans. Appl. Environ. Microbiol. 45:1247-1253.

Carr, G., and S. J. Ferguson. 1988. Nitric oxide reductase of Paracoccus denitrificans. Biochem. Soc. Trans. 16:187-188.

Carr, G. J., M. D. Page, and S. J. Ferguson. 1989. The energy-conserving nitric-oxide-reductase system in Paracoccus denitrificans. Distinction from the nitrite reductase that catalyses synthesis of nitric oxide and evidence from trapping experiments for nitric oxide as a free intermediate during denitrification. Eur. J. Biochem. 179:683-692.

Chung, C. W., and V. A. Najjar. 1956. Cofactor requirements for enzymatic de nitrification. II. Nitric oxide reductase J. Biol. Chem. 218:627-632.

Cox, C. D., Jr., and W. J. Payne. 1973. Separation of soluble denitrifying enzymes and cytochromes from Pseudomonas perfectomarinus. Can. J. Microbiol. 19:861-872.

Cox, C. D., Jr., W. J. Payne, and D. V. DerVartanian. 1971. Electron paramagnetic resonance studies on the nature of hemoproteins in nitrite and nitric oxide reduction. Biochim. Biophys. Acta 253:290-294.

Coyle, C. L., W. G. Zumft, P. M. H. Kroneck, H. Körner, and W. Jakob. 1985. Nitrous oxide reductase from denitrifying Pseudomonas perfectomarina. Purification and properties of a novel multicopper enzyme. Eur. J. Biochem. 153:459-467.

Daniel, R. M., A. W. Limmer, K. W. Steele, and I. M. Smith. 1982. Anaerobic growth, nitrate reduction and denitrification in 46 Rhizobium strains. J. Gen. Microbiol. 128:1811-1815.

Dhesi, R., and R. Timkovich. 1984. Patterns of inhibition for bacterial nitrite reductase. Biochem. Biophys. Res. Commun. 123:966-972.

Dooley, D. M., R. S. Moog, and W. G. Zumft. 1987. Characterization of the copper sites in Pseudomonas perfectomarina nitrous oxide reductase by resonance Raman spectroscopy. J. Am. Chem. Soc. 109:6730-6735.

Ferguson, S. J. 1988. The redox reactions of the nitrogen and sulphur cycles, p. 1-29. In J. A. Cole, and S. J. Ferguson (ed.) The nitrogen and sulphur cycles. Cambridge University Press, Cambridge.

Fewson, C. A., and J. D. Nicholas. 1961. Nitric oxide reductase from Pseudomonas aeruginosa. Biochem. J. 78:9-10p.

Firestone, M. K., R. B. Firestone, and J. M. Tiedje. 1979. Nitric oxide as an intermediate in denitrification: evidence from nitrogen-13 isotope exchange. Biochem. Biophys. Res. Commun. 91:10-16.

Freitag, A., M. Rudert, and E. Bock. 1987. Growth of Nitrobacter by dissimilatoric nitrate reduction. FEMS Microbiol. Lett. 48:105-109.

Friedrich, M. J., L. C. DeVeaux, and R. J. Kadner. 1986. Nucleotide sequence of the btuCED genes involved in vitamin B_{12} transport in Escherichia coli and homology with components of periplasmic-binding protein-dependent transport systems. J. Bacteriol. 167:928-934.

Frunzke, K., and W. G. Zumft. 1986. Inhibition of nitrous-oxide respiration by nitric oxide in the denitrifying bacterium Pseudomonas perfectomarina. Biochim. Biophys. Acta 852:119-125.

Gamble, T. N., M. R. Betlach, and J. M. Tiedje. 1977. Numerically dominant denitrifying bacteria from world soils. Appl. Environ. Microbiol. 33:926-939.

Garber, E. A. E., D. Castignetti, and T. C. Hollocher. 1982. Proton translocation and proline uptake associated with reduction of nitric oxide by denitrifying Paracoccus denitrificans. Biochem. Biophys. Res. Commun. 107:1504-1507.

Garcia, J.-L., F. Pichinoty, M. Mandel, and B. Greenway. 1977. A new denitrifying saprophyte related to Pseudomonas pickettii. Ann. Microbiol. (Inst. Pasteur) 128A:229-237.

Grant, M. A., S. E. Cronin, and L. I. Hochstein. 1984. Solubilization and resolution of the membrane-bound nitrite reductase from Paracoccus halodenitrificans into nitrite and nitric oxide reductases. Arch. Microbiol. 140:183-186.

Greenberg, E. P., and G. E. Becker. 1977. Nitrous oxide as end product of denitrification by strains of fluorescent pseudomonads. Can. J. Microbiol. 23:903-907.

Hart, L. T., A. D. Larson, and C. S. McCleskey. 1965. Denitrification by Corynebacterium nephridii. J. Bacteriol. 89:1104-1108.

Heiss, B., K. Frunzke, and W. G. Zumft. 1989. Formation of the N-N bond from nitric oxide by a membrane-bound cytochrome bc complex of nitrate-respiring (denitrifying) Pseudomonas stutzeri. J. Bacteriol. 171:3288-3297.

Higgins, C. F., I. D. Hiles, G. P. C. Salmond, D. R. Gill, J. A. Downie, I. J. Evans, I. B. Holland, L. Gray, S. D. Buckels, A. W. Bell, and M. A. Hermodson. 1986. A family of related ATP-binding subunits coupled to many distinct biological processes in bacteria. Nature (London) 323:448-450.

Higgins, C. F., M. P. Gallagher, M. L. Mimmack, and S. R. Pearce. 1988. A family of closely related ATP-binding subunits from prokaryotic and eukaryotic cells. BioEssays 8:111-116.

Hochstein, L. I., and G. A. Tomlinson. 1988. The enzymes associated with denitrification. Annu. Rev. Microbiol. 42:231-261.

Hoglen, J., and T. C. Hollocher. 1986. Purification of nitric oxide reductase from Paracoccus denitrificans. Fed. Proc. 45:1604.

Hollocher, T. C. 1982. The pathway of nitrogen and reductive enzymes of denitrification. Antonie van Leeuwenhoek J. Microbiol. Serol. 48:531-544.

Hynes, R. K., A.-L. Ding, and L. M. Nelson. 1985. Denitrification by Rhizobium fredii. FEMS Microbiol. Lett. 30:183-186.

Itoh, M., S. Mizukami, K. Matsuura, and T. Satoh. 1989. Involvement of cytochrome bc_1 complex and cytochrome c_2 in the electron-transfer pathway for NO reduction in a photodenitrifier, Rhodobacter sphaeroides f. s. denitrificans. FEBS Lett. 244:81-84.

Iwasaki, H., T. Saigo, and T. Matsubara. 1980. Copper as a controlling factor of anaerobic growth under N_2O and biosynthesis of N_2O reductase in denitrifying bacteria. Plant Cell Physiol. 21:1573-1584.

Iwasaki, H., and H. Terai. 1982. Analysis of N_2 and N_2O produced during growth of denitrifying bacteria in copper-depleted and -supplemented media. J. Gen. Appl. Microbiol. 28:189-93.

Jensen, B. B., and R. H. Burris. 1986. N_2O as a substrate and as a competitive inhibitor of nitrogenase. Biochemistry 25:1083-1088.

Jin, H., H. Thomann, C. L. Coyle, and W. G. Zumft. 1989. Copper coordination in nitrous oxide reductase from Pseudomonas stutzeri. J. Am. Chem. Soc. 111:4262-4269.

Kaplan, W. A., and S. C. Wofsy. 1985. The biogeochemistry of nitrous oxide: a review. Adv. Aquat. Microbiol. 3:181-206.

Koike, I., and A. Hattori. 1975. Energy yield of denitrification: an estimate from growth yield in continuous culture of Pseudomonas denitrificans under nitrate-, nitrite-, and nitrous oxide-limited conditions. J. Gen. Microbiol. 88:11-19.

Körner, H., K. Frunzke, K. Döhler, and W. G. Zumft. 1987. Immunochemical patterns of distribution of nitrous oxide reductase and nitrite reductase (cytochrome cd$_1$) among denitrifying pseudomonads. Arch. Microbiol. 148:20-24.

Körner, H., and W. G. Zumft. 1989. Expression of denitrification enzymes in response to the dissolved oxygen level and respiratory substrate in continuous culture of Pseudomonas stutzeri. Appl. Environ. Microbiol. 55:1670-1676.

Krieg, N. R. 1976. Biology of the chemoheterotrophic spirilla. Bacteriol. Rev. 40:55-115.

Kroneck, P. M. H., W. A. Antholine, J. Riester, and W. G. Zumft. 1988. The cupric site in nitrous oxide reductase contains a mixed-valence (Cu(II),Cu(I)) binuclear center: a multifrequency electron paramagnetic resonance investigation. FEBS Lett. 242:70-74.

Kroneck, P. M. H., W. A. Antholine, J. Riester, and W. G. Zumft. 1989. The nature of the cupric site in nitrous oxide reductase and of Cu_A in cytochrome c oxidase. FEBS Lett. 248:212-213.

Lee, H. S., R. E. W. Hancock, and J. L. Ingraham. 1989. Properties of a Pseudomonas stutzeri outer membrane channel-forming protein (NosA) required for production of copper-containing N_2O reductase. J. Bacteriol. 171:2096-2100.

Li, P. M., J. Gelles, S. I. Chan, R. J. Sullivan, and R. A. Scott. 1987. Extended X-ray absorption fine structure of copper in Cu_A-depleted, p-(hydroxymercuri)benzoate-modified, and native cytochrome c oxidase. Biochemistry 26:2091-2095.

Malkin, R., and B. G. Malmström. 1970. The state and function of copper in biological systems. Adv. Enzymol. 33:177-244.

Mancinelli, R. L., S. Cronin, and L. I. Hochstein. 1986. The purification and properties of a cd-cytochrome nitrite reductase from Paracoccus halodenitrificans. Arch. Microbiol. 145:202-208.

Matsubara, T. 1971. Studies on denitrification. XIII. Some properties of the N_2O-anaerobically grown cell. J. Biochem. 69:991-1001.

Matsubara, T., K. Frunzke, and W. G. Zumft. 1982. Modulation by copper of the products of nitrite respiration in Pseudomonas perfectomarinus. J. Bacteriol. 149:816-823.

Matsubara, T., and H. Iwasaki. 1971. Enzymatic steps of dissimilatory nitrite reduction in Alcaligenes faecalis. J. Biochem. 69:859-868.

Matsubara, T., and H. Iwasaki. 1972. Nitric oxide-reducing activity of Alcaligenes faecalis cytochrome cd. J. Biochem. 72:57-64.

Matsubara, T., and M. Sano. 1985. Isolation and some properties of a novel violet copper protein from a denitrifying bacterium, Alcaligenes sp. Chem. Lett. 1053-1056.

Matsubara, T., and W. G. Zumft. 1982. Identification of a copper protein as part of the nitrous oxide-reducing system in nitrite-respiring (denitrifying) pseudomonads. Arch. Microbiol. 132:322-328.

McEwan, A. G., A. J. Greenfield, H. G. Wetzstein, J. B. Jackson, and S. J. Ferguson. 1985. Nitrous oxide reduction by members of the family Rhodospirillaceae and the nitrous oxide reductase of Rhodopseudomonas capsulata. J. Bacteriol. 164:823-830.

Michalski, W. P., D. H. Hein, and D. J. D. Nicholas. 1986. Purification and characterization of nitrous oxide reductase from Rhodopseudomonas sphaeroides f.sp. denitrificans. Biochim. Biophys. Acta 872:50-60.

Minagawa, N., and W. G. Zumft. 1988. Cadmium-copper antagonism in the activation of periplasmic nitrous oxide reductase of copper-deficient cells from Pseudomonas stutzeri. Biol. Metals 1:117-122.

Miyata, M., T. Matsubara, and T. Mori. 1969. Studies on denitrification. XI. Some properties of nitric oxide reductase. J. Biochem. 66:759-765.

Mokhele, K., Y. J. Tang, M. A. Clark, and J. L. Ingraham. 1987. A Pseudomonas stutzeri outer membrane protein inserts copper into N_2O reductase. J. Bacteriol. 169:5721-5726.

Payne, W. J., M. A. Grant, J. Shapleigh, and P. Hoffman. 1982. Nitrogen oxide reduction in Wolinella succinogenes and Campylobacter species. J. Bacteriol. 152:915-918.

Pichinoty, F., J. Bigliardi-Rouvier, M. Mandel, B. Greenway, G. Méténier, and J.-L. Garcia. 1976. The isolation and properties of a denitrifying bacterium of the genus Flavobacterium. Antonie van Leeuwenhoek J. Microbiol. Serol. 42:349-354.

Pichinoty, F., M. Mandel, and J.-L. Garcia. 1977a. Étude de six souches de Agrobacterium tumefaciens et A. radiobacter. Ann. Microbiol. (Inst. Pasteur) 128A:303-310.

Pichinoty, F., M. Mandel, and J.-L. Garcia. 1979. The properties of novel mesophilic denitrifying Bacillus cultures found in tropical soils. J. Gen. Microbiol. 115:419-430.

Pichinoty, F., M. Mandel, B. Greenway, and J.-L. Garcia. 1977b. Étude de 14 bactéries dénitrifiantes appartenant au groupe Pseudomonas stutzeri isolées du sol par culture d'enrichissement en présence d'oxyde nitreux. Ann. Microbiol. (Inst. Pasteur) 128A:75-87.

Poth, M. 1986. Dinitrogen production from nitrite by a Nitrosomonas isolate. Appl. Environ. Microbiol. 52:957-959.

Poth, M., and D. D. Focht. 1985. ^{15}N kinetic analysis of N_2O production by Nitrosomonas europaea: an examination of nitrifier denitrification. Appl. Environ. Microbiol. 49:1134-1141.

Riester, J., W. G. Zumft, and P. M. H. Kroneck. 1989. Nitrous oxide reductase from Pseudomonas stutzeri. Redox properties and spectroscopic characterization of different forms of the multicopper enzyme. Eur. J. Biochem. 178:751-762.

Robertson, L. A., and J. G. Kuenen. 1983. Thiosphaera pantotropha gen. nov. sp. nov., a facultatively anaerobic, facultatively autotrophic sulphur bacterium. J. Gen. Microbiol. 129:2847-2855.

Römermann, D., and B. Friedrich. 1985. Denitrification by Alcaligenes eutrophus is plasmid dependent. J. Bacteriol. 162:852-854.

Scott, R. A., W. G. Zumft, C. L. Coyle, and D. M. Dooley. 1989. Pseudomonas stutzeri N_2O reductase contains Cu_A-type sites. Proc. Natl. Acad. Sci. USA 86:4082-4086.

Shapleigh, J. P., K. J. P. Davies, and W. J. Payne. 1987. Detergent inhibition of nitric-oxide reductase activity. Biochim. Biophys. Acta 911:334-340.

Shapleigh, J. P., and W. J. Payne. 1985. Nitric oxide-dependent proton translocation in various denitrifiers. J. Bacteriol. 163:837-840.

Snyder, S. W., D. A. Bazylinski, and T. C. Hollocher. 1987. Loss of N_2O reductase activity as an explanation for poor growth of Pseudomonas aeruginosa on N_2O. Appl. Environ. Microbiol. 53:2045-2049.

Snyder, S. W., and T. C. Hollocher. 1984. Nitrous oxide reductase and the 120,000 MW copper protein of N_2-producing denitrifying bacteria are different entities. Biochem. Biophys. Res. Commun. 119:588-592.

Snyder, S. W., and T. C. Hollocher. 1987. Purification and some characteristics of nitrous oxide reductase from Paracoccus denitrificans. J. Biol. Chem. 262:6515-6525.

Stanier, R. Y., N. J. Palleroni, and M. Doudoroff. 1966. The aerobic pseudomonads: a taxonomic study. J. Gen. Microbiol. 43:159-271.

Stouthamer, A. H. 1988a. Dissimilatory reduction of oxidized nitrogen compounds, p. 245-303. In A. J. B. Zehnder (ed.) Biology of anaerobic microorganisms. John Wiley & Sons, Inc., New York.

Stouthamer, A. H. 1988b. Bioenergetics and yields with electron acceptors other than oxygen, p. 345-437. In L. E. Erickson and D. Y.-C. Fung (ed.) Handbook on anaerobic fermentations. Marcel Dekker, Inc., New York.

Teraguchi, S., and T. C. Hollocher. 1989. Purification and some characteristics of a cytochrome c-containing nitrous oxide reductase from Wolinella succinogenes. J. Biol. Chem. 264:1972-1979.

Tibelius, K. H., and R. Knowles. 1984. Uptake hydrogenase activity in denitrifying Azospirillum brasilense grown anaerobically with nitrous oxide or nitrate. J. Bacteriol. 157:84-88.

Tiedje, J. M. 1988. Ecology of denitrification and dissimilatory nitrate reduction to ammonium, p. 179-244. In A. J. B. Zehnder (ed.), Biology of anaerobic microorganisms. John Wiley & Sons, Inc. New York.

Urata, K., and T. Satoh. 1985. Mechanism of nitrite reduction to nitrous oxide in a photodenitrifier, Rhodopseudomonas sphaeroides forma sp. denitrificans. Biochim. Biophys. Acta 841:201-207.

Urata, K., K. Shimada, and T. Satoh. 1982. Periplasmic location of nitrous oxide reductase in a photodenitrifier, Rhodopseudomonas sphaeroides forma sp. denitrificans. Plant Cell Physiol. 23:1121-1124.

Vaughn, S., and B. K. Burgess. 1989. Nitrite: a new substrate for nitrogenase. Biochemistry 28:419-424.

Viebrock, A., and W. G. Zumft. 1987. Physical mapping of transposon Tn5 insertions defines a gene cluster functional in nitrous oxide respiration by Pseudomonas stutzeri. J. Bacteriol. 169:4577-4580.

Viebrock, A., and W. G. Zumft. 1988. Molecular cloning, heterologous expression, and primary structure of the structural gene for the copper enzyme nitrous oxide reductase from denitrifying Pseudomonas stutzeri. J. Bacteriol. 170:4658-4668.

Weeg-Aerssens, E., J. M. Tiedje, and B. A. Averill. 1988. Evidence from isotope labeling studies for a sequential mechanism for dissimilatory nitrite reduction. J. Am. Chem. Soc. 110:6851-6856.

Wijler, J., and C. C. Delwiche. 1954. Investigation on the denitrifying process in soil. Plant Soil 5:155-169.

Wood, A. P., and D. P. Kelly. 1983. Autotrophic, mixotrophic and heterotrphic growth with denitrification by Thiobacillus A2 under anaerobic conditions. FEMS Microbiol. Lett. 16:363-370.

Yoshimura, T., H. Iwasaki, S. Shidara, S. Suzuki, A. Nakahara, and T. Matsubara. 1988. Nitric oxide complex of cytochrome c' in cells of denitrifying bacteria. J. Biochem. 103:1016-1019.

Yoshinari, T. 1980. N_2O reduction by Vibrio succinogenes. Appl. Environ. Microbiol. 39:81-84.

Zumft, W. G., C. L. Coyle, and K. Frunzke. 1985a. The effect of oxygen on chromatographic behavior and properties of nitrous oxide reductase. FEBS Lett. 183:240-244.

Zumft, W. G., K. Döhler, and H. Körner. 1985b. Isolation and characterization of transposon Tn5-induced mutants of Pseudomonas perfectomarina defective in nitrous oxide respiration. J. Bacteriol. 163:918-924.

Zumft, W. G., K. Döhler, H. Körner, S. Löchelt, A. Viebrock, and K. Frunzke. 1988a. Defects in cytochrome cd_1-dependent nitrite respiration of transposon Tn5-induced mutants from Pseudomonas stutzeri. Arch. Microbiol. 149:492-498.

Zumft, W. G., and K. Frunzke. 1982. Discrimination of ascorbate-dependent nonenzymatic and enzymatic, membrane-bound reduction of nitric oxide in denitrifying Pseudomonas perfectomarinus. Biochim. Biophys. Acta 681:459-468.

Zumft W. G., and T. Matsubara. 1982. A novel kind of multi-copper protein as terminal oxidoreductase of nitrous oxide respiration in Pseudomonas perfectomarinus. FEBS Lett. 148:107-112.

Zumft, W. G., and J. M. Vega. 1979. Reduction of nitrite to nitrous oxide by a cytoplasmic membrane fraction from the marine denitrifier Pseudomonas perfectomarinus. Biochim. Biophys. Acta 548:484-499.

Zumft, W. G., A. Viebrock, and H. Körner. 1988b. Biochemical and physiological aspects of denitrification, p. 245-279. In J. A. Cole, and S. J. Ferguson (ed.) The nitrogen and sulphur cycles. Cambridge University Press, Cambridge.

PHYSIOLOGY, BIOCHEMISTRY AND GENETICS OF NITRATE DISSIMILATION TO AMMONIA

J.A. Cole

Department of Biochemistry
University of Birmingham
Birmingham B15 2TT, U.K.

INTRODUCTION

Many textbooks, symposium publications and even edited papers in leading journals present denitrification as the only dissimilatory pathway for bacterial NO_3^- reduction. In fact, the rapid, dissimilatory reduction to NH_4^+ by fermentative bacteria was documented many years ago (see, for example, Woods, 1938). As the limited literature on NO_3^- dissimilation to NH_4^+ published before 1988 has recently been reviewed extensively (Cole, 1988; 1989), this article will focus on recent developments which confirm or conflict with the previous conclusions. Although the practical importance of denitrification by essentially respiratory bacteria is beyond doubt, Figure 1 presents it - possibly for the first time - in the context of the emerging diversity of enzymes which have evolved to dissimilate NO_3^- and NO_2^- in different bacterial groups.

Virtually all bacteria that are able to reduce NO_3^- are also able to reduce NO_2^-, but it is sometimes difficult to detect the nitrite reductase activity in the laboratory. The converse is certainly not correct, however, because NO_2^- reduction can be detected in some species that are devoid of nitrate reductase activity (Forsythe et al., 1987). Note, however, that reports that eukaryotic microbes can respire NO_3^- have so far been unconfirmed (see, for a recent example, Finlay, Span & Harman, 1983). We must also wait to learn whether any denitrifying bacteria can also dissimilate NO_3^-, NO or N_2O to NH_4^+. As a further example of our current ignorance, little is known about the electron donors, prosthetic groups and structures of nitrate and nitrite reductases in the majority of bacteria that dissimilate NO_3^- or NO_2^-. Fig. 1 must therefore inevitably be an over-simplification based on too few data.

METABOLIC SIGNIFICANCE OF NITRITE REDUCTION TO AMMONIA

Nitrite is a key intermediate in both denitrification and NO_3^- reduction to NH_4^+. Because NO_3^- dissimilation is usually a rapid process, some bacteria catalyse a rapid and massive accumulation of NO_2^- which, in some but not all bacteria, is enhanced because nitrite reductase synthesis is repressed or NO_2^-

Denitrification in Soil and Sediment, Edited by N. P. Revsbech and J. Sørensen, Plenum Press, New York, 1990

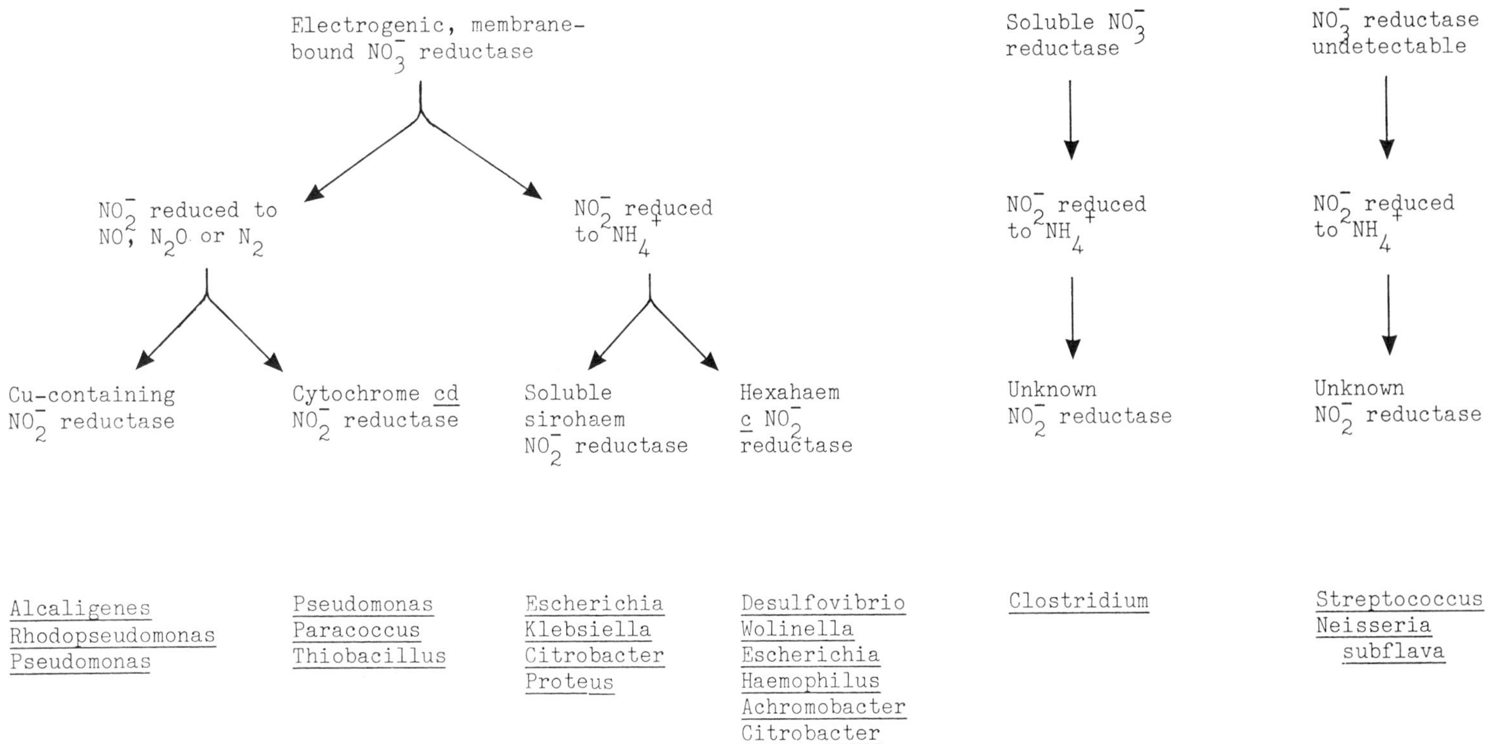

Figure 1. Sub-groups of denitrifying bacteria and bacteria that reduce nitrate to ammonia based on biochemical differences in the enzymes involved in nitrate or nitrite reduction. The genera listed at the bottom of the figure are typical representatives of each sub-group.

reduction is inhibited by NO_3^-. At least nine consequences of the reduction of this NO_2^- to NH_4^+ have been recognised (Cole, 1989): many of them are inter-dependent and some apply only to a few bacterial genera. They are that the NH_4^+ produced is a useful source of nitrogen for growth, oxidized pyridine nucleotides which are in short supply during anaerobic growth are regenerated, fermentation products are oxidized, the pH of the environment increases, a toxic chemical is removed either from the extra-cellular or intra-cellular environment, energy is conserved either by substrate level phosphorylation or by the generation of an electrochemical proton gradient, and finally anaerobic growth becomes possible on non-fermentable carbon sources such as glycerol.

The end products of NO_3^- reduction can be correlated, to a first approximation, with two other aspects of anaerobic bacterial metabolism. First, most non-fermentative organisms such as the pseudomonads, chemolithotrophs, and photosynthetic bacteria that can use NO_3^- as an alternative electron acceptor during anaerobic growth are denitrifiers. In contrast, most fermentative bacteria reduce NO_3^- to NH_4^+. Secondly, reducing equivalents for nitrate reduction by respiratory organisms tend to be generated by the same reactions, for example the tricarboxylic acid cycle, that generates NADH for O_2 reduction during aerobic growth (Wimpenny & Warmsley, 1968). In contrast, fermentative organisms like the enteric bacteria metabolise pyruvate preferentially to formate during anaerobic growth irrespective whether NO_3^- is present or absent. Formate is then an excellent electron donor for NO_3^- reduction (Wimpenny & Cole, 1967). Exceptions to these broad generalisations are already becoming apparent. Nishimura, Kamihara & Fukui (1979) were the first to demonstrate formate-dependent denitrification by a pseudomonad, but it is not clear how such bacteria encounter this substrate. Conversely, Samuelson (1985) documented the rapid dissimilation of NO_3^- to NH_4^+ by a Pseudomonas putrefaciens strain.

ECOLOGICAL ASPECTS OF NITRATE DISSIMILATION TO AMMONIA

The theme that has emerged so far is that denitrification is the dominant process in environments that favour a respiratory mode of life but NO_3^- dissimilation to NH_4^+ should predominate in more anaerobic ecosystems. Available data fully support such expectations.

Sørensen (1978) reported that the denitrification capacity of a coastal marine sediment decreased rapidly with depth, but the potential for NO_3^- reduction to NH_4^+ in the anaerobic layers was better maintained. The importance of NO_3^- reduction to NH_4^+ was further indicated by similar data from Koike & Hattori (1978a) and Cole & Brown (1980). The isolation of members of the Enterobacteriaceae from such sediments was at first surprising, but provided a possible explanation for the accumulation of high concentrations of NH_4^+. Subsequently, Rõnner & Samuelsson (1985) demonstrated that potential rates of denitrification were high in the anaerobic deep water below 80 m of the Baltic. Dissimilation of nitrate to NH_4^+ was stimulated more than denitrification by supplementation with both glucose and $^{15}NO_3^-$, the ^{15}N isotope being readily detected in both $^{15}N_2$ and $^{15}NH_4^+$. Ammonia was also the dominant product of NO_3^- reduction in a carbon-rich digested sludge (Kaspar & Tiedje, 1981) and in yet more marine sediments (Koike & Hattori, 1978a,b; Samuelsson & Rönner, 1982). In contrast, only about 10% of the NO_3^- added to the sediment of a eutrophic lake was converted to NH_4^+ and the typical figure for soils is of the order of 1% to 5% (Kaspar & Tiedje, 1982; Caskey & Tiedje, 1979). In an even more recent study, both Agrobacterium radiobacter and Enterobacter amnigenus

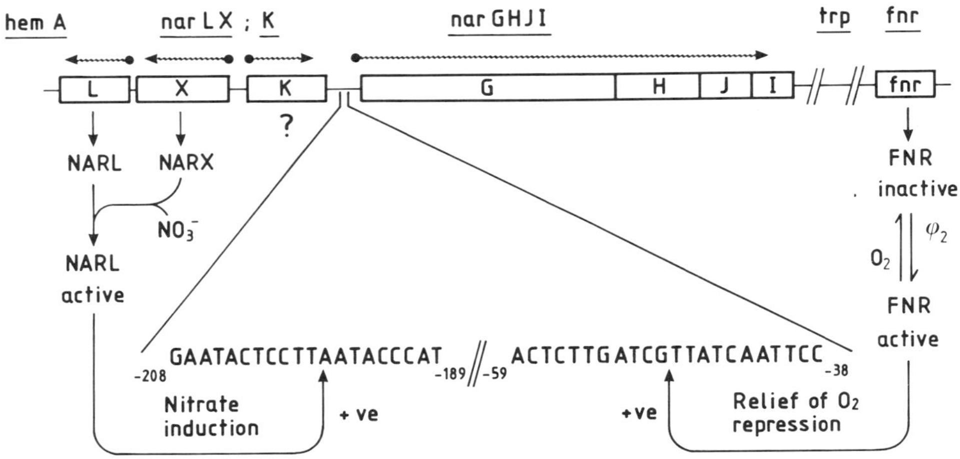

Fig. 2. The nitrate reductase operon of E. coli and its regulatory elements.

were isolated from soil in which the concentration of NH_4^+ was inexplicably high: biochemical analysis elegantly revealed that the Erwinia was reducing NO_3^- rapidly to NH_4^+ (Fazzolari et al., 1989).

Studies of bacteria isolated from the human stomach or from the rumen are totally consistent with the view that NO_3^- reduction to NH_4^+ is the dominant process in a carbon-rich environment inhabited by fermentative bacteria. Lewis (1951) established that any NO_3^- which enters the highly-reducing environment of the rumen is reduced quantitatively to NH_4^+ ; H_2, formate or glucose are potential electron donors for this process (Jones,1972). Similar results were obtained by Forsythe et al. (1987) who reported that all of the nitrate-reducing bacteria isolated from human gastric juice dissimilated NO_2^- to NH_4^+ , the only exception being Neisseria subflava which reduces nitrite slowly to gaseous products.

DIVERSITY OF BACTERIAL NITRATE REDUCTASES

Nitrate reductases involved in electron transfer phosphorylation have been isolated from a wide range of bacteria which continues to increase rapidly. The older literature was reviewed by Haddock & Jones (1977), Ingledew & Poole (1984) and Cole (1988); most recent data can be found in abstracts by Bielkie & Ketchum (1989) and Lang & Hochstein (1989).

The paradigm for these enzymes is the four sub-unit nitrate reductase isolated from Escherichia coli which has been grown anaerobically in the presence of NO_3^-. The presence of the fourth sub-unit is inferred from the translated DNA sequence of the narG operon which encodes the three previously-purified polypeptides (Figure 2). The molybdenum cofactor is attached to the largest sub-unit (product of the narG gene) and the smallest (product of the narI gene) is a cytochrome b specifically required for NO_3^- reduction. Nitrate reductase preparations frequently lack this cytochrome after extensive purification, a fact which has led to considerable controversy (see, for example,

Table 1. Enzymes that dissimilate nitrite to ammonia

Source	Cell location	Molecular weight	Prosthetic group	Reference
E. coli	cytoplasm	180 kDa	sirohaem, FAD [Fe-S] cluster(s)	Jackson et al., 1981
Vibrio fischeri	periplasm	58 kDa	6 x haem c	Liu et al., 1988
Wolinella succinogenes	periplasm	63 kDa	6 x haem c	Schroder et al., 1985
E. coli	periplasm	69 kDa	6 x haem c	Kajie & Anraku, 1986
Desulfovibrio desulfuricans	periplasm	66 kDa	6 x haem c	Liu & Peck, 1981

Knook et al., 1973; MacGregor et al., 1974; MacGregor, 1975). The product of the narH gene is post-translationally modified: its function appears to be to position the other sub-units correctly in the cytoplasmic membrane. The function of the fourth sub-unit, the narJ gene product, has not yet been defined (Li & DeMoss, 1988).

Perhaps the most striking differences between the E. coli enzyme and those described more recently are the wide variations in the sizes of the smaller sub-units. The largest of them so far, 49-52 kDa, (presumably corresponding to the 19.5 kDa sub-unit in the E. coli enzyme) were isolated from Xanthomonas maltophila and Klebsiella aerogenes (Bielkie & Ketchum, 1989; van't Riet & Planta, 1975). The significance of this variation, if any, is unknown.

Nitrate reduction by obligately fermentative bacteria

Obligate anaerobes such as the clostridia reduce nitrate slowly to NH_4^+ but apparently cannot use non-fermentable carbon sources such as glycerol for growth in the presence of NO_3^- (Woods, 1938; Hasan & Hall, 1975). Nitrate is reduced too slowly by clostridia for it to increase the growth rate significantly, typical rates of 100 nmol h^{-1} (mg dry weight)$^{-1}$ being about 1% of the rate of NO_3^- reduction by a derepressed culture of E. coli (Hasan & Hall, 1975; Cole, 1988). Nevertheless, the growth yields per mole of fermentable carbohydrate consumed are about 20% greater in the presence of NO_3^- than in its absence. This increase in yield is the consequence of additional substrate level phosphorylation during the production of acetate instead of more reduced products such as H_2, ethanol and butyrate (Hasan & Hall, 1975).

In contrast to the electrogenic, membrane-bound nitrate reductases of denitrifying and facultatively anaerobic bacteria, nitrate reductase from Clostridium perfringens is a soluble, non-haem iron-sulphur protein which functions independently of a cytochrome-containing electron transfer chain (Seki-Chiba & Ishimoto, 1977; Seki et al., 1987). Ferredoxin appears to donate electrons to this single sub-unit enzyme (Mr 90 000) which again contains the typical bacterial molybdenum cofactor, molybdopterin. A much larger monomeric nitrate reductase was isolated from another obligate anaerobe, Mitsuokella

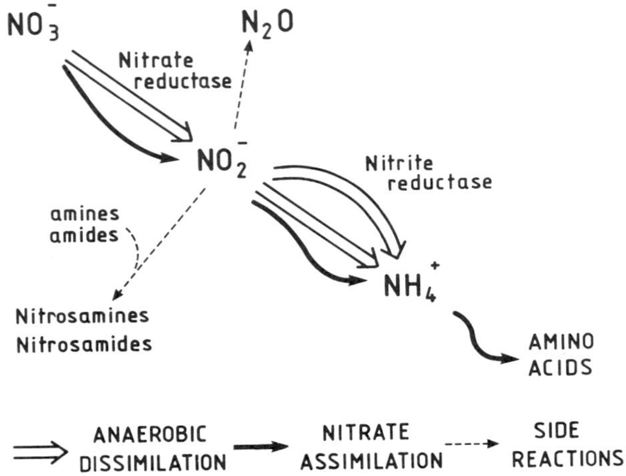

Fig. 3. Major pathways for nitrate and nitrite reduction by E. coli strains.

multiacidus (Yamamoto et al., 1986): the electron donor for this soluble enzyme (Mr 160 kDa) is also ferredoxin.

PHYSIOLOGY AND BIOCHEMISTRY OF NITRITE DISSIMILATION TO NH_4^+

So far only 5 enzymes that dissimilate NO_2^- to NH_4^+ have been purified to homogeneity (Table 1). Four of them are clearly related both structurally and functionally, being hexa-haem c-type cytochromes located in the bacterial periplasm. The fifth enzyme is more like the assimilatory nitrite reductases of plants and fungi: it is a soluble, cytoplasmic protein which accepts electrons from NADH and transfers them via FAD and non-haem iron-sulphur centres to sirohaem (Jackson et al., 1981; 1982a,b). Both types of enzyme occur widely in the Enterobacteriaceae, but formate-dependent NO_2^- reduction by the periplasmic c-type cytochrome has not yet been detected in Proteus or Klebsiella species (Abou-Jaoude, Chippaux & Pascal, 1979).

NITRITE METABOLISM OF ESCHERICHIA COLI

The principal pathways for NO_3^- and NO_2^- reduction by various E. coli strains are shown in Figure 3. It must be stressed, however, that some of these activities are difficult to detect or even absent from laboratory strains. For example, K-12 strains grow very poorly on minimal agar in which NO_3^- is the only source of nitrogen. In contrast, NO_3^- assimilation is readily detected in fresh clinical isolates. By far the most active nitrate reductase is the electrogenic, membrane-bound enzyme discussed above.

Anaerobic cultures of most E. coli strains reduce NO_2^- rapidly to NH_4^+. NADH-dependent NO_2^- reduction by the cytoplasmic NIRB enzyme provides the host with a potential nitrogen source for growth, regenerates NAD^+ from NADH, increases the pH of the environment and removes the toxic product of nitrate reduction from the cytoplasm. In contrast, the periplasmic formate-dependent nitrite reductase is linked to a membrane-bound electron transfer chain. This effectively removes two potentially toxic species, extracellular NO_2^- and intracellular formate, with the additional advantage of generating a proton electrochemical gradient. There is also a much slower process in which NO_2^- is reduced to N_2O.

The Soluble, NADH-dependent Nitrite Reductase

Zarowny & Sanwal (1963) were the first to report that, during anaerobic growth in the presence of NO_2^-, some strains of E.coli synthesize a very active, soluble and NADH-dependent nitrite reductase (EC 1.6.6.4). The enzyme is extremely unstable in vitro, so many of its properties were only revealed subsequently. For example, Kemp & Atkinson (1966) reported the very low Km for NO_2^- and, contrary to Zarowny & Sanwal (1963), identified NH_4^+ as the product of NO_2^- reduction.

Anaerobic cultures of certain chlorate-resistant mutants are constitutive for the synthesis of this nitrite reductase even in rich media unsupplemented with NO_2^- (Coleman et al., 1978a). Nitrite reductase constitutes up to 4% of the soluble protein in such cultures from which it has been purified almost to homogeneity by two cycles of gel filtration and anion exchange chromatography following breakage in a French pressure cell, high speed centrifugation and ammonium sulphate fractionation (Jackson et al., 1981a). The purified enzyme is apparently a homodimer of 90 kDa subunits, each of which contains either two [2Fe-2S] or one [4Fe-4S] iron-sulphur clusters (deduced from the Fe content of the purified protein and ESR spectroscopy). By comparing the amino acid sequence of the N-terminus of the purified protein with the nucleotide sequence of the nirB gene, nirB has been indentified as the structural gene for the enzyme (Macdonald & Cole, 1985; Jayaraman et al., 1987). Both the FAD and sirohaem moieties are readily lost from the purified protein, even during gel filtration. The amber coloured enzyme can be stored in the dark at -20 °C in Tris buffer, ascorbate, EDTA, FAD, NO_2^- and 20% (w/v) glycerol without significant loss of activity and without further precautions to prevent inactivation by oxygen. All of the chemicals listed above contribute towards stability during storage. NADH is the only effective electron donor, but alternative electron acceptors to NO_2^- include NH_2OH, SO_3^{2-}, mammalian cytochrome c and a variety of dyes (Jackson et al., 1982a).

Perhaps the most unusual property of the enzyme is its requirement for a reaction product, NAD^+, for activity (Coleman et al., 1978b). By analogy with Massey & Veeger's (1961) classical studies with lipoamide dehydrogenase, the role of NAD^+ is suggested to be to prevent the FAD moiety being fully reduced to a catalytically inactive form. Results of a kinetic analysis for the reduction of the alternative electron acceptors have been reported (Jackson et al., 1981b, 1982).

Cytochrome c_{552} and the Electrogenic Formate-nitrite Oxidoreductase Complex

Following the discovery by Postgate (1956) of cytochrome c_3 in an obligately anaerobic sulphate-reducing bacterium, Gray et al. (1963) discovered a similar, low redox potential c-type cytochrome in anaerobic cultures of E. coli. By analogy with the presumed role of cytochrome c_3 in hydrogen metabolism, it was suggested that cytochrome c_{552} might also be one of the unidentified components, x_1 or x_2, of the formate hydrogenlyase complex (Peck & Gest, 1957). This became unlikely when it was realised that cytochrome c_{552} synthesis correlates far better with nitrite reductase activity than with hydrogenase activity. As the reduced cytochrome is rapidly oxidised by NO_2^-, both Fujita & Sato (1966b) and Cole (1968) suggested that it was more likely to be a nitrite reductase than an electron donor to hydrogenase.

Further work by Fujita & Sato (1967), subsequently confirmed by Cole (1968), established that although 98% of the NADH-dependent nitrite reductase activity was retained when bacteria were subjected to osmotic shock, the majority of the cytochrome $\underline{c}_{552}$ was released with other periplasmic proteins. This loss coincided with a loss of NO_2^--dependent CO_2 evolution from glucose or an endogenous substrate, possibly formate (Fujita & Sato, 1967). The purification of NADH-dependent nitrite reductase with high specific activity but free from cytochrome $\underline{c}_{552}$ established that the cytochrome was not necessary for NO_2^- reduction by NADH.

Evidence that the NIRB protein and cytochrome $\underline{c}_{552}$ are components of independent pathways for NO_2^- reduction came from the characterisation of the formate-dependent nitrite reductase activity by Chippaux and his colleagues (Abou-Jaoude et al., 1979a,b), the isolation of nirB mutants that lack NADH-dependent nitrite reductase but retain formate-dependent nitrite reductase activity, and from the characterization of a similar, hexahaem cytochrome $\underline{c}$ which is the terminal nitrite reductase of formate-dependent nitrite reduction by Desulfovibrio strains (Steenkamp & Peck, 1980; 1981). Furthermore, Motteram et al. (1981) and Pope & Cole (1982) demonstrated that although no proton electrochemical gradient was established during NADH-dependent nitrite reduction, the formate-dependent activity was electrogenic. In summary, four lines of evidence indicate that cytochrome $\underline{c}_{552}$ is the terminal enzyme of a trans-membrane formate-nitrite oxidoreductase complex:

(a) the cytochrome is rapidly oxidised by nitrite and will catalyse benzyl viologen-dependent nitrite reduction;

(b) synthesis of the cytochrome, both within and between strains, correlates with formate-nitrite oxidoreductase activity;

(c) like the periplasmic nitrite reductase from Desulfovibrio, it is a hexahaem cytochrome $\underline{c}$;

(d) NO_2^--dependent gas evolution is restored when sphaeroplasts are complemented by high concentrations of cytochrome $\underline{c}_{552}$.

Note, however, that although each of the above correlations is consistent with the assumed role for the cytochrome, rigorous proof that it is the terminal nitrite reductase must await the cloning and sequence analysis of the structural gene and demonstration that the cloned gene is both necessary and sufficient to restore NO_2^- reduction by formate to Nrf^- mutants. Such cytochrome-deficient mutants have recently been isolated in the author's laboratory.

Components of the formate-nitrite oxidoreductase other than cytochrome $\underline{c}_{552}$ remain to be identified. Nitrite reduction is inhibited by cyanide but not by azide. It is partially sensitive to hydroxyquinoline-N-oxides which inhibit some $\underline{b}$-type cytochromes. A mutant deficient in both ubiquinone and menaquinone retained 70% of the activity of the wildtype, but as this mutant was leaky for quinone synthesis, it is still not clear whether quinones are essential components of this specialized electron transfer chain (Abou-Jaoude et al., 1979b).

Nitrate Assimilation by Escherichia coli

Throughout the 1950's and 1960's there was considerable controversy concerning whether a single enzyme or two independent enzymes from E. coli and other enteric bacteria were required for nitrate assimilation and NO_3^- respiration. By 1970 the consensus view favoured a single enzyme being involved in both processes, although the possibility was recognised that the product of the structural gene for this enzyme (the narG gene product) might associate with different auxillary proteins in the two processes. The discovery by Ingraham and his colleagues that biochemically and genetically distinct enzymes are required for NO_3^- assimilation and respiration by Pseudomonas aeruginosa is therefore particularly important (Jeter et al., 1984). A report that there are two homologous but independent genes for nitrate reductase in E.coli has recently been confirmed, (Bonnefoy et al., 1987), so it is possible - indeed likely - that there are also independent pathways for NO_2^- assimilation and dissimilation to NH_4^+ in E. coli. Nothing is yet known about the biochemistry and genetics of NO_2^- assimilation by E. coli.

Other 'Nitrite Reductases' of Escherichia coli

At least four additional NO_2^- reducing activities in E. coli have been detected or postulated. The best characterised is due to the broad substrate specificity of the NADPH-dependent sulphite reductase which is required for cysteine biosynthesis during growth in minimal media (Kemp et al., 1963), Another is due to non-specific reduction of NO_2^- to N_2O oxide by nitrate reductase (Fig. 3; Smith, 1983) and a third is probably an artifact created by combining a diaphorase activity with FAD and hexahaem cytochrome c_{552} (Liu et al.,1981). Little is known about NO_2^- reduction using ethanol and pyruvate as electron donors (Pope & Cole, 1984).

GENETIC REGULATION OF NADH-DEPENDENT NITRITE REDUCTION

The structural gene for the NADH-dependent nitrite reductase is located at minute 74 on the E. coli chromosome. The nirB operon has been cloned and sequenced (Figure 4). The operon consists of at least three genes, nirB, nirC and cysG with a possible fourth gene, nirD, located between nirB and nirC. Computer-aided analysis of the derived amino acid sequence of NIRB revealed an FAD binding domain at the N terminus of the protein followed immediately by NAD binding, central and interface domains. Next there is a tandem repeat sequence containing the motif -Cys-X-X-Cys- which is a prime candidate for the location of one or more iron-sulphur custers. Presumably the sirohaem binding site is located in the C-terminal region which was not similar to any of the proteins in the database used to locate other prosthetic group domains.

Immediately after the translation stop codon for the nitrite reductase apoprotein is the sequence 5'-GAGGA-3' immediately preceding an ATG and open reading frame which might encode a fourth polypeptide. This in turn is followed by an inverted repeat sequence typical of rho-independent transcription terminators and an A-T-rich untranslated region. The nirC gene is known to be transcribed and translated co-ordinately with nirB: it encodes a highly hydrophobic protein with regions of amino-acid sequence which are similar to regions in cytochrome oxidase polypeptide 1. The function of the NIRC protein is unknown.

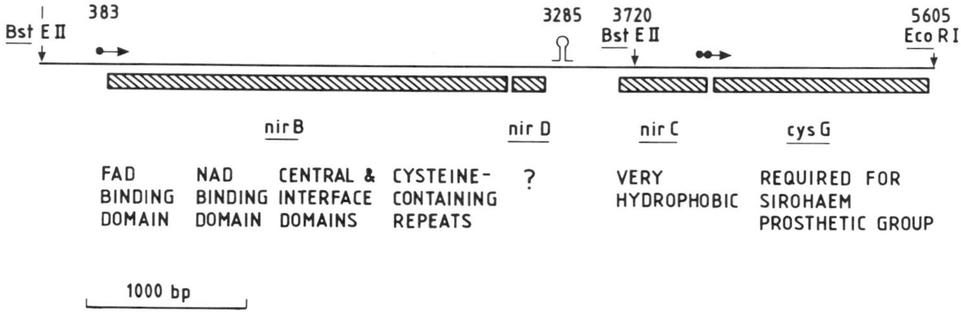

Fig. 4. The nirB operon of E. coli K-12.

The final gene in the nirB operon is cysG which codes for a protein essential for the synthesis of the sirohaem prosthetic group of nitrite reductase. The CYSG protein is probably the methylase which converts uroporphyrinogen III into factor 2, the precursor of sirohaem. Although cysG is transcribed from the nirB promoter during anaerobic growth, it is expressed aerobically from its own promoter to ensure that sirohaem is available for SO_3^{2-} reduction to S^{2-} in the cysteine biosynthetic pathway even when the nirB promoter is inactivated by O_2 (Peakman, Cole & Busby, unpublished results).

Synthesis of the NADH-dependent nitrite reductase is regulated by dual control mechanisms, NO_2^- induction during anaerobic growth but total repression during aerobic growth. Operon fusion studies have established that both of these responses are mediated at the level of transcription (Griffiths & Cole, 1987). Although, as mentioned before, some chlorate resistant mutants are constitutive for nitrite reductase synthesis during anaerobic growth in the absence of NO_2^-, such mutants remain sensitive to oxygen repression.

HOW OXYGEN REGULATES NITRITE REDUCTION IN E. COLI

The nirB promoter requires the FNR protein, product of the fnr gene, to be expressed in response to O_2 starvation. Indeed, the first nitrite reductase mutants to be described were the nirA mutants of Cole & Ward (1973) which were subsequently shown to be identical to the fnr mutants described by Lambden and Guest (1976) (Newman & Cole, 1978). Many other aspects of anaerobic metabolism are also regulated by the FNR protein; examples include the induction of dimethylsulfoxide, nitrate, fumarate, and trimethylamine-N-oxide reductases as well as cytochrome c_{552} synthesis and anaerobic repression of the respiratory enzymes NADH dehydrogenase and cytochrome oxidase o.

The fnr gene is autoregulated, but only during anaerobic growth (Spiro & Guest, 1987b). This suggests that the FNR protein is inactive in aerated cultures, but becomes activated, possibly as a result of a conformational change, during anaerobic growth. The FNR protein shows considerable homology with the cyclic AMP binding protein, CRP (product of the crp gene), which responds to conditions of carbon starvation by forming a complex with cAMP. This complex then activates genes required for the metabolism of secondary carbon sources. Like CRP, the FNR protein contains a helix-turn-helix motif typical of many DNA binding proteins (Pabo & Sauer, 1984), but there are also the following

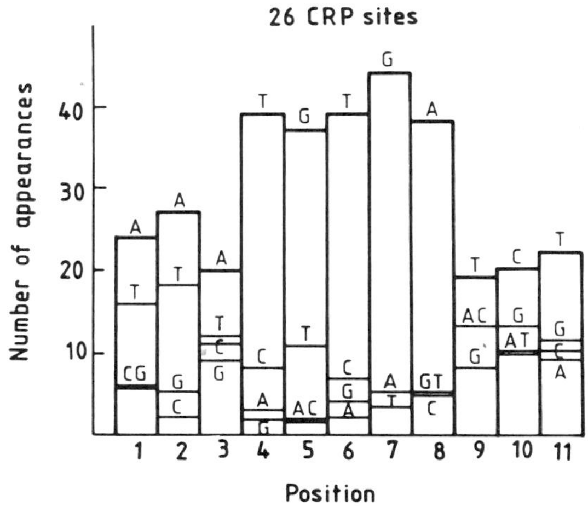

Fig. 5. Frequency of appearance of bases at CRP binding sites. The histogram shows the number of times each base appears at each of the 11 positions in the 52 CRP half-sites of 26 CRP-dependent promoters (each half site binds one CRP monomer).

three major differences between these two transcriptional activator proteins:

(1) FNR lacks key amino acid residues which enable CRP to interact with its signal molecule cAMP: presumably this is how FNR and CRP respond to different physiological signals (anaerobiosis and an elevated cAMP concentration due to starvation, respectively).

(2) The amino acid terminus of FNR includes a cysteine-rich region not present in CRP which might sense the redox state of the cell.

(3) The putative DNA binding domains of FNR and CRP are different at three key amino acid residues, presumably accounting for their specificity for different promoters.

Spiro & Guest (1987a) replaced these three key amino acids in the presumed DNA binding helix of FNR with those found in CRP. The modified FNR protein now activated the CRP-dependent lac operon of E. coli in response to a switch from aerobic to anaerobic growth. This provided the first clear evidence that FNR is indeed a DNA binding protein.

The CRP protein is a sequence dependent DNA binding protein. This was initally deduced by noticing that the binding site for CRP usually includes the motif 5'-TGTGA-3' separated by 6 bases from an imperfect inverted repeat motif, 5'-TCACA-3'. By comparing the DNA sequences of 26 CRP-dependent promoters, Bell et al. (1989), concluded that a typical CRP site extends for at least 22 bases consisting of two 11bp inverted repeats. By scoring the frequency of occurrence of each of the 4 bases at each position in these promoters, an 11 bp consensus sequence for each subunit of the CRP dimer was deduced (Fig. 5).

We have recently identified a similar possible FNR binding site in the nirB promoter. First, results of a deduction analysis (Figure 6) established that bases

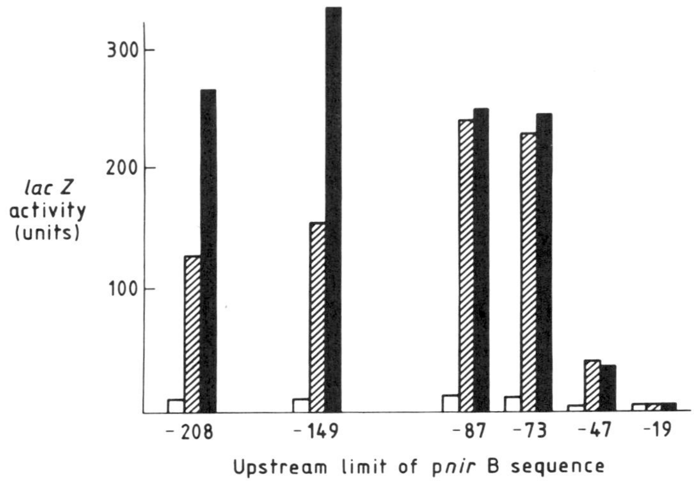

Fig. 6. Deletion analysis of the nirB promoter. Histogram showing the ß-galactosidase activity due to transcription from the truncated promoter fragments after aerobic (open boxes) or anaerobic growth in the presence (solid boxes) or absence (cross-hatched boxes) of NO_2^-.

between -147 and -87 with respect to the transcription start point are required for NO_2^- induction, and sequence downstream from base -55 is required for FNR-dependent activation during anaerobic growth (Jayaraman et al.,1988). This region between position -55 and the transcription start site includes the sequence 5'-TTGAT-N4-ATCAA-3' which again is a perfect hyphenated inverted repeat similar to the consensus CRP binding site. Two lines of evidence strongly suggested that this region includes a target site at which the FNR protein activates nitrite reductase transcription during anaerobic growth.

(a) A hybrid promoter was constructed in vitro in which the CRP site of the galactose promoter P1 was replaced by the putative FNR binding site of the nirB promoter. This hybrid promoter was actively transcribed only during anaerobic growth and was dependent upon the FNR protein (Jayaraman et al., 1988).

(b) Deletions and mutations in this region of dyad symmetry abolish or severely diminish both promoter strength and anaerobic induction (Jayaraman et al., 1989).

A consensus sequence for an FNR binding site was then deduced by comparing the published DNA sequences of 7 FNR dependent promoters (Jayaraman et al., 1989). Once again a hyphenated inverted repeat was suggested which differed significantly from the corresponding CRP consensus sequence at position 5 where a G in the CRP sequence is replaced by a T in the FNR sequence. Thus in a 22bp site, a C at position 18 of the CRP site would be replaced by an A in the corresponding FNR site (Figure 7). To determine whether these differences were sufficient to alter the sensitivity of a promoter from CRP-dependent activation to FNR-dependent activation, a set of four synthetic promoters was constructed in which a naturally occurring (and therefore "imperfect") CRP site was replaced by a consensus CRP site, or by derivatives of the consensus CRP site in which the G at position 5 was replaced by a T, the C

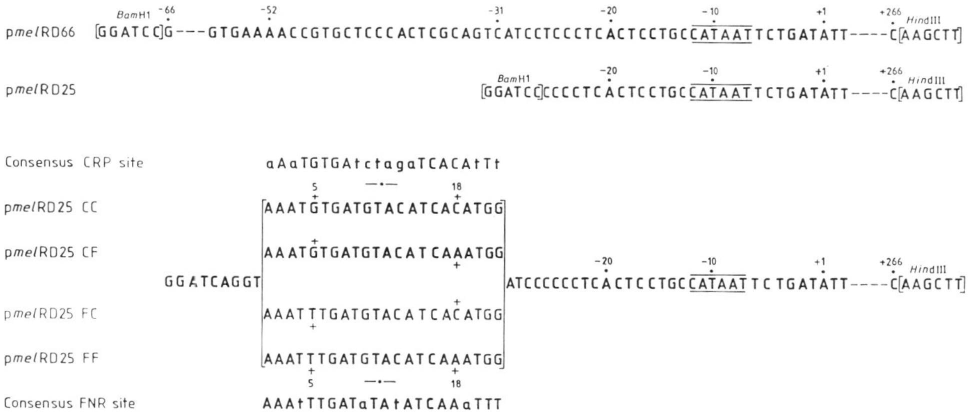

Fig. 7. Cloning of DNA sequences which determine promoter selection by the E. coli transcription activator proteins, CRP and FNR.

at position 18 was replaced by an A, or both positions 5 and 18 were altered to construct the consensus FNR binding sequence (Bell et al., 1989; Figure 7). The melR promoter, which is totally dependent upon the CRP protein, was used for these experiments and the N terminus of the melR gene was fused to the lacZ gene of E. coli to provide a convenient way of assaying promoter activity. Promoter activity which was dependent upon the CRP protein was determined by transforming CRP^+ FNR^- bacteria with recombinant plasmids carrying the hybrid promoters and assaying ß-galactosidase activity after growing the transformants aerobically in minimal medium under conditions of minimal catabolite repression. Conversely, FNR-dependent promoter activity was assayed using transformants of a CRP^- FNR^+ host which had been grown anaerobically in rich medium supplemented with glucose as carbon source.

From the results shown in Figure 8a it can be seen that the synthetic promoter containing the consensus CRP site, 'CC', was both more active and more dependent on the CRP protein than the wild-type melR promoter. The activity of the hybrid promoters in which one half-site was identical to the CRP consensus but the other was an FNR site ('CF' and 'FC') were similar to that of the naturally occurring melR promoter, but the 'FF' promoter was totally inactive. We conclude that substitution of just 2 bases in the 22 base consensus CRP site is sufficient to abolish promoter activation by the CRP protein. The crucial question, however, was whether these changes were sufficient to enable the FNR protein to activate this promoter during anaerobic growth.

Results of experiments to determine the ability of the FNR protein to activate the same set of synthetic promoters during anaerobic growth are shown in Figure 8b. Both the naturally occurring and consensus 'CC' promoters were totally inactive in the absence of a functional CRP protein, as expected. In contrast, the promoter with the 'FF' sequence was now extremely active; furthermore, this activity was totally dependent on a functional FNR protein. The activities of the hybrid promoters 'CF' and 'FC' and their dependence on FNR were intermediate between those of the 'FF' and 'CC' constructs.

Two significant conclusions can be drawn from these experiments. First, the FNR protein activates genes required for anaerobic electron transfer reactions,

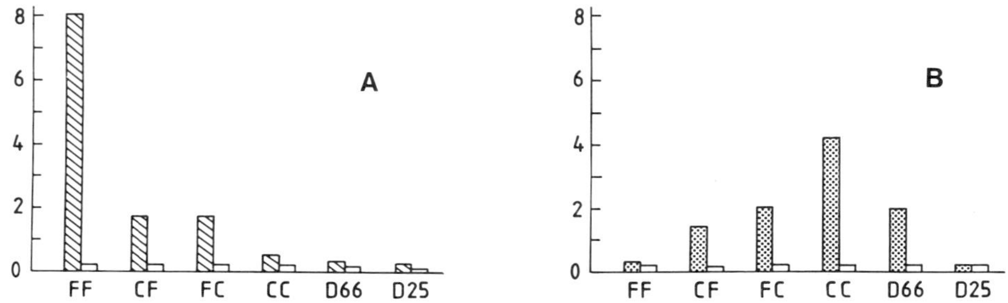

Fig. 8. Identification of bases involved in the discrimination between FNR and CRP. The activities of the synthetic promoters bases on melR (Fig. 7) were compared in Crp$^+$ Fnr$^-$ and Crp$^-$ Fnr$^-$ hosts (Fig. 8A) after aerobic growth with fructose; or in Crp$^-$ Fnr$^+$ and Crp$^-$ Fnr$^-$ hosts after anaerobic growth with glucose (8B). Promoter activities are standard ß-galactosidase units, x 10^{-3}.

including NO_3^- and NO_2^- reduction to NH_4^+, by activating transcription of the structural genes of the required enzymes. Secondly, FNR is a sequence-dependent DNA binding protein which, like the CRP protein, apparently binds as a dimer, each monomer requiring a consensus half-site to activate transcription.

NITRITE INDUCTION AND WHY THERE ARE TWO PATHWAYS FOR NITRITE DISSIMILATION

It was stated previously that nitrite reductase synthesis, like nitrate reductase synthesis, is regulated by two control mechanisms, oxygen repression and substrate induction during anaerobic growth. Induction of nitrate reductase synthesis by NO_3^- requires a second transcription activator protein, NARL, which was recently shown to be part of a two-component regulatory system (Stewart et al., 1989; Drummond et al., 1986). The presence of NO_3^- is sensed by NARX, the product of the narX gene (Figure 2). NARX then activates NARL, converting it into a transcription activator which is essential for inducing the narGHJI operon to full expression (Stewart, 1982; but for a more cautious interpretation, see Stewart et al, 1989). The NARL protein also functions as a repressor of fumarate reductase synthesis, but only when the more powerful electron acceptor, NO_3^-, is available. The NARL protein therefore achieves the hierarchical control of anaerobic respiratory metabolism by ensuring that the energetically most favorable electron acceptor is used preferentially.

Nitrite is the product of NO_3^- reduction: one might argue that because intracellular NO_2^- is toxic, nirB expression should be regulated coordinately with narGHJI, this regulation being mediated by the NARL protein. An alternative view is that because NO_2^- is a less powerful oxidant than NO_3^-, nirB transcription should be repressed by NARL. Data in Table 2 show that the former is correct: transcription of the nirB operon is stimulated rather than repressed by a functional NARL protein. This stimulation is, however, far less than the activation of the nar operon by NARL.

The data in Table 2 also establish that NARL is not essential for NO_2^- induction of the nirB operon. The deletion analysis of the nirB promoter (Fig. 6)

Table 2. Activation of the nirB promoter by the NARL protein

Strain	Activity of the nirB promoter after growing the bacteria with		
	O_2	$Ø_2$	$Ø_2+NO_2^-$
JCB 321 narL$^+$ Chl^r	60	1020	1220
JCB 321 narL Chl^r	30	400	480
JCB 321 narL$^+$ Chl^s	30	190	650
JCB 321 narL Chl^s	15	115	290

Numbers are ß -galactosidase activities (units / mg dry weight) of chromosomal nirB::lacZ operon fusion strains. $Ø_2$ = anaerobic incubation.

revealed that DNA sequence upstream from the FNR binding site is required for NO_2^- induction. Induction by NO_2^- is also lost by some chlorate-resistant mutants, but the mechanism of NO_2^- activation remains to be determined. One possibility is clearly that the molybdenum cofactor or one of its precursors is a co-repressor of nirB but, if so, why has the repressor gene not been identified?

Regulation of nitrite reduction by formate

Synthesis of the formate-dependent nitrite reductase and cytochrome c_{552} are also regulated by oxygen repression and NO_2^- induction during anaerobic growth (Page & Cole, unpublished). Indeed, the dependence of Nrf expression on NO_2^- is far greater than for Nir. As Nrf activity results in the generation of a membrane potential, it was suggested that although NARL was an activator of Nir, it would repress the expression of the Nrf genes. The results in Table 3 fully support this suggestion. Furthermore they provide for the first time an explanation of why there are two independent pathways in some bacteria for NO_2^- dissimilation to NH_4^+. The primary role of the NADH-dependent pathway is to reduce intracellular NO_2^- generated during NO_3^- reduction: by contrast, the Nrf pathway functions as a secondary energy-generating pathway and is therefore

Table 3. Effect of a narL mutation on nitrite reduction by formate and cytochrome c_{552} synthesis.

Strain	Nitrite reduction by formate nmol min^{-1} (mg dry wt.)$^{-1}$	Cytochrome c_{552} pmol (mg protein)$^{-1}$
narL$^+$	0.32	86
narL	1.5	570

Bacteria were grown anaerobically with 0.4% glycerol and 20 mM KNO_3

subject to NO_3^- repression mediated by the NARL-NARX regulatory system (Table 3). It will be interesting to see how widely these two types of NO_2^- dissimilation occur in organisms isolated from natural environments.

SUMMARY

Two types of dissimilatory NO_3^- reduction by anaerobic bacteria have been recognised. Fermentative bacteria predominantly reduce NO_3^- to NH_4^+; in contrast, most respiratory bacteria are denitrifiers. Both types are usually repressed during aerobic growth. Further sub-groups have been defined on the basis of the types of NO_3^- and NO_2^- reductases synthesised. Denitrifiers dominate environments in which the N:C ratio is high; conversely, NO_3^- is primarily reduced to NH_4^+ in carbon-rich niches.

Enzymes which catalyse NO_2^- dissimilation to NH_4^+ are of two types. The hexa-haem c-type cytochromes located in the periplasm are the terminal components of electron transfer chains in which formate is the electron donor. The second type, so far found only in enteric bacteria, are cytoplasmic flavoproteins with sirohaem as the prosthetic group. Synthesis of both types of NO_2^- reductase is regulated by dual control mechanisms at the level of initiation of transcription of the structural genes. Oxygen prevents the synthesis of both NO_3^- and NO_2^- reductases by inactivating the global regulator protein, FNR, which is a transcription activator very similar to the cyclic AMP receptor protein required for the relief of catabolite repression. The availability of NO_3^- is sensed by a two-component regulatory system, NARX-NARL; the NARX protein is a NO_3^- sensor which interacts with the DNA binding protein, NARL.

ACKNOWLEDGEMENTS

The author is grateful to the many colleagues cited in the text who have contributed to the studies of nitrite reduction by E. coli. Particular thanks go to Steve Busby who has designed and co-supervised the molecular genetic experiments on the nirB operon, and to A. Bell, K. Gaston, L. Griffiths, L. Page, T. Peakman and C. Webster for providing data from recent unpublished experiments.

REFERENCES

Abou-Jaoude, A., Chippaux, M., & Pascal, M.-C., 1979a, Formate-nitrite reduction in Escherichia coli K12. 1. Physiological study of the system, Eur. J. Biochem., 95:309.

Abou-Jaoude, A., Pascal, M.C. & Chippaux, M., 1979b, Formate-nitrite reduction in Escherichia coli K12. 2. Identification of components involved in the electron transfer, Eur. J. Biochem., 95:315.

Bell, A.I., Gaston, K.L., Cole, J.A. & Busby, S.J.W, 1989, Cloning of binding sequences for the Escherichia coli transcription activators, FNR and CRP; location of bases involved i discrimination between FNR and CRP. Nucl. Acids Res. 17:3865.

Bielkie, A., & Ketchum, P.A., 1989, Purification of nitrate reductase from Xanthomonas maltophila. Abstract K-121. Abstracts of the Annual Meeting of the American Society for Microbiology. p.265.

Bonnefoy, V., Burini, J.F., Giordano, G., Pascal, M.-C. & Chippaux, M., 1987, Presence in the "silent" terminus region of Escherichia coli K12 chromosome of cryptic gene(s) encoding a new nitrate reductase, Mol. Microbiol., 1:143.

Caskey, W.H., & Tiedje, J.M., 1979, Evidence for clostridia as agents of dissimilatory reduction of nitrate to ammonium in soil, Soil Sci. Soc. Am. J., 43:931.

Cole, J.A., 1968. Cytochrome $\underline{c}_{552}$ and nitrite reduction in Escherichia coli, Biochimica et Biophysica Acta., 162:356.

Cole, J.A., 1988, Assimilatory and dissimilatory reduction of nitrate to ammonia. In: "The Nitrogen and Sulphur Cycles" Eds. Cole, J.A., & Ferguson, S.J., Symposium 42, The Society for General Microbiology, Cambridge University Press, U.K.

Cole, J.A., 1989, Physiology, biochemistry and genetics of nitrite reduction by Escherichia coli, In: "Molecular and Genetic Aspects of Nitrate Assimilation" Eds. Wray, J.L., & Kinghorn, J.R., Oxford University Press, U.K.

Cole, J.A., & Brown, C.M., 1980, Nitrite reduction to ammonium by fermentative bacteria: a short circuit in the biological nitrogen cycle, FEMS Microbiol. Lett., 7:65.

Cole, J.A., & Ward, F.B., 1973, Nitrite reductase-deficient mutants of Escherichia coli K12, J. Gen. Microbiol., 76:21.

Coleman, K.J., Cornish-Bowden, A., & COLE, J.A., 1978a, Purification and properties of nitrite reductase from Escherichia coli K12, Biochem. J. 175, 483.

Coleman, K.J., Cornish-Bowden, A., & Cole J.A., 1978b, Activation of nitrite reductase from Escherichia coli K12 by oxidized nicotine-adenine dinucleotide, Biochem. J., 175:495.

Drummund, M.H., Whitty, P.W., & Wootton, J.C., 1986, Sequence and domain relationships of ntrC and nifA from Klebsiella pneumoniae; homologies to other regulatory proteins, EMBO J., 5:441.

Fazzolari Correa, E., Mariotti, A., & Germon, J.C., 1989, Dissimilatory nitrate reduction into ammonium by Enterobacter amnigenus: ability and competition with denitrification with Agrobacterium radiobacter, Abstract N-83, Abstracts of the Annual Meeting of the American Society for Microbiology, p.299.

Forsythe, S.J., Dolby, J.M., Webster, A.D.B., & Cole, J.A., 1988, Nitrate- and nitrite-reducing bacteria in the achlorhydric stomach, J. Med. Microbiol., 25: 253.

Fujita, T., & Sato, R., 1966a, Studies on soluble cytochromes in Enterobacteriaceae III. Localization of cytochrome $\underline{c}_{552}$ in the surface layer of cells, J. Biochem., 60:568.

Fujita, T., & Sato, R., 1966b, Studies on soluble cytochromes in Enterobacteriaceae IV. Possible involvement of cytochrome $\underline{c}_{552}$ in anaerobic nitrite metabolism, The Journal of Biochemistry, 60:691.

Fujita, T., & Sato, R., 1967, Studies on soluble cytochromes in Enterobacteriaceae V. Nitrite-dependent gas evolution in cells containing cytochrome $\underline{c}_{552}$, J. Biochem., 62:230.

Gray, C.T., Wimpenny, J.W.T., Hughes, D.E., & Ranlett, M., 1963, A soluble c-type cytochrome from anaerobically grown Escherichia coli and various Enterobacteriaceae, Biochim. Biophys. Acta., 67:157.

Griffiths, L., & Cole, J.A., 1987, Lack of redox control of the anaerobically-induced $nirB^+$ gene of Escherichia coli K-12. Arch. Microbiol., 147:364.

Haddock, B.A., & Jones, C.W., 1977, Bacterial respiration, Bacteriol. Rev., 41: 47.

Hasan, S.M., & Hall, J.B., 1975, The physiological function of nitrate reduction in Clostridium perfringens, J. Gen. Microbiol., 87:120.

Ingledew, W.J., & Poole, R.K., 1984, The respiratory chains of Escherichia coli. Microbiol. Rev., 48:222.

Jackson, R.H., Cornish-Bowden, A., & Cole, J.A., 1981a, Prosthetic groups of the NADH-dependent nitrite reductase from Escherichia coli K12, Bioch. J., 193:861.

Jackson, R.H., Cole, J.A., & Cornish-Bowden, A., 1981b, The steady-state kinetics of the NADH-dependent nitrite reductase from Escherichia coli K12: nitrite and hydroxylamine reduction, Biochem. J., 199: 171.

Jackson, R.H., Cole, J.A., & Cornish-Bowden, A., 1982, The steady state kinetics of the NADH-dependent nitrite reductase from Escherichia coli: the reduction of single-electron acceptors, Biochem. J., 203: 505.

Jayaraman, P.S., Peakman, T.C., Busby, S.J.W., Quincey, R.V., & Cole, J.A., 1987, Location and sequence of the promoter of the gene for the NADH-dependent nitrite reductase of Escherichia coli and its regulation by oxygen, the FNR protein and nitrite, J. Mol. Biol., 196:781.

Jayaraman, P.-S., Gaston, K.L., Cole, J.A., & Busby, S.J.W., 1988, The nirB promoter of Escherichia coli: location of nucleotide sequences essential for regulation by oxygen, the FNR protein and nitrite, Mol. Microbiol., 2:527.

Jayaraman, P.-S., Cole, J.A., & Busby, S.J.W., 1989, Mutational analysis of the sequence at the FNR-dependent nirB promoter in Escherichia coli, Nucl. Acids Res., 17:135.

Jeter, R.M., Sias, S.R., & Ingraham, J.L., 1984, Chromosomal location and function of genes affecting Pseudomonas aeruginosa nitrite assimilation, J. Bacterol., 157, 673.

Jones, G.A., 1972, Dissimilatory metabolism of nitrate by the rumen microbiota, Can. J. Microbiol., 18:1783.

Kajie, S., & Anraku, Y., 1986, Purification of a hexaheme cytochrome c_{552} from Escherichia coli K12 and its properties as a nitrite reductase, Eur. J. Biochem., 154:457.

Kaspar, H.F., & Tiedje, J.M., 1982, Dissimilatory reduction of nitrate and nitrite in the bovine rumen: nitrous oxide production and effect of acetylene, Appl. Environ. Microbiol., 41:705.

Kemp, J.D., & Atkinson, D.E., 1966, Nitrite reductase of Escherichia coli specific for reduced nicotinamide adenine dinucleotide, J. Bacteriol., 92:628.

Kemp, J.D., Atkinson, D.E., Ehret, A., & Lazzarimi, R.A., 1963, Evidence for the identity of the nicotinamide adenine dinucleotide phosphate-specific sulphite and nitrite reductases of Escherichia coli, J. Biol. Chem., 238:3466.

Koike, I., & Hattori, A., 1978a, Denitrification and ammonia formation in anaerobic coastal sediments. Appl. Environ. Microbiol., 35:278.

Koike, I., & Hattori, A., 1978b, Simultaneous determination of nitrification andnitrate reduction in coastal sediments by a ^{15}N dilution technique, Appl. Environ. Microbiol., 35:853.

Lambden, P.R., & Guest, J.R., 1976, Mutants of Escherichia coli K12 unable to use fumarate as an electron acceptor, J. Gen. Microbiol., 97:145.

Lang, F. & Hochstein, L.I., 1989, Purification and properties of a nitrate reductase from Halobacterium denitrificans. Abstract K-122. Abstracts of the Annual Meeting of the American Society for Microbiology, p.265.

Lazzarini, R.A., & Atkinson, D.E., 1961, A TPN specific nitrite reductase from E. coli, J. Biol. Chem., 236:3330.

Lewis, D., 1951, The metabolism of nitrate and nitrite in sheep.2. Hydrogen donators in nitrate reduction by rumen microorganisms in vitro, Biochem.J., 49:149.

Liu, M.-C., Baker, B.W., Liu, M.-Y., & Dao, T.N., 1988, Purification of Vibrio fischeri nitrite reductase and its characterisation as a hexaheme c-type cytochrome, Arch. Biochem. Biophys., 262:259.

Liu, M.-C., & Peck, H.D., 1981, The isolation of a hexaheme cytochrome from Desulfovibrio desulfuricans and its identification as a new type of nitrite reductase, J. Biol. Chem., 256:13159.

Macdonald, H., & Cole, J.A., 1985, Molecular cloning and functional analysis of the cysG and nirB genes of Escherichia coli K12, two closely-linked genes required for NADH-dependent nitrite reductase activity, Mol. Gen.Genet. 200:328.

Massey, V., & Veeger, C., 1961, Studies on the reaction mechanisms of lipoyl dehydrogenase, Biochim. Biophys. Acta., 48:33.

Motteram, P.A.S., Mc Carthy, J.E.G., Ferguson, S.J., Jackson, J.B., & Cole, J.A., 1981, Energy conservation during the formate-dependent reduction of nitrite by Escherichia coli, FEMS Microbiol. Letts., 12:317.

Newman, B.M., & Cole, J.A., 1978, The chromosomal location and pleiotropic effects of mutations in the $nirA^+$ gene of Escherichia coli K12: the essential role of $nirA^+$ in nitrite reduction and in other anaerobic redox reactions, J. Gen. Microbiol., 106:1.

Pabo, C.O., & Sauer, R.T., 1984, Protein-DNA recognition, Ann. Rev. Biochem., 53:293.

Peck, H.D., & Gest, H., 1957, Formic dehydrogenase and the hydrogenlyase enzyme complex in Coli-Aerogenes bacteria, J. Bacteriol., 73:706.

Pope, N.R., & Cole, J.A., 1982, Generation of a membrane potential by one of two independent pathways for nitrite reduction by Escherichia coli, J. Gen. Microbiol., 128:319.

Pope, N.R., & Cole, J.A., 1984, Pyruvate and ethanol as electron donors for nitrite reduction by Escherichia coli K12, J. Gen. Microbiol., 130:1279.

Postgate, J.R., 1954, Presence of cytochrome in an obligate anaerobe, Biochem. J., 56:xi

Rönner, V., & Sørensen, F., 1985, Denitrification rates in the low-oxygen waters of the stratified Baltic proper, Appl. Environ. Microbiol., 50:801.

Samuelsson, M.-O., & Rönner, V., 1982, Ammonium production by dissimilatory nitrate reducers isolated from Baltic sea water, as indicated by ^{15}N study, Appl. Environ. Microbiol., 44:1241.

Schroder, I., Robertson, A.M., Bokranz, M., Unden, G., Bocher, R., & Kroger, A., 1985, The membranous nitrite reductase involved in the electron transport of Wolinella succinogenes, Arch. Microbiol., 140:380.

Seki, S., Hattori, Y., Hasegawa, T., Haraguchi, H., & Ishimoto, M., 1987, Studies on nitrate reductase of Clostridium perfringens. IV. Identification of metals, molybdenum cofactor and iron-sulfur cluster, J. Biochem., 101:503.

Seki-Chiba, S., & Ishimoto, M., 1977, Studies on nitrate reductase of Clostridium perfringens: I. Purification, some properties, and effect of tungstate on its formation, J. Biochem., 82:1663.

Smith, M., 1983, Nitrous oxide production by Escherichia coli is correlated with nitrate reductase activity, Appl. Environ. Microbiol., 45:1545.

Sørensen, J., 1978, Capacity for denitrification and reduction of nitrate to ammonia in a coastal marine sediment, Appl. Environ. Microbiol., 35:301.

Spiro, S., & Guest, J.R., 1987a, Activation of the lac operon of Escherichia coli by a mutant FNR protein, Mol. Microbiol., 1:53.

Spiro,S., & Guest, J.R., 1987b, Regulation and over-production of the fnr gene of Escherichia coli, J. Gen. Microbiol., 133:3279.

Steenkemp, D.J., & Peck, H.D., 1980, The association of hydrogenase and dithionite reductase of Desulfovibrio desulfuricans, Biochem. Biophys. Res. Commun., 94:41.

Steenkemp, D.J., & Peck, H.D., 1981, Proton translocation associated with nitrite respiration in Desulfovibrio desulfuricans, J. Biol. Chem., 256:5450.

Stewart, V., 1982, Requirement of Fnr and Narl functions for nitrate reductase expression in Escherichia coli K12, J. Bacteriol., 151:1320.

Stewart, V., Parales, J., & Merkel, S.M., 1989, Structure of genes narL and narX of the nar (Nitrate Reductase) locus in Escherichia coli K-12, J. Bacteriol., 171: 2229.

Van't Riet, J., & Planta, R.J., 1975, Purification, structure and properties of the respiratory nitrate reductase of Klebsiella aerogenes, Biochim. Biophys. Acta., 379:81.

Wimpenny, J.W.T., & Cole, J.A., 1967, The regulation of metabolism of facultative bacteria.III. The effect of nitrate. Biochim. Biophys. Acta., 148:233.

Wimpenny, J.W.T., & Warmsley, A.M.H., 1968, The effect of nitrate on Krebs cycle enzymes in various bacteria, Biochim. Biophys. Acta, 156:297.

Woods, D.D., 1938, The reduction of nitrate to ammonia by Clostridium welchii, Biochem. J., 32:2000.

Yamamoto, I., Shimizu, H., Tsuji,T., & Ishimoto, M., 1986, Purification and properties of nitrate reductase from Mitsuokella multiacidus, J. Biochem, 99:961.

Zarowny, D.P., & Sanwal, B.D., 1963, Characterization of an NADH-specific nitrite reductase from E. coli K12, Can. J. Microbiol., 9:531.

ROLE OF ENVIRONMENTAL FACTORS IN REGULATING NITRATE RESPIRATION IN INTERTIDAL SEDIMENTS

Rodney A. Herbert[1] and David B. Nedwell[2]

[1]Department of Biological Sciences, University of Dundee
Dundee DD1 4HN
[2]Department of Biology, University of Essex
Colchester CO4 3SQ
U.K.

INTRODUCTION

The microbial decomposition of moribund organic matter involves the consumption of an equivalent quantity of mineral oxidant, either directly as in the case of respiratory metabolism, or indirectly as a consequence of fermentation. Oxidants used in respiratory metabolism include O_2, NO_3^- and NO_2^-, manganese and ferric oxides, SO_4^- and CO_2. Organic matter decomposition within sediments results in the depletion of these oxidants. As a consequence microbial processes which take place in the surface layers of sediments are not only intimately related to the quality and quantity of organic carbon present but to the concentration and availability of potential inorganic electron acceptors. Aerobic respiration, which takes place in the surface layers of sediments, results in the rapid depletion of O_2 and alternative electron acceptors, if present, are then successively utilised by different physiological groups of microorganisms thus creating more reduced conditions (Kaplan, 1979). In the absence of O_2 many aerobic and facultatively anaerobic bacteria are able to carry out NO_3^- respiration using NO_3^- as terminal electron acceptor (Payne, 1973; Herbert, 1982). In NO_3^- respiration the end-product(s) of reduction depend (Abd. Aziz and Nedwell, 1986) upon the microorganism involved, growth conditions and may be either NO_2^-, NH_4^+, N_2O or N_2. Whilst NO_3^- respiration to NO_2^- is a fairly widespread property of chemosynthetic bacteria the further reduction of NO_2^- is carried out by fewer species. When the end-products are gaseous the process is termed denitrification and this represents a net loss of nitrogen from an ecosystem. In contrast a number of fermentative bacteria have been shown to reduce NO_2^- to NH_4^+ with the resultant conservation of nitrogen in a potentially useful form (Hadjepetrou and Stouthamer, 1965; Dunn et al., 1978; Cole and Brown, 1982; Macfarlane and Herbert, 1982). This process termed 'NO_3^- ammonification' has been shown to occur widely in freshwater, estuarine and marine environments.

In this review we discuss the methods for isolating and enumerating NO_3^- respiring bacteria from marine, estuarine and salt marsh sediments and the factors regulating NO_3^- respiration <u>in situ</u>.

Denitrification in Soil and Sediment, Edited by N. P. Revsbech and J. Sørensen, Plenum Press, New York, 1990

ISOLATION OF NO_3^- RESPIRING BACTERIA

Whilst a large number of studies have been undertaken to measure denitrification and more recently 'NO_3^- ammonification' rates in estuarine, marine and saltmarsh sediments (Koike and Hattori, 1976; Oremland et al., 1984; Macfarlane and Herbert, 1984; Koike and Sørensen, 1988) relatively little attention has been focussed on isolating the microbial populations involved. The fundamental difficulty facing microbial ecologists is how to isolate those microbial populations which are important in carrying out a particular transformation and not those members of the resident microflora which grow most rapidly under the imposed culture conditions. Direct plating techniques (agar plates), whilst the simplest system to use, suffer from several disadvantages notably that the nutrient levels employed are unrealistically high in order to promote the growth of large populations of bacteria and there is no removal of toxic end-products of metabolism. Additionally, there is no control of growth rate and hence microorganisms growing most rapidly under the imposed conditions predominate. In an attempt to overcome these inherent difficulties Jannasch (1967) pioneered the use of continuous culture methods to isolate microorganisms. The advantage of open systems is that nutrient levels approaching those found in natural environments can be used, metabolic end- products do not accumulate and growth rates can be precisely controlled by altering the dilution rate. Direct plating methods may give information on the identity of the total NO_3^- reducing population present without providing any indication of which bacteria are most active. Conversely, chemostat enrichment systems yield no data on the composition of the overall populations present but give valuable information regarding those bacteria best adapted to the prevailing physio-chemical conditions of NO_3^- availability, temperature and salinity.

In experiments involving direct plating techniques for isolating NO_3^- respiring bacteria two media were used viz:- Giltay agar contained asparagine, citrate, 0.2M NaCl and KNO_3 whilst "Nitrate" agar comprised nutrient agar (Difco), 0.2M NaCl and KNO_3. Data presented in Table 1 and 2 show the identification to genus level of 100 bacterial isolates selected randomly from each media type after inoculation with appropriate dilutions of estuarine sediment and incubation under a H_2/CO_2 atmosphere at 15°C for 7 days. Nitrate respiring bacteria isolated on Giltay agar (Table 1) were all Gram-negative, catalase positive bacilli or cocco-bacilli. The dominant organisms were identified as *Pseudomonas* spp with *Aeromonas/Vibrio* spp, *Acinetobacter* spp and members of the enterobacteriaceae accounting for approximately equal numbers of the remaining isolates. In contrast, on Nitrate agar the situation was reversed with fermentative types (*Aeromonas/Vibrio* spp and enterobacteria) predominating over the pseudomonads (Table 2). These data underline the importance of the choice of carbon source used in different selective media and how it can influence the dominance of different groups of NO_3^- respiring bacteria. Citrate, the carbon source in Giltay's medium selects for oxidative bacteria belonging to the genera *Pseudomonas* and *Acinetobacter*, whereas Nitrate agar is more nutrient rich and is less selective. The significance of these observations is that when these media were employed for the enumeration of NO_3^- respiring bacteria substantially higher counts of NO_3^- respirers were always recorded on Nitrate agar confirming that this medium was less selective (Dunn et al., 1980). In a more extensive study of NO_3^- respiring bacteria populations in Scottish east-coast estuarine sediments Macfarlane and Herbert (1984) obtained similar

Table 1. Bacteria isolated from Kingoodie Bay sediments on Giltay's agar and incubated anaerobically at 15°C (100 colonies selected at random from plate counts)

Number of Bacteria	Catalase	Oxidase	Oxidative - O Fermentative - F	Identification
44	+	+	O	<u>Pseudomonas</u> sp.
20	+	+	F	<u>Aeromonas/Vibrio</u> sp.
18	+	-	O	<u>Acinetobacter</u> sp.
14	+	-	F	Enterobacteria
4	-	-	-	No growth

results. In an attempt to avoid selection based on competition for NO_3^- Nedwell (1975) used peptone-sea water agar to isolate heterotrophic bacteria from two sediment sites in a tropical estuary. One site was eutrophic with high NO_3^- concentrations whereas the other had low NO_3^- concentrations in the overlying water. The isolates (434 in total) were tested for their abilities to use respire NO_3^- anaerobically. In both sites about 40% of the total heterotrophic isolates reduced NO_3^- to NO_2^-, and about 30% in both sites were NO_3^--ammonifiers.

The proportion of denitrifiers rose from 2% in the unpolluted site to 11% in the polluted site, suggesting selection of denitrifiers in the presence of high NO_3^- concentrations. In laboratory experiments the ratio of organic carbon : NO_3^- was shown to be important in determining the outcome of competition between denitrifiers and NO_3^- ammonifiers (Rehr and Klemme, 1989) but in field situations the amount of available organic carbon, and hence the true organic carbon : NO_3^- ratio, is usually unknown.

As an alternative approach to direct isolation methods we have carried out a series of continuous culture enrichments to isolate NO_3^- respiring bacteria under different salinity regimes (freshwater, half strength seawater and full strength seawater) and different carbon sources. The rationale and full experimental details of the methods employed are described in Dunn et al., (1980). The enrichments were run aerobically for 10-14 days to select against obligate anaerobes and

Table 2. Bacteria isolated from Kingoodie Bay sediments on nutrient agar + 1% w/v KNO_3 incubated anaerobically at 15°C (100 colonies selected at random from plate counts)

Number of Bacteria	Catalase	Oxidase	Oxidative - O Fermentative - F	Identification
21	+	+	O	<u>Pseudomonas</u> sp.
48	+	+	F	<u>Aeromonas/Vibrio</u> sp.
4	+	-	O	<u>Acinetobacter</u> sp.
24	+	-	F	Enterobacteria
3	-	-	-	No growth

Table 3. Bacteria isolated from Kingoodie Bay sediments by enrichment in continuous culture at different salinities and on different C- sources

Salinity (o/oo)	Carbon Source			
	Glycerol		Acetate	
0	Enterobacteria	77%	Acinetobacter sp.	87%
	Pseudomona sp.	22%	Pseudomona sp.	12%
	Pseudomonas sp.	1%	Enterobacteria	1%
17	Enterobacteria	67%	Enterobacteria	85%
	Aeromonas/Vibrio	33%	Pseudomonas sp.	15%
35	Enterobacteria	76%	Pseudomonas sp.	82%
	Enterobacteria	17%	Enterobacteria	12%
	Acinetobacter sp.	4%	Pseudomonas sp.	6%
	Enterobacteria	3%		

then anaerobically under a gas phase of high purity for a further 7 days. Data presented in Table 3 summarises the results obtained.

The final bacterial populations consisted of facultatively anaerobic bacteria able to grow anaerobically on glycerol or acetate as C-sources in the presence of NO_3^-. The dominant populations in the glycerol enrichments, irrespective of salinity, were members of the Enterobacteriaceae. More detailed identification showed that these bacteria were predominantly Klebsiella spp, but occasionally strains of E. coli were isolated. In contrast to the direct plating method Aeromonas/Vibrio sp were rarely encountered and it was only in the ½ strength seawater enrichment that they formed a significant component of the final population. In acetate enrichments pseudomonads became increasingly dominant with increasing salinity and under fully marine conditions constituted 88% of the final population.

Chemostat enrichments have also been successfully used by King and Nedwell (1984) to examine the seasonal changes in NO_3^- respiring populations in the River Colne. These workers used anaerobic carbon-limited chemostat enrichments with glucose as the organic substrate and NO_3^- in the growth medium, at 10°C to select for psychrotrophic NO_3^- respirers and at 25°C for mesophilic strains. Psychrotrophic and mesophilic populations were successfully isolated by this means with temperature characteristics similar to the populations selected in situ (King and Nedwell, 1984). When representative isolates were identified there was a preponderance of Pseudomonas spp at 10°C and of Vibrio spp and enterobacteria at 25°C.

COMPETITION FOR NITRATE

Field studies

Wheatland et al., (1959) demonstrated that the majority of NO_3^- added to a mixture of Thames mud and sewage was denitrified to nitrogen gas. The potential for NO_3^- reduction, particularly where there are high NO_3^- concentrations arising from agricultural run-off pollution, is considerable as NO_3^- reduction in sediments is almost invariably NO_3^--limited under natural conditions. Increase of either NO_3^- or organic matter (substrate concentration) tends to enhance NO_3^- reduction rates in aquatic sediments (van Kessel, 1978) indicating that NO_3^- reduction can be simultaneously NO_3^- and substrate limited.

For NO_3^- to be reduced in the anaerobic layers of sediments, it has to be transported into the sediment from the overlying water. The rate of transport of NO_3^- into the sediment is a function of the concentration gradient across the sediment-water interface, and usually exhibits a hyperbolic response to NO_3^- concentration in the overlying water (Nedwell, 1975; 1982). As transport rates of NO_3^- into the sediment tend to be rapid, the trends in dissolved NO_3^- concentrations within the pore waters of the surface layers of the sediment tend to reflect those of the overlying water. For example, along the estuary of the River Colne, in Essex, U.K., there are distinct gradients of NO_3^-, NO_2^- and NH_4^+ in the estuarine waters because of input of treated sewage effluent in the upper reaches (King and Nedwell, 1984; 1985). At any one site the concentrations of NO_3^- and NO_2^- in the pore waters of the surficial sediment are lower than in the overlying water, but changes between sites reflect the gradient in the water along the estuary. In contrast pore water NH_4^+ is higher than in the overlying water and there is no trend along the estuary, suggesting that it is principally derived within the sediment from ammonification of organic matter and is independent of the pollution gradient.

The pore water concentrations of NO_3^- in the sediments reflect NO_3^- concentrations in the overlying water. Concentration gradients in estuarine water therefore may kinetically affect NO_3^- reduction rates in the sediments. Nedwell (1975) showed that the dissolved NO_3^- concentration influenced the end-products and rate of NO_3^- reduction, the rate of denitrification and the proportion of the NO_3^- denitrified both increasing with increased dissolved NO_3^- concentration. At concentrations >0.3 mM NO_3^- was reduced to gaseous products, but denitrification was negligible at lower NO_3^- concentrations and NH_4^+ was the major end product of NO_3^- reduction. This observation in a tropical mangrove-lined estuary has since been confirmed in a number of other studies. At the NO_3^- concentrations of unpolluted salt marsh sediment denitrification accounts for <10% of NO_3^- reduction. Less than 1% of the ammonified nitrogen liberated from mineralised organic matter in a salt marsh sediment was nitrified and subsequently denitrified (Abd. Aziz and Nedwell, 1986).

Along the Colne estuary the first order rate constant for relative NO_3^- reduction increased along the concentration gradient, showing that the sedimentary microflora responded to the high NO_3^- levels in the water and became physiologically adapted to the extent that the community's capacity to reduce NO_3^- increased (King and Nedwell, 1987). On top of this faster rate of NO_3^- reduction the proportion of NO_3^- denitrified also increased (King and Nedwell, 1985; 1987). A number of possible reasons could be put forward to explain this observation. It could be that the denitrifiers were physiologically better adapted to compete for NO_3^- (and NO_2^-?) at high concentrations than the bacteria bringing about NO_3^- ammonification: and _vice versa_ at low concentrations. Indeed, in recent work Rehr and Klemme (1989) demonstrated with pure cultures of the denitrifying _Pseudomonas stutzeri_ and NO_3^- ammonifying _Citrobacter_ freundii and Klebsiella spp. that the C/N ratio required for denitrification was much lower than that required for NO_3^- ammonification. This may result in differential selection of these two groups of NO_3^--reducers under different C:N ratios in the field. Alternatively, the ecological observation could be explained if certain bacteria were able to change their NO_3^- metabolism in response to changes in the substrate:NO_3^- ratio. Nonetheless, the greater capacity for denitrification at higher NO_3^- concentrations argues for some expansion of the denitrifying component of the sedimentary microbial community.

Laboratory studies

In order to account for the recorded difference in dominance of bacterial types isolated by direct plating and continuous culture methods, and for differences in end products of NO_3^- reduction in the field, competition experiments were carried out using selected strains of nitrate respiring bacteria. The experiments were carried out in 1 litre all glass chemostats with a working volume of 750 ml using half strength Evans' medium (Evans et al., 1970) with either glycerol or acetate as C-source and nitrate as sole N-source. The growth temperature was 20°C and a dilution rate of 0.1 h^{-1} Data presented in Figure 1 show the outcome of competition between Klebsiella K312 and Vibrio V48 when grown on NO_3^- under C-limitation. Under these growth conditions both bacteria undergo cyclical oscillations in cell numbers with the Klebsiella sp dominant. Spent media analysis show NO_2^- is the principal end-product of NO_3^- reduction. In contrast, under NO_3^- limited conditions (Figure 2) Klebsiella K312 outcompeted Vibrio V48 and NH_4^+ was the major end-product of NO_3^- dissimilation. These data are consistent with the results obtained for chemostat enrichments performed under NO_3^- limiting conditions using Kingoodie Bay sediments (Table 3).

When populations of Klebsiella K312 were grown with Pseudomonas P388 under C- limitation the Klebsiella spp always outcompeted the pseudomonad (Figure 3) whereas under NO_3^- limiting conditions the reverse was true (Figure 4). Nitrite was the principal end-product

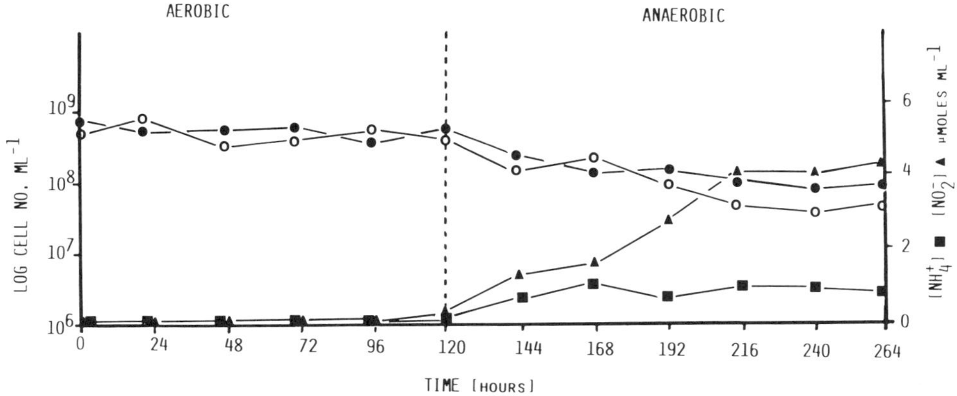

Figure 1. Competition experiment involving Klebsiella K312 (● - ●) and Vibrio V48 (○ - ○) grown aerobically and then anaerobically under glycerol limiting conditions (10mM KNO_3, 25mM glycerol, D = 0.1 h^{-1}, 20°C)

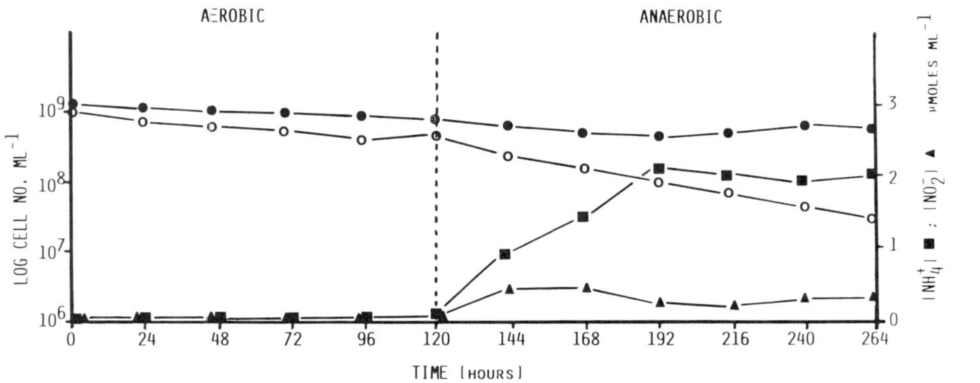

Figure 2. Competition experiment involving Klebsiella K312 (●-●) and Vibrio V48 (○-○) grown aerobically and then anaerobically under nitrate limiting conditions (5mM KNO_3, 50mM glycerol, D = 0.1 h^{-1}, 20°C)

of NO_3^- respiration whereas under NO_3^--limitation NH_4^+ accumulated in the spent media. These data are in agreement with the field data of King and Nedwell (1985; 1987) who demonstrated in the Colne estuary that as the NO_3^- concentration increased denitrification became the dominant process. Similarly, Smith and co-workers (1982) demonstrated that NO_3^- dissimilation to NH_4^+ in a saltmarsh sediment decreased from over 50% to 4% of the total NO_3^- reduced when NO_3^- concentrations were increased. These observations may reflect the different competitive success of denitrifiers or NO_3^--ammonifiers, and hence selection of the NO_3^--reducing communities, under different C:N ratios in the field (Rehr and Klemme, 1989).

In competition experiments involving Vibrio V48 and Pseudomonas P388 using acetate as C-source (Figure 5) the pseudomonad outcompeted the Vibrio which confirm the data obtained with Giltay's medium that low

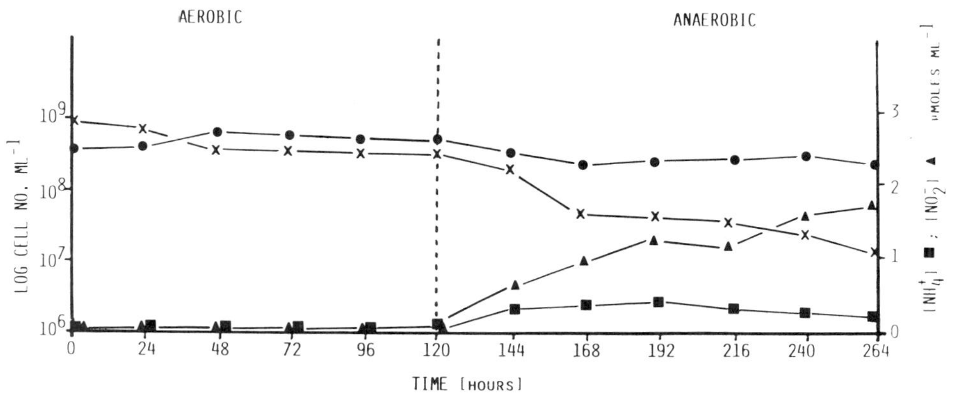

Figure 3. Competition experiment involving Klebsiella K312 (●-●) and Pseudomonas P388 (×-×) grown aerobically and then anaerobically under glycerol limiting conditions (10mM KNO_3, 25mM glycerol, D = 0.1 h^{-1}, 20°C).

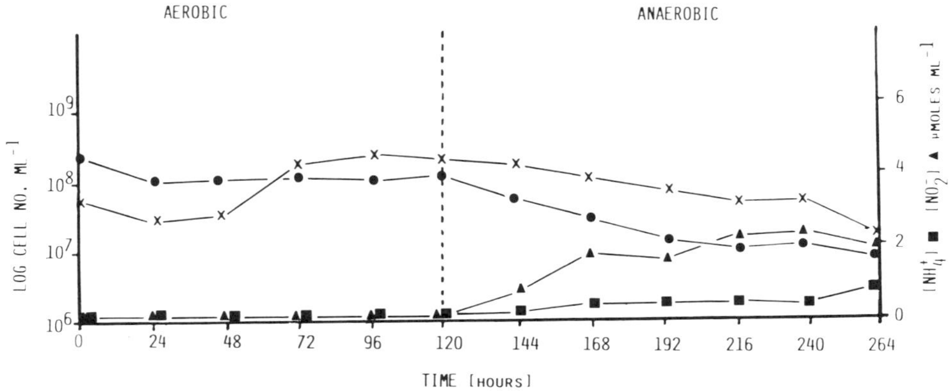

Figure 4. Competition experiment involving Klebsiella K312 (● - ●) and Pseudomonas P388 (× - ×) grown aerobically and then anaerobically under nitrate limiting conditions (5mM KNO_3, 50mM glycerol, D = 0.1 h^{-1}, 20°C).

molecular weight organic acids such as citrate and acetae are selective for pseudomonads.

Until recently it has been proposed that bacteria such as Pseudomonas spp with an oxidative mode of metabolism were responsible for denitrification while those such as Vibrio or Aeromonas with a predominantly fermentative type of metabolism were responsible for NO_3^- ammonification. This difference has become less clear cut with the reports of bacteria such as Pseudomonas putrefaciens which are able to reduce NO_3^- either to gaseous products or to NH_4^+ depending upon the substrate:NO_3^- ratio in their environment (Samuelson 1985; Samuelson and Ronner, 1982; Samuelson et al., 1988). Smith (1982) has similarily demonstrated that the reduction of NO_3^- by Citrobacter sp was also influenced by the glucose:NO_3^- ratio. At high ratios NH_4^+ is produced; a ratio near 4 yielded both NO_2^- and NH_4^+ whilst ratios < 4 led to NO_2^- production. Thus, physiological adaptation of the existing NO_3^--reducing community could conceivably explain the change in end products of NO_3^- reduction at different NO_3^- concentrations. However, the report by Sørensen (1978) that the capacity for denitrification down a sediment profile remained constant, while that for reduction of NO_3^- to NH_4^+ declined with depth may indicated that the two possible pathways of reduction of NO_3^- were indeed carried by distinct groups of microorganisms. Alternatively, the results may indicated that below a threshold C:N ratio denitrification is limited (although the potential on NO_3^- addition is maintained) whereas ammonification declines with increased depth as available organic carbon diminishes.

EFFECT OF TEMPERATURE UPON NITRATE REDUCERS

The effects of changing environmental temperature upon NO_3^- respiring bacteria are manifested in several ways. There may be direct effects upon the bacterial populations either directly on rates of enzyme activity, or because of the selective action upon the bacterial community of a change in environmental temperature. Additionally there are indirect effects of temperature upon NO_3^-

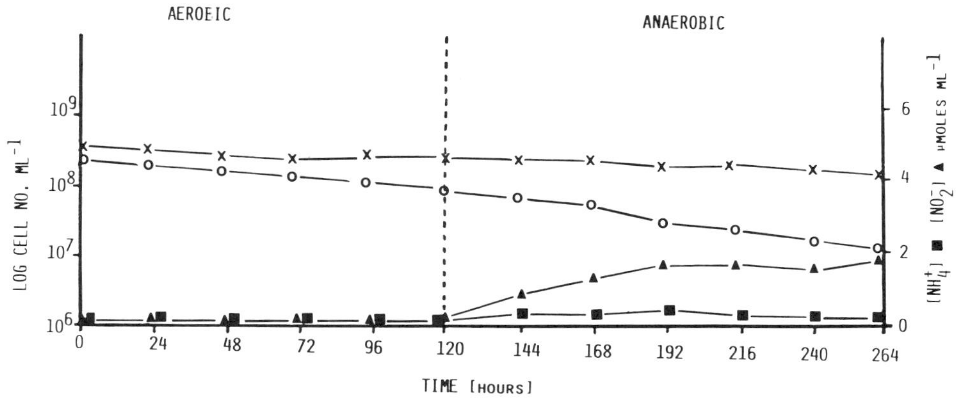

Figure 5. Competition experiment involving <u>Pseudomonas</u> P388 (● - ●) and <u>Vibrio</u> V48 (× - ×) grown aerobically and then anaerobically under nitrate excess conditions (10mM KNO_3, 50mM sodium acetate, D = 0.1 h^{-1}, 20°C).

reduction rates via other factors such as the depth of the oxidised layer of sediment within which nitrification can occur (Billen, 1982). As NO_3^- reduction in unpolluted aquatic sediments is usually NO_3^--limited, any factor which reduces nitrification rates will also regulate NO_3^- respiration rates (Nedwell et al., 1983).

Nitrate reducing bacteria comprise a heterogeneous group of microorganisms and it is therefore not surprising that temperature may exert a selective pressure on which groups are dominant <u>in situ</u> at particular times of the year. Kaplan et al., (1979) sparged salt marsh sediment with helium and then used N_2 gas accumulation as a measure of denitrification rate. These workers compared N_2 evolution from sediments at temperatures of 5, 10 and 20°C at different times of the year and postulated that seasonal temperatures selected for at least two denitrifying populations, one of which was cold sensitive. King and Nedwell (1984) showed that in the River Colne temperature exerted a distinct seasonal selection of the indigenous NO_3^- reducing community. Their studies examined the potential for NO_3^- reduction by the microbial community at different temperatures along a gradient block incubator. They showed that a mesophilic component of the NO_3^- respiring microflora was always present with an optimum growth temperature of 28°C, although it was proportionately less significant during the winter than in the summer. With the onset of lower temperatures during the autumn, a second, lower temperature optimum for NO_3^- reduction developed. These data indicate the appearance of a second population of cold-adapted (psychrotrophic) NO_3^- respirers which were physiologically better adapted to reduce NO_3^- at low temperatures than the mesophilic community. These observations were further substantiated by the increasingly lower temperature optimum for this emerging psychrotrophic population which reached a minimum value of 12.7°C in May. This represented the maximum psychrotrophic adaptation of the NO_3^- reducing community to low temperatures, and occurred 2 months after the lowest environmental temperatures were recorded. These results clearly demonstrate the long lag period (2 months) involved before the selective pressure of low temperature was reflected in a change of the physiological characteristics of the NO_3^- respiring microflora.

Further work showed that the response of the NO_3^- reducers to temperature described in the field could be reproduced in laboratory experiments. Sediment was held in containers in the laboratory to exclude air and incubated at 5, 15, 25 and 35°C for 4 weeks. After 4 weeks adaptation the sediment was made into a slurry with 1mM NO_3^- added, and NO_3^- reduction potential measured at different temperatures along a temperature gradient block incubator as described by King and Nedwell (1984). This experiment was first carried out during October (*in situ* temperature 11°C) when the sediment showed maximum adaptation to high summer temperature and the psychrotrophic population was not detectable; and again in April (*in situ* temperature 9°C) when the sediment bacteria were maximally adapted to low temperature and the psychrotrophic population was best developed. In both cases, as a control, sediment was also stored at the *in situ* temperature when it was collected. In both experiments the temperature profile of NO_3^- reduction activity was modified in response to the storage temperature, even within 4 weeks (Fig. 6; for clarity only data from sediment stored at 5, 35°C and controls at *in situ* temperatures are shown). The summer sediment taken in October did not initially exhibit a low psychrotrophic temperature optimum, yet after 4 weeks storage at 5°C a distinct optimum at 12.4°C had developed in addition to the mesophilic optimum at 27.8°C. In the sediment stored at 15°C also a low temperature optimum at 14.5°C developed. Nitrate reduction activity at the lower optimum (as described by the first order rate constant for NO_3^- reduction) was greater in the sediment stored at 5°C than at 15°C, indicating greater psychrotrophic selection at the lower temperature (Table 4). In contrast, the rate constant for NO_3^- reduction at the higher mesophilic optimum was greatest in sediment stored at 35°C, and declined with storage temperature. This showed that the mesophilic population was strongly selected for by higher temperatures, and actually declined from the *in situ* control value at all storage temperatures other than 35°C. This illustrates the strong mesophilic adaptation of the summer-adapted sediment NO_3^--reducing community.

In contrast, the winter-adapted sediment taken during April showed a distinct psychrotrophic optimum at 16.3°C (Fig. 6), but after 4 weeks incubation the sediment microflora had adapted (Table 5). At 5°C storage the psychrotrophic optimum was at 12.3°C and the rate constant for nitrate reduction at 13°C was at its greatest value. At the higher incubation temperatures (25 and 35°C) the mesophilic population was selected to the extent that the rate constant for nitrate reduction at 29°C more than doubled over that of the control sediment.

As described in a previous section of this chapter King and Nedwell (1984) used chemostat enrichments to isolate the representative members of the psychrotrophic and mesophilic microbial communities. *Pseudomonas* spp were the predominant psychrotrophs isolated at 10°C whereas *Vibrio* spp and enterobacteria were the principal mesophiles obtained at 25°C. Pseudomonads have an oxidative mode of metabolism and are usually associated with the ability to denitrify whereas *Vibrio* and enterobacteria have a fermentative metabolism of the type associated with NO_3^- reduction to NH_4^+. The implication of the observed adaptation of the sedimentary microflora is that temperature not only selects for different populations of NO_3^- respiring bacteria but indirectly also influences the end-products of NO_3^- reduction. If a predominantly oxidative NO_3^--reducing microflora develops at low winter temperatures, this will tend to favour denitrification at the

Table 4. The effect of storage temperature on bacterial nitrate reduction in summer (October) sediment after 4 weeks incubation

STORAGE TEMPERATURE (°C)	POSITION OF HIGH TEMPERATURE PEAK (°C)	VALUE OF K AT HIGH PEAK (h^{-1})	POSITION OF LOW TEMPERATURE PEAK (°C)	VALUE K AT LOW PEAK (h^{-1})
CONTROL	28.4	0.170	NONE	-
5	27.8	0.118	12.3	0.121
15	28.0	0.141	14.5	0.109
25	27.0	0.156	NONE	-
35	38.3	0.159	NONE	-

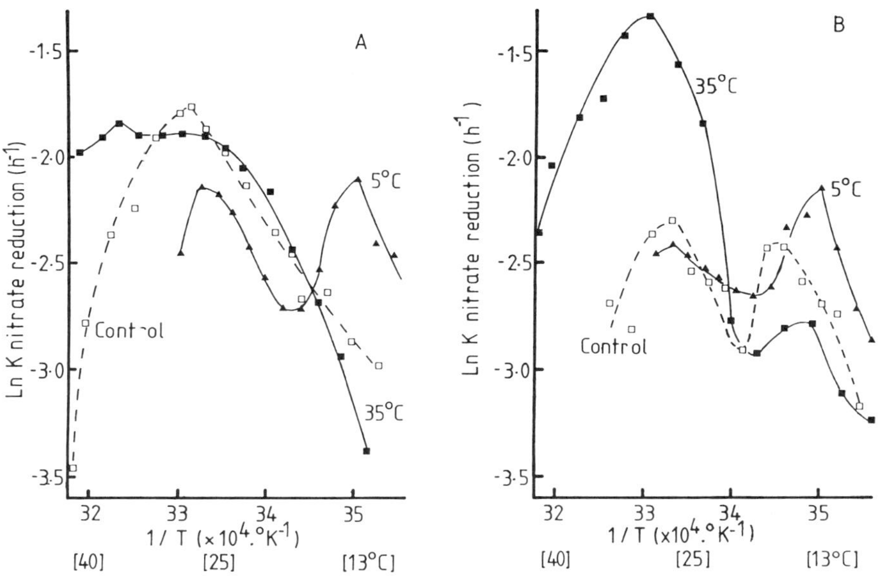

Figure 6. Influence of incubation temperature upon nitrate reduction rate of microflora in sediments taken A, during October (summer adapted) and B during April (winter adapted)

= control sediment, incubated at in situ temperature (October 11°C, April, 9°C) when taken,
= sediment incubated at 5°C for 4 weeks
= sediment incubated at 35°C for 4 weeks

Table 5. The effect of storage temperature on bacterial nitrate reduction in winter (April) sediment after 4 weeks incubation

STORAGE TEMPERATURE (°C)	POSITION OF HIGH TEMPERATURE PEAK (°C)	VALUE OF K AT HIGH PEAK (h^{-1})	POSITION OF LOW TEMPERATURE PEAK (°C)	VALUE K AT LOW PEAK (h^{-1})
CONTROL	27.8	0.100	16.3	0.095
5	27.0	0.090	12.3	0.116
15	26.5	0.120	14.3	0.093
25	28.8	0.259	14.7	0.061
35	29.5	0.264	13.9	0.063

expense of NO_3^- - ammonification. In contrast, at higher summer temperatures NO_3^- - ammonification by fermentative microorganisms will tend to be favoured. These data are in agreement with the observations of Jørgensen & Bertelsen (cited by Koike and Sørensen, 1988) that denitrification potential increases during the 3 winter months; and also with those of Jørgensen and Sørensen (1988) that NO_3^- ammonification was maximal in a Danish coastal sediment during summer, but denitrification was maximal during winter.

SUMMARY

In the absence of molecular oxygen, a diverse range of facultatively anaerobic bacteria are able to continue respiratory activity using nitrate as an alternative terminal e^- acceptor. The initial step in nitrate respiration is the reduction of nitrate to nitrite but after this step the products vary depending upon the bacterial species involved and the prevailing environmental conditions. Denitrifying bacteria such as _Pseudomonas aeruginosa_ and _Paracoccus denitrificans_ are able to reduce nitrite to gaseous products (N_2O and N_2) whereas fermentative bacteria such as _E. coli_, _K. aerogenes_ K312 and _Achromobacter fischeri_ reduce nitrite to ammonia. Direct plating methods for the isolation of nitrate respiring bacteria select for the growth of the fastest growing members which may not necessarily be important _in situ_. Using direct plating methods on nutrient rich media _Aeromonas/Vibrio_ spp. were numerically dominant whereas chemostat enrichments yielded primarily oxidative bacteria e.g. _Pseudomonas_ and _Acinetobacter_ spp. (acetate as C-source) or fermentative types e.g. _Klebsiella_ and _Vibrio_ spp. (glycerol as C-source). Field studies have demonstrated that NO_3^- availability exerts a major influence on the end-products of nitrate respiration. At low NO_3^- concentrations nitrate dissimilation to NH_4^+ is the predominant process whereas denitrificaiton is dominant at high NO_3^- concentrations. Temperature also exerts an influence on the distribution of nitrate respiring bacteria in salt marsh sediments. At low temperatures (10°C) _Pseudomonas_ spp. predominated whereas at 25°C _Vibrio_ spp. were dominant. These data are discussed in relation to the role of environmental factors which regulate nitrate respiration in marine, estuarine and salt marsh sediment.

REFERENCES

Abd. Aziz, A.S., and Nedwell, D.B., 1986, The Nitrogen Cycle of an East Coast, U.K. Saltmarsh: II Nitrogen Fixation, Nitrification, Denitrification, Tidal Exchange, Estuar. Coastal Shelf Sci., 22, 689.

Billen, G., 1982, Modelling the processes of organic matter degradation and nutrient recycling in sedimentary systems, In: "Sediment Microbiology," pp. 1-52, D.B. Nedwell and C.M. Brown, eds., Academic Press, London.

Cole, J.A., and Brown, C.M. 1982, Nitrate reduction to ammonia by fermentative bacteria, FEMS Microbiol, Lett., 7: 65.

Dunn, G.M., Herbert, R.A., and Brown, C.M., 1978, Physiology of denitrifying bacteria from tidal mudflats in the River Tay, In: "Physiology and Behaviour of Marine Organisms," pp. 135-140, D.S. McLusky and A.J. Berry, eds., Pergamon Press, Oxford.

Dunn, G.M., Wardell, J.M., Herbert, R.A., and Brown, C.M., 1980, Enrichment, enumeration and characterisation of nitrate reducing bacteria present in sediments of the River Tay estuary, Proc. Roy. Soc. Edin. 78B, 47.

Evans, C.G.T., Herbert, D., and Tempest, D.W., 1970, The continuous culture of microorganisms, In: "Methods of Microbiology," Vol. 2, pp. 277-327, Academic Press, London.

Hadjepetrou, L.P., and Stouthamer, A.H., 1965, Energy production during nitrate respiration by Aerobacter aerogenes, J. Gen. Microbiol., 38, 29.

Herbert, R.A., 1982, Nitrate dissimilation in marine and estuarine sediments, In: "Sediment Microbiology," pp. 53-71, D.B. Nedwell and C.M. Brown, eds., Academic Press, London.

Jannasch, H.W., 1967, Enrichments of aquatic bacteria in continuous culture, Arch. Mikrobiol., 59, 165.

Jørgensen, K.S. and J. Sørensen, 1988, Two annual maxima of nitrate reduction and denitrification in estuarine sediment (Norsminde fjord, Denmark). Mar. Ecol. Prog. Ser. 48: 147-154.

Kaplan, W.A., Valiela, I., and Teal, J.M., 1979, Denitrification in a marsh ecosystem, Limnol. Oceanogr., 24, 726.

Keith, S.M., Russ, M.A., Macfarlane, G.T., and Herbert, R.A., 1987, The ecology and physiology of anaerobic bacteria isolated from Tay estuary sediments. Proc. Roy. Soc. Edin. 92B, 323.

King, D., and Nedwell, D.B., 1984, Changes in the nitrate reducing community of an anerobic saltmarsh sediment in response to seasonal selection by temperature J. Gen. Microbiol, 130, 2935.

King, D., and Nedwell, D.B., 1985, The influence of nitrate concentration upon the endproducts of nitrate dissimilation by bacteria in anaerobic salt marsh sediments, FEMS Microbiol. Lett., 31, 23.

King, D., and Nedwell, D.B., 1987, The adaptation of nitrate reducing bacterial communities in estuarine sediments in response to overlying nitrate load, FEMS Microb. Ecol. 45, 15.

Koike, I., and Hattori, A., 1978, Denitrification and ammonia formation in anaerobic coastal sediments, Appl. Environ. Microbiol., 35, 278.

Koike, I., and Sørensen, J., 1988, Nitrate reduction and denitrification in marine sediments, In: "Nitrogen cycling in coastal marine environments," T.H. Blackburn & J. Sørensen, eds., pp. 251-273, J. Wiley & Sons, Chichester.

Macfarlane, G.T., and Herbert, R.A., 1982, Nitrate dissimilation by Vibrio spp isolated from estuarine sediments, J. Gen. Microbiol., 128, 2463.

Macfarlane, G.T., and Herbert, R.A., 1984, Dissimilatory nitrate reduction and nitrification in estuarine sediments J. Gen. Microbiol., 130, 2301.

Nedwell, D.B., 1975, Inorganic nitrogen metabolism in a eutrophicated tropical mangrove estuary., Wat. Res., 9, 221.

Nedwell, D.B., 1982, Exchange in nitrate and the products of nitrate reduction between seawater and sediment from a U.K. salt marsh, Estuar. Coastal Shelf Sci., 14, 557.

Nedwell, D.B., Hall, S.E., Andersson, A., Hagstrom, A., and Lindstrom, E.B., 1983, Seasonal changes in the distribution and exchange of inorganic nitrogen between sediments and water in the northern Baltic (Gulf of Bothnia), Estuar. Coastal Shelf Sci., 17, 169.

Oremland, R.S., Umberger, C., Culbertson, C.W., and Smith, R.L., 1984, Denitrification in San Francisco sediments, Appl. Environ. Microbiol., 47, 1106.

Payne, W.J., 1973, Reduction of nitrogenous oxides by microorganisms, Bact. Rev. 37, 409.

Rehr, B., and Klemme, J.-H., 1989, Competition for nitrate between denitrifying *Pseudomonas stutzeri* and nitrate ammonifying enterobacteria. FEMS Microbiol. Ecol., 62, 51.

Samuelson, M.O., 1985, Dissimilatory nitrate reduction to nitrite, nitrous oxide and ammonium by Pseudomonas putrefaciens, Appl. Environ. Microbiol. 50, 812.

Samuelson, M.O. and Ronner, U. 1982, Ammonium production by dissimilatory nitrate reducers. Appl. Environ. Microbiol., 44, 1241.

Samuelson, M.O., Cadez, P., and Gustafsson, L. 1988, Heat production by the denitrifying bacterium, *Pseudomonas fluorescens* and the dissimilatory ammonium producing bacterium *Pseudomonas putrefaciens* during anaerobic growth with nitrate as electron acceptor, Appl. Environ. Microbiol.,54, 2220-2225.

Smith, M., 1982, Dissimilatory nitrate reduction of NO_2^- to NH_4^+ and N_2O by a soil citrobacter sp., Appl. Environ. Microbiol., 43, 854.

Smith, C.J., De Laune, R.D., and Patrick, W.H.Jr., 1982, Nitrate reduction in *Spartina alterniflora* marsh soil. Soil. Sci. Soc. Am. J., 46, 748-750.

Sørensen J., 1978, Capacity for denitrification and reduction of nitrate to ammonia in a coastal marine sediment Appl. Environ. Microbiol., 35, 301.

van Kessel, J.F., 1978, Gas production in aquatic sediments in the presence and absence of nitrate. Water Res. 12, 291.

Wheatland, A.B., Barnett, M.J., and Bruce, A.M., 1959, Some observations of denitrification in rivers and estuaries. Journal of the Institute of Sewage Purification 2, 149.

PHYSIOLOGICAL AND ECOLOGICAL ASPECTS OF AEROBIC DENITRIFICATION, A LINK WITH HETEROTROPHIC NITRIFICATION?

Lesley A. Robertson and J.G. Kuenen

Kluyver Laboratory for Biotechnology
Delft University of Technology
Delft
The Netherlands

INTRODUCTION

For about a century, the existence of aerobic denitrification was a subject of debate, if not controversy (Robertson & Kuenen, 1984a). However, as available technology (e.g. sensitive oxygen - analyzing equipment, chemostats, mass spectrometers) has improved, it has been possible to establish that a number of strains do indeed denitrify while simultaneously respiring oxygen. This is, of course, not true of all denitrifiers. _Thiomicrospira denitrificans_ appears to be the most specialized as it can only tolerate oxygen in limiting concentrations. Other species, including most _Paracoccus denitrificans_ strains, can grow under fully aerobic conditions, but only denitrify in the virtual absence of oxygen. These species fit the classical definition of denitrifiers. A third group is made up of strains which simultaneously utilize oxygen and denitrify - the aerobic denitrifiers. The members of this group do not show a uniform response to oxygen, denitrification by some of them is inhibited by relatively low amounts of oxygen (e.g. _Hyphomicrobium_ X; Meiberg et al., 1980) while others continue to denitrify even at air saturation (e.g. _Thiosphaera pantotropha_). It is not unlikely that, as with electron donors, the environmental conditions and nutrient supplies determine whether specialist or versatile denitrifiers occur in any given ecosystem.

This paper aims to document some of the more recent reports of aerobic denitrification and to discuss some of the ecological and physiological pressures which might determine whether aerobic or specialist denitrifiers are selected in any situation.

THIOSPHAERA PANTOTROPHA

We first became interested in the possibility of aerobic denitrification six years ago when when we found that a new isolate, _Thiosphaera pantotropha_ possessed constitutive denitrifying enzymes which were active, even in cultures grown at dissolved O_2 concentrations

Denitrification in Soil cnd Sediment, Edited by N. P. Revsbech and J. Sørensen, Plenum Press, New York, 1990

80-90% of air saturation (Robertson & Kuenen, 1983a; 1984b; Robertson et al., 1986). In batch culture, the simultaneous use of two electron acceptors conferred a higher growth rate than when either electron acceptor was supplied separately. During these experiments, it was also found that _Tsa. pantotropha_ can oxidize ammonia to nitrite, provided that it is supplied with an organic substrate. Preliminary experiments using a mass spectrometer with labelled and unlabelled inorganic nitrogen compounds confirmed that N_2 rather than N_2O is produced by _Tsa. pantotropha_ during denitrification. Moreover, cells supplied with $^{15}NH_4Cl$ and metabolizing endogenous reserves generated ^{15} $^{15}N_2$ (E.W.J. van Niel, L.A. Robertson & R.P. Cox, unpublished results).

The oxidation of ammonia to nitrite or nitrate in the presence of an organic substrate is known as heterotrophic nitrification (Verstraete, 1975; Killham, 1986) and has, for a long time, been dismissed as of little ecological importance because heterotrophic nitrifiers accumulate very little nitrite or nitrate. However, the activity of heterotrophic nitrifiers such as _Tsa. pantotropha_, which simultaneously denitrify the nitrite they produce, can be seriously underestimated unless complete nitrogen balances are made. Because nitrification and denitrification appear to be closely linked in _Tsa. pantotropha_ as well as some of the other aerobic denitrifiers, the two phenomena will be discussed together.

It became very clear early in our studies that _Tsa. pantotropha_ adjusted its behaviour to accommodate even small changes in its environment (e.g. raising the dissolved oxygen concentration from 80% air saturation to 90% air saturation can add an hour to the doubling time), and most subsequent experiments were therefore carried out in chemostats with pH and dissolved oxygen control. By running steady state, acetate-limited cultures at different dilution (growth) rates but with the dissolved oxygen concentration controlled at 80% air saturation, several factors immediately became apparent (Table 1):

1: The presence of nitrate or nitrite in the medium reduced the rate of nitrification, except when low amounts (5 mM) of nitrite were supplied, and the organism was able to denitrify all of it, thus maintaining an effective concentration of 0 in the culture medium.

2: Both the nitrification and denitrification rates increased as the growth (dilution) rate increased.

3: Thiosulphate inhibited nitrification.

4: Heterotrophic nitrification cannot be treated as a form of mixotrophy (the metabolism of an organic substrate and an inorganic energy supplement) because not only did the cultures not gain extra energy for growth from the reaction, but biomass yields from cultures with the highest nitrification rates were lower than any of the others, and only 60% of the yields expected.

Cultures run at different dissolved oxygen concentrations revealed that both the nitrification and denitrification rates increased as the dissolved oxygen fell, reaching a maximum at 25-30% of air saturation. Below this, the cultures not provided with nitrate or nitrite accumulated poly β-hydroxybutyrate (PHB) and started forming biolayers on all available surfaces, a general response of this organism to oxygen limitation.

Table 1. Correlation of chemostat yields (mg/L) with nitrification (NH_4^+ oxidation) and denitrification (reduction of NO_3^- supplied and NO_2^- supplied or produced from nitrification) rates ($nmol^{-1}$ min^{-1} mg $protein^{-1}$). All cultures were substrate limited, and supplied with NH_4^+. Dissolved O_2=80% air saturation. D=0.04 h^{-1}. * indicates yield lower than expected. (Data from Robertson et al., 1988).

Additive	Nitrification	Denitrification	Yield
-	43	43	81*
NO_3^-	12	107	103
NO_2^- (limiting)	48	85	80*
NO_2^- (saturating)	25	98	115
$S_2O_3^{2-}$	21	21	145
$S_2O_3^{2-}$ / NO_3^-	6	36	120

In order to attempt to explain the behaviour of Tsa. pantotropha, and to help in the design of further experiments, a model based on the physiological data and preliminary cytochrome determinations was constructed (Fig. 1). A basic assumption is that, for an unknown reason, there is a bottleneck in the flow of electrons to oxygen via cytochrome aa_3, thus limiting the rate at which NADH can be reoxidized by this route. This bottleneck may be at the cytochrome c level. The somewhat greater degree of reduction in the cytochrome chain that such a

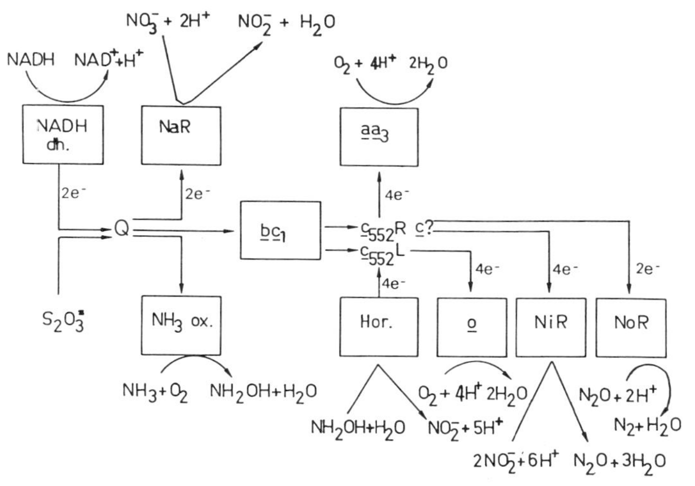

Figure 1. Schematic representation of the model used to describe the various potential routes for electron transport in Tsa. pantotropha. Cytochromes are underlined; NaR = nitrate reductase; NH_3 ox. = NH_3 monooxygenase; Hor. = hydroxylamine oxidoreductase; NiR = nitrite reductase; NoR = nitrous oxide reductase; c? indicates one or more additional cytochromes c; e = electrons; $S_2O_3^{2-}$ = thiosulphate oxidizing pathway (Robertson, 1988).

Table 2. Nitrogen balances from Tsa. pantotropha cultures growing on acetate with nitrate (7.5 mM) as the sole source of nitrogen. D = 0.04 h^{-1}. a = PHB synthesized; b = ammonia (7.5 mM) + nitrate (40 mM) cultures shown for comparison.

Dissolved oxygen (% air)	Nitrate lost (mM)	Nitrite formed (mM)	Protein (g mol^{-1} acetate)
80	1.3	0.03	5.9
50	0.8	0.05	4.3
33	1.8	0.45	3.6
23	6.3	0.02	2.7[a]
80	0.5	0.03	5.5
40	3.5	0.05	5.5
20	14.7	0.10	2.8
80[b]	13.4	0.02	5.2
0[b]	24.0	0.10	2.5

bottleneck would cause should then allow electrons to flow to the denitrifying pathway, thus allowing a faster rate of NADH oxidation. For simplicity, ammonia monoxygenase is shown as a part of the sequence. It should be noted that NAD(P)H rather than electron flow is involved in this reaction.

The ammonia monooxygenase of Tsa. pantotropha is NAD(P)H dependent (Robertson & Kuenen, 1988) and thus provides an additional means of regenerating reducing power, if necessary. However, since the use of this pathway does not result in energy generation (and may even cost the cell energy), denitrification is obviously more favourable. A certain amount of energy may, of course, be gleaned from the denitrification of nitrite produced during nitrification. The apparently inhibitory effect of nitrate or nitrite on nitrification may thus be two-fold. Firstly,

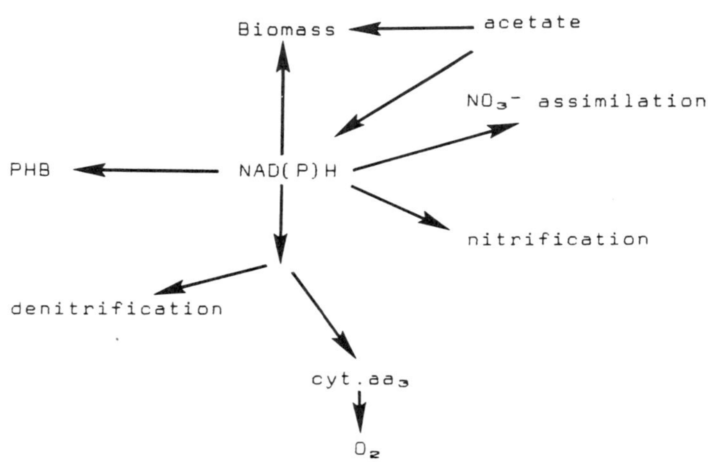

Figure 2. Simplified scheme showing the various possible options for NAD(P)H utilization available to Tsa. pantotropha (Robertson, 1988).

both ammonia and hydroxylamine oxidation are inhibited by nitrite (Robertson & Kuenen, 1988) and, secondly, there is less need for nitrification when denitrification can occur. Similarly, thiosulphate probably has a double effect. As with the autotrophic nitrifiers, it inhibits ammonia monooxygenase (Robertson & Kuenen, 1987), but growth on thiosulphate also results in the synthesis of additional cytochromes c (Robertson, 1988) which may thus overcome our postulated bottleneck. Two other methods of dealing with its redox problems are available to *Tsa. pantotropha*. Firstly, the assimilatory nitrite reductase requires NADH, and thus during nitrate or nitrite assimilation in the absence of ammonium any surplus NADH could be consumed. Aerobic denitrification rates in cultures where nitrate was the sole source of nitrogen were indeed found to be only 10% of those obtained when ammonia was also present (Table 2). The final option may be summarized as "if all else fails, make PHB". Figure 2 summarizes the different ways by which *Tsa. pantotropha* can oxidize NADH.

NITRIFICATION AND DENITRIFICATION BY OTHER STRAINS

It was obviously of great interest to discover whether *Tsa pantotropha* was unique, or whether other bacteria shared its capacity for heterotrophic nitrification and aerobic denitrification. In 1984, Castignetti & Hollocher described their finding that a number of denitrifying soil bacteria were also capable of heterotrophic nitrification. However, many of these strains did not accumulate very much nitrite, and it therefore seemed possible that they were also denitrifying. Several strains were therefore checked in batch culture and, as with *Tsa. pantotropha*, were found to grow more rapidly with nitrate and oxygen than on a single electron acceptor (Table 3). Like *Tsa. pantotropha*, and unlike the *Pa. denitrificans* control strain, the protein yield of these batch cultures was lower in the oxygen/nitrate cultures, and a substantial amount of nitrate has disappeared. Chemostat experiments with two of these strains showed that both were capable of aerobic denitrification, but that their behaviour was slightly different from that of *Tsa. pantotropha* and from each other. *Alcaligenes* sp. nitrified and denitrified efficiently at lower dissolved oxygen concentrations, but accumulated nitrification and denitrification intermediates in increasing quantities as the dissolved oxygen was increased (figure 3). A strain from the heterogenous group known as "*Pseudomonas denitrificans*", on the other hand, was as insensitive to dissolved oxygen as *Tsa. pantotropha*, but needed to induce its dissimilatory nitrate reductase. The remainder of its denitrifying pathway was constitutive. Unlike *Tsa. pantotropha*, nitrate in the medium did not reduce the rate of nitrification by "*Ps. denitrificans*".

There were two other major points of similarity between these nitrifier/denitrifiers. All of them possessed the copper-based nitrite reductase rather than cytochrome cd (Kuenen & Robertson, 1987; Robertson et al., 1989a; 1989b) and, until recently, it appeared that all aerobic denitrifiers had this enzyme. This will be discussed further below. With one exception, all of them were able to induce a significant thiosulphate - oxidizing capacity after batch culture on a mixture of acetate and thiosulphate (Table 4). In fact, "*Ps. denitrificans*" was able to grow autotrophically on thiosulphate in a chemostat, the other strains have yet to be checked in this way.

It is not yet clear whether the possession of such diverse abilities as nitrification and thiosulphate oxidation are coincidental

Table 3. Comparison of the maximum specific growth rates (μ_{max}), protein concentrations and nitrate reduction obtained from aerobic or anaerobic batch cultures of bacteria known to be capable of heterotrophic nitrification. Ammonia was the nitrogen source. The dissolved oxygen concentration was kept above 80% of air saturation. (Adapted from Robertson et al., 1989A). Pseudomonas sp. LMD 84.60 is the strain formerly known as Ps. denitrificans. Pa. denitrificans LMD 22.21 was the negative control.

Organism	μ_{max} (h^{-1})			Protein (mg/l)		NO_3^-
	O_2	O_2/NO_3^-	NO_3^-	O_2	O_2/NO_3^-	mM
Pseudomonas sp. LMD 84.60	0.1	0.41	0.15	78	60	5.0
A. faecalis LMD 84.59	0.17	0.25	0.07	30	14	4.1
Ps. aureofaciens LMD 37.26	0.19	0.21	0.07	66	66	5.0
T. pantotropha LMD 82.5	0.28	0.34	0.25	81	60	5.5
Pa. denitrificans LMD 22.21	0.28	0.28	nd.	92	88	<1.0

nd. = not determined

or, in some way, linked. That not all thiosulphate-oxidizing bacteria fall into this group is illustrated by the fact that our control strain of Pa. denitrificans can grow autotrophically on thiosulphate, but does not nitrify to any significant degree and is not an aerobic denitrifier. Two indirect pieces of evidence suggest that there might be a link. Firstly, the preliminary cytochrome studies with Tsa. pantotropha indicated that a particular cytochrome c is common to both nitrification and thiosulphate oxidation. Secondly, the apparent ecological niche for a versatile denitrifier (and all of the aerobic denitrifiers thus far tested have also been heterotrophic nitrifiers) is where the dissolved

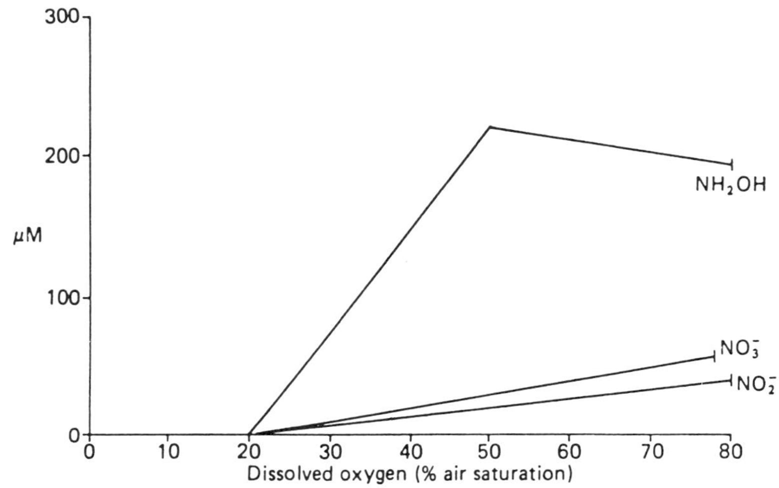

Figure 3. The gradual accumulation of oxidized nitrogen compounds in acetate-limited chemostat cultures of Alcaligenes sp. as the dissolved oxygen was increased. Ammonium was the sole source of nitrogen. (Kuenen & Robertson, 1987).

Table 4. Ammonia disappearance and thiosulphate oxidation rates obtained with aerobic, heterotrophic batch cultures. NH_3 was the sole source of nitrogen, and 5 mM acetate and 10 mM thiosulphate were supplied as substrates. The NH_3 disappearance figures have been corrected for assimilated nitrogen. The _Pseudomonas_ sp. was grown in a chemostat (20 mM acetate and 5 mM thiosulphate). * = N-limited, nd = not determined.

Organism	NH_3 (mM)	Oxygen uptake rate (nmol/min/mg protein)
Pseudomonas sp. LMD 84.60	nd	615
_A_. _faecalis_ LMD 84.59	-3.6*	598
_A_. _faecalis_ S6 LMD 60.48	-2.7	0
_Ps_. _aureofaciens_ LMD 37.26	-2.9	652
_Ps_. _fluorescens_ LMD 60.46	-4.6	109
_Ps_. _fluorescens_ LMD 60.44	-3.3	128
_Ps_. _fluorescens_ LMD 60.48	0	0
_Ps_. _fluorescens_ LMD 60.71	-1.2	0

oxygen concentration is limiting, or fluctuates. This adequately describes life at aerobic/anaerobic interfaces, where reduced sulphur compounds can also be expected to occur.

REPORTS OF AEROBIC DENITRIFICATION IN THE LITERATURE

It is neither practical or desirable to review the historical arguments about aerobic denitrification here, the interested reader will find this subject extensively covered in Robertson & Kuenen (1984a). However, since this review was published, other accounts of aerobic denitrification have appeared, and it is appropriate to consider them here.

The _Nitrosomonas_ species described by Poth (1986) was shown to denitrify at a very low rate (less than 1 nmol min^{-1} mg $protein^{-1}$) under "oxygen stress" (concentration not reported). This strain, like _Aquaspirillum magnetotacticum_ (Bazylinski & Blakemore, 1983), requires oxygen for growth, and might thus be described as obligately aerobic denitrifiers. However, it should be remembered that all denitrifiers are able to denitrify under oxygen limitation.

In a recent paper, Strand et al. (1988) examined aerobic denitrification by chemostat-grown cultures of a non-flocculating strain of _Zooglea ramigera_ at a range of dissolved oxygen concentrations. As with _Tsa_. _pantotropha_ the rate of denitrification by _Z_. _ramigera_ achieved its maximum at an oxygen concentration somewhat lower than 50% of air saturation. Strand et al.. (1988) also measured ammonia levels in their cultures and it appears from the data provided that ammonia was disappearing from their cultures at a rate about half that found with similar cultures of _Tsa_. _pantotropha._ It would therefore appear that this strain also combines heterotrophic nitrification and aerobic denitrification.

As mentioned above, modern techniques, especially the use of ^{15}N-labelled compounds and mass spectrometry, have recently been used (Degn et al., 1985; Lloyd et al., 1986), to analyse the products of denitrification (aerobic or otherwise) in more detail. Lloyd et al. (1987) used this technique to investigate the persistence of denitrification in anaerobically-grown batch cultures when incubated under aerobic conditions, and came to the conclusion that the phenomenon is very widespread among bacteria. They found that strains of *Paracoccus denitrificans*, *Ps. aeruginosa*, *Ps. stutzeri*, *Propionibacterium thoenii* and newly isolated *Pseudomonas* species all continued to denitrify, even when the dissolved oxygen concentration was equivalent to air saturation. In most of the strains, the rate of N_2 evolution was low (0.13-0.40 nmol min^{-1} mg $protein^{-1}$), but *Ps. stutzeri* and *Pr. thoenii* produced N_2 at significantly higher rates (3.42-5.61 nmol min^{-1} mg $protein^{-1}$). This is, of course, still considerably lower (10-100 times, depending on the conditions) than the denitrification rates found with chemostat cultures of *Tsa. pantotropha*. All of the strains also produced small amounts of N_2O. Davies et al. (1989) then studied two of these strains, *Pa. denitrificans* (NCIB 8944) and *Ps. aeruginosa* (PA0129) in greater detail, growing the cells both aerobically and anaerobically, and then measuring denitrification under aerobic and anaerobic conditions. As might be expected, the anaerobically grown, anaerobically tested suspensions gave the highest denitrification rates. As the dissolved oxygen increased, N_2O began to appear as well as N_2. In order to completely inhibit N_2 production, extremely high (360 μM dissolved oxygen concentrations were required, and even then N_2O production continued. The position of *Pa. denitrificans* as a typical classical denitrifier is further complicated by results published by Kawakami et al. (1985). The authors showed that N_2O production in the presence of acetylene by anaerobically grown cells was only partially inhibited by oxygen, even at 50% of air saturation. N_2O production continued at reduced levels for periods of at least 4 hours. Moreover, it took approximately 1 hour for inhibition by oxygen to reach its maximum. In acetate-limited, steady state chemostat culture, the strain of *Pa. denitrificans* (LMD 22.21) used as a reference culture during our studies did not denitrify nitrate supplied in the medium aerobically, although the nitrogen balance made for the culture indicated that it was nitrifying (and denitrifying the product) at a low rate (6 nmol min^{-1} mg $protein^{-1}$, compared with 18 nmol min^{-1} mg $protein^{-1}$ obtained with *Tsa. pantotropha* under the same conditions). These observed differences in the denitrifying abilities of *Pa. denitrificans* are probably due to strain variation, perhaps resulting from the original enrichment or maintenance conditions. We have recently gained evidence that a variant of *Tsa. pantotropha* appeared in cultures which were not maintained on nitrate-containing medium. This variant had lost much of its ability to denitrify externally supplied nitrite or nitrate under aerobic conditions, but continued to simultaneously nitrify and then denitrify the product.

The sequential inhibition of the denitrifying enzymes by increasing amounts of oxygen was shown for *Pa. halodenitrificans* (Hochstein et al., 1984). Chemostat cultures and gas chromatography were used to demonstrate that as oxygen was supplied to the cultures, the amount of N_2 produced by the cells fell and the N_2O level increased. With 5% oxygen in the influent gas (and the dissolved oxygen undetectable), only N_2O was produced. With an oxygen supply greater than 5%, dissolved oxygen became detectable, the N_2 production began to fall, and nitrite began to accumulate. At air saturation, N_2O production had stopped and only trace amounts of nitrite were observed.

Our original hypothesis that the separation between classical and aerobic denitrifiers might be dependent on which nitrite reductase (cytochrome cd or the copper-based enzyme) was present may need to be revised in view of the observations with the two Paracoccus strains described above. Further evidence that this idea is not completely correct came from the work of Korner & Zumft (1989) in which they examined the synthesis of the denitrifying enzymes in chemostat cultures of Ps. stutzeri (which has cytochrome cd) under different oxygen regimes. When supplied with nitrate in the medium, Ps. stutzeri contained significant amounts of nitrate and N_2O reductase at 50% air saturation, and the N_2O reductase was even detectable at air saturation. Nitrite reductase began to appear at about 25% air saturation, and reached its maximum just before the onset of complete anaerobiosis. It now seems more likely that the capacity for aerobic denitrification is determined by the redox state of the cytochrome chain, and the various factors which affect this.

Aerobic denitrification (of nitrite) was recently shown for Rhizobium "hedysari" strain HCNT1 (Casella et al., 1988). It seems that relatively low (10-75 μM) concentrations of nitrite inhibited oxygen uptake by bacteroids of this strain. Oxygen uptake began once the nitrite had been reduced. However, rather than having a respiratory function, the authors suggest that this form of "aerobic denitrification" is a means of protecting nitrogenase (which is also very sensitive to nitrite) from nitrite generated by the nitrate reductase of the host plant.

Observations in the field and in natural samples were made by Trevors (1985), Trevors & Starodub (1987) and Rönner & Sörensson (1985). They showed denitrification in the presence of reduced oxygen concentrations in soil samples, freshwater sediments and water from the deeper areas of the Baltic, respectively. In all three cases, denitrification was measured as N_2O production after acetylene treatment. Trevors & Starodub (1987) showed that the initial availability of nitrate was more important for the onset of denitrification than the degree of anoxia. During studies of N_2O and NO production (without acetylene) by common soil bacteria, Anderson & Levine (1986) found that Rhizobium japonicum and Pseudomonas fluorescens only continued to generate these gases up to a dissolved oxygen concentration of 5% air saturation, but that their strain of Alcaligenes faecalis continued to do so at air saturation and, in other words, must be an aerobic denitrifier. Stimulation of denitrification by the presence of oxygen was shown by Abou Seada & Ottow (1985) for Aeromonas "denitrificans", Azospirillum lipoferum and Bacillus licheniformis. They attributed this phenomenon to oxygen- stimulated mineralisation rates, and considered that the rate of mineralisation in soil was a more important controlling factor in denitrification than anaerobiosis. Unfortunately, only headspace analysis of the oxygen concentration was used during these experiments, and it is therefore not possible to relate the results to those from Trevors and Rönner & Sörensson, mentioned above.

It is clear that a number of denitrifying bacteria are capable of simultaneously respiring oxygen and nitrate (or another nitrogen oxide), the phenomenon known as aerobic denitrification. As has been shown, especially with the Pa. halodenitrificans experiments (Hochstein et al., 1984), most denitrifying bacteria are able to do so "aerobically". The important difference is the concentration of oxygen at which either denitrification activity or the synthesis of the denitrifying enzymes is reduced or shut down completely, a concentration which can vary enormously between different organisms. From a physiological point of view the important question is whether or not oxygen is saturating the terminal cytochrome oxidase. This generally occurs at concentrations in the order of 1-2% of air saturation (2-4 micromoles/litre). If the level of dissolved oxygen is above this threshold, the term "aerobic denitrification" is justified. From an ecological or environmental point of view, aerobic denitrification means that the process occurs at detectable levels of air saturation (in common practice, around 5% of of air saturation).

Can aerobic denitrification be rationalized, or even justified, in ecological terms? The obvious conceptual problem is that, because of the lower biomass yield of denitrifying cultures, it seems inefficient for an organism to co-respire oxygen and nitrate. In biochemical terms this might indeed be a disadvantage and, at first glance, it would appear that evolutionary pressure would act against the maintenance of such a trait in a species. The same could be said of heterotrophic nitrification, where a high nitrification rate is linked to a low yield. This reasoning depends on the implicit assumption that these species exist in a niche in which steady state aerobic and anaerobic conditions exist, or where aerobiosis and anaerobiosis alternate over periods favourable for maximal yield. However, these conditions are usually not fulfilled in nature. The specific growth rate under the relevant conditions and the ability of the organism to respond efficiently to a changing environment (i.e. the reactivity of the organism (Kuenen & Beudeker, 1982; Leegwater, 1983)) are of as great importance to the survival of the organism as biomass yield.

A number of recent ecophysiological studies have considered these aspects of electron donor specificity and/or alternating mixed substrate utilization (Rittenberg, 1969; Gottschal & Kuenen, 1980; Kuenen & Gottschal, 1982; Kuenen & Robertson, 1984A; 1984B). It seems that some parallels can be drawn between these experiments and electron acceptor utilization. Certainly, a species which did not need to re-induce its denitrifying system after exposure to oxygen, or which could continue to denitrify while taking advantage of the extra energy provided by limiting amounts of oxygen, would have a selective advantage over other, less versatile species. In addition, a denitrifier living in an

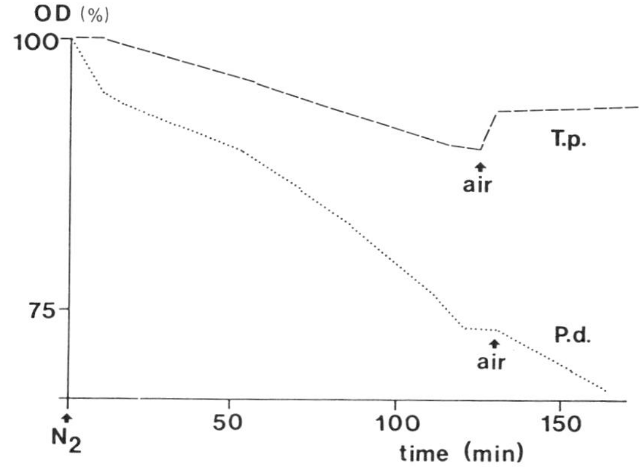

Figure 4. The outcome of switching steady-state, acetate-limited chemostat cultures of <u>Tsa</u>. <u>pantotropha</u> (T.p.) and <u>Pa</u>. <u>denitrificans</u> (P.d.) from aerobiosis (80% air saturation) to anaerobiosis (Robertson, 1988).

environment where long periods of anaerobic conditions are interspersed by short periods of aerobiosis would be at a disadvantage if it inactivated its denitrifying system too readily, because this would result in a lack of sufficient denitrifying capacity when anaerobic conditions were restored.

Another of the more obvious situations in which a versatile denitrifier might have a selective advantage over a specialist is under fluctuating oxygen levels. An experiment during which cultures of Tsa. pantotropha and of a specialist strain of Pa. denitrificans were grown to steady state at 80% air in an acetate-limiting medium with ammonia and nitrate showed that Tsa. pantotropha adjusted rapidly to the sudden onset of anaerobiosis, while Pa. denitrificans washed out with the dilution rate. When the air supply was equally suddenly restored, Tsa. pantotropha adjusted smoothly again, but Pa. denitrificans continued to wash out (figure 4). This was found to be due to the accumulation of up to 14 mM nitrite in the medium. 10 mM was sufficient to inhibit the oxygen uptake rate of these cultures by 50%. It is clear that Pa. denitrificans induced its nitrate reductase first, and was unable to readjust to the return of oxygen. If the anaerobic period was extended, the nitrite level eventually fell, and the culture was able to recover. These results emphasize the advantages which flexibility with regard to electron acceptors can confer under changing conditions. Tsa. pantotropha was isolated from a waste water treatment system receiving nitrate as well as intermittent amounts of oxygen. Other aerobic (i.e. constitutive) denitrifiers have been isolated from wastewater treatment systems (e.g. Krul, 1976; Krul & Veeningen, 1977; Meiberg et al., 1980; Robertson & Kuenen, 1983), and their activities may account for some of the nitrogen losses reported to occur in aerobic treatment plants. They may even explain the occurrence of levels of denitrifying enzymes in well aerated soils (Tiedje et al., 1982).

SUMMARY

Although "aerobic denitrification" has been a subject of some controversy for over a century, and had largely come to be discredited, reports of the simultaneous use of oxygen and denitrification by different species of bacteria have recently become more common. Research with some strains (e.g. Thiosphaera pantotropha) has revealed that bacteria simultaneously using two electron acceptors (i.e. nitrate and oxygen) grow more rapidly than when only a single electron acceptor is available. Further tests have indicated that there might be a link between aerobic denitrification and a form of nitrification which requires rather than generates energy, and is therefore known as heterotrophic nitrification. Heterotrophic nitrification (when defined as the oxidation of reduced nitrogen compounds during heterotrophic metabolism) is relatively common, being found in bacteria from different genera as well as fungi and even tissue from rat livers. Bacterial heterotrophic nitrification has tended to be under-estimated as, in many cases, the heterotrophic nitrifiers were simultaneously converting some of the nitrite (or other nitrogen oxide) to gaseous products (i.e. denitrifying). The relative importance of these strains within the nitrogen cycle may thus require some reconsideration. This paper reviews recent research into heterotrophic nitrification and aerobic denitrification, and presents a preliminary model which, if verified, will provide at least a partial explanation for the simultaneous occurrence of nitrification and denitrification in some bacteria. Finally, the ecological impact of bacteria capable of simultaneous nitrification and denitrification is considered.

REFERENCES

Abou Seada, M.N.I. and Ottow, J.C.G.,1985, Effect of increasing oxygen concentration on total denitrification and nitrous oxide release from soil by different bacteria. Biol. Fert. Soils 1:31-38.

Anderson, I.C. and Levine, J.S., 1986, Relative rates of NO and N_2O production by nitrifiers, denitrifiers and nitrate respirers. Appl. Env. Microbiol. 51:938-945.

Bazylinski D.A. and Blakemore R.P., 1983, Denitrification and assimilatory nitrate reduction in Aquaspirillum magnetotacticum. Appl. Env. Microbiol. 46:1118-1124.

Casella, S., Shapleigh, J.P., Lupi, F. and Payne, W.J., 1988, Nitrite reduction in bacteroids of Rhizobium "hedysari" strain HCNT1. Arch. Microbiol. 149:384-388.

Castignetti, D., and Hollocher, T.C., 1984, Heterotrophic nitrification among denitrifiers. Appl. Env. Microbiol. 47:620-623.

Davies, K.J.P., Lloyd, D. and Boddy, L., 1989, The effect of oxygenon denitrification in Paracoccus denitrificans and Pseudomonas aeruginosa. J. Gen. Microbiol. 135:2445-2451.

Degn, H., Cox, R.P. and LLoyd, D., 1985, Continuous measurement of dissolved gases in biochemical systems with the quadrupole massspectrometer. Methods Biochem. Anal. 31:166-194.

Gottschal, J.C. and Kuenen, J.G., 1980, Selective enrichment of facultatively chemolithotrophic Thiobacilli and related organisms in continuous culture. FEMS Microbiol. Letts. 7:241-247.

Hochstein L.I., Betlach M. and Kritikos G., 1984, The effect of oxygen on denitrification during steady-state growth of Paracoccus halodenitrificans. Arch. Microbiol. 137:74-78.

Kawakami, Y., Pacaud, B. and Nishimura, H., 1985, Inhibition of denitrification by oxygen in Paracoccus denitrificans. J. Ferment. Technol. 63:437-442.

Killham, K., 1986, Heterotrophic nitrification. IN Nitrification J.I. Prosser (editor) IRL Press. pp.117-126.

Korner, H. and Zumft, W.G., 1989, Expression of denitrification enzymes in response to the dissolved oxygen level and respiratory substrate in continuous culture of Pseudomonas stutzeri. Appl. Env. Microbiol. 55:1670-1676.

Krul, J.M., 1976, Dissimilatory nitrate and nitrite reduction under aerobic conditions by an aerobically and anaerobically grown Alcaligines sp. and by active sludge. J. Appl. Bacteriol. 40:245-260.

Krul, J.M. and Veeningen, R., 1977, The synthesis of the dissimilatory nitrate reductase under aerobic conditions in a number of denitrifying bacteria isolated from active sludge and drinking water. Water Res. 11:39-43.

Kuenen, J.G. and Beudeker, R.F., 1982, Microbiology of Thiobacilli and other sulphur oxidizing autotrophs, mixotrophs and heterotrophs. Phil. Trans. Royal Soc. London. Ser. B. 298:473-497.

Kuenen, J.G. and Gottschal, G.C., 1982, Competition among chemolithotrophs and methylotrophs and their interactions with heterotrophic bacteria. IN Microbial Interactions and Communities vol 1. A.T. Bull and J.H. Slater. (editors) Academic Press. pp.153-187.

Kuenen, J.G. and Robertson, L.A., 1984A. Interactions between obligately and facutatively chemolithotrophic sulphur bacteria. IN Continuous Culture 8 Biotechnology, medicine and the environment. A.C.R. Dean, D.C. Ellwood and C.G.T. Evans (editors) Ellis Horwood Ltd. pp.139-158.

Kuenen, J.G. and Robertson, L.A., 1984B. Competition among

chemolithotrophic bacteria under aerobic and anaerobic conditions. IN Current Perspectives in Microbial Ecology. M.J. Klug and C.A. Reddy (editors, American Society for Microbiology. pp.306-313.

Kuenen, J.G. and Robertson, L.A., 1987, Ecology of nitrification and denitrification. IN The Nitrogen and Sulphur Cycles J.A. Cole and S. Ferguson (editors) Cambridge University Press. pp.162-218.

Leegwater, M.P.M., 1983, Microbial reactivity: its relevance to growth in natural and artificial environments. PhD Thesis, University of Amsterdam.

Lloyd D., Boddy, L. and Davies, K.J.P., 1987, Persistance of bacterial denitrification capacity under aerobic conditions: the rule rather than the exception. FEMS Microbiol. Ecol. 45:185-190.

Lloyd, D., Davies, K.J.P. and Boddy, L., 1986, Mass spectrometry as an ecological tool for in situ measurement of dissolved gases in sediment systems. FEMS Microbiol. Ecol. 38:11-17.

MacFarlane, G.T. and Herbert, R.A. ,1984, Dissimilatory nitrate reduction and nitrification in estuarine sediments. J. Gen. Microbiol. 130:2301-2308.

Meiberg, J.B.M., Bruinenberg P.M., Harder W. ,1980, Effect of dissolved oxygen tension on the metabolism of methylated amines in Hyphomicrobium X in the absence and presence of nitrate: evidence for "aerobic denitrification". J. Gen. Microbiol. 120:453-463.

Poth, M., 1986, Dinitrogen production from nitrite by a Nitrosomonas isolate. Appl. Env. Microbiol. 52:957-959.

Rittenberg S.C., 1969, The roles of exogenous organic matter in the physiology of chemolithotrophic bacteria. Adv. Microb. Physiol. 3: 151-196.

Robertson, L.A., 1988, Aerobic denitrification and heterotrophic nitrification in Thiosphaera pantotropha and other bacteria. PhD thesis. Delft University of Technology.

Robertson, L.A. and Kuenen, J.G., 1983a, Thiosphaera pantotropha gen. nov. sp. nov., a facultatively anaerobic, facultatively autotrophic sulphur bacterium. J. Gen. Microbiol. 129:2847-2855.

Robertson, L.A. and Kuenen, J.G., 1983b, Anaerobic and aerobic denitrification by sulphide oxidizing bacteria from waste water. IN Anaerobic Waste Water Treatment W.J. van den Brink (editor) pp.3-12.

Robertson, L.A. and Kuenen, J.G., 1984a, Aerobic denitrification - old wine in new bottles? Antonie van Leeuwenhoek 50:525-544.

Robertson, L.A. and Kuenen, K.G., 1984b, Aerobic denitrification: a controversy revived. Arch. Microbiol. 139:351-354.

Robertson, L.A. and Kuenen, J.G., 1988, Heterotrophic nitrification in Thiosphaera pantotropha - oxygen uptake and enzyme studies. J. Gen. Microbiol. 134:857-863.

Robertson, L.A., Cornelisse, R., de Vos, P., Hadioetomo, R. and Kuenen, J.G., 1989a, Aerobic denitrification in various heterotrophic nitrifiers. Antonie van Leeuwenhoek 56:289-299.

Robertson, L.A., Cornelisse, R., Zheng, R. and Kuenen, J.G., 1989b, The effect of thiosulphate and other inhibitors of autotrophic nitrification on heterotrophic nitrifiers. Antonie van Leeuwenhoek 56: 301-309.

Robertson, L.A., van Kleeff, B.H.A. and Kuenen, J.G., 1986, A microcomputer-based method for semi-continuous monitoring of biological activities J. Microbiol. Methods 5:237-242.

Robertson, L.A., van Niel, E.W.J., Torremans, R.A.M. and Kuenen, J.G., 1988, Simultaneous nitrification and denitrification in aerobic chemostat cultures of Thiosphaera pantotropha. Appl. Env. Microbiol. 54:2812-2818.

Rönner, U. and Sörensen, F., 1985, Denitrification rates in the low-

oxygen waters of the stratified Baltic proper. Appl. Env. Microbiol. 50:801-806.
Strand, S.E., McDonnell, A.J. and Unz, R.F., 1988, Oxygen and nitrate reduction kinetics of a nonflocculating strain of *Zooglea ramigera*. Antonie van Leeuwenhoek 54:245-255.
Tiedje, J.M., Sexstone, A.J., Myrold, D.D. and Robinson, J.A., 1982, Denitrification: ecological niches, competition and survival. Antonie van Leeuwenhoek 48:569-583.
Trevors, J.T., 1985, The influence of oxygen concentrations on denitrification in soil. Appl. Microbiol. Biotechnol. 23:152-155.
Trevors, J.T. and Starodub, M.E., 1987, Effect of oxygen concentration on denitrification in freshwater sediment. J. Basic Microbiol. 27:387-391.
Verstraete, W., 1975. Heterotrophic nitrification in soils and aqueous media. Izvestija Akademii Nauk SSSR Ser. Biol. 4:541-558.

FLUX OF NO_x BETWEEN SOIL AND ATMOSPHERE: IMPORTANCE AND SOIL MICROBIAL METABOLISM

Ralf Conrad

Fakultät für Biologie, Universität Konstanz
P.O.Box 5560, D-7750 Konstanz
F.R.G.

INTRODUCTION

Nitric oxide and nitrogen dioxide ($NO + NO_2 = NO_x$) are trace gases which occur only in amounts of less than 1 ppbv in the clean atmosphere. A comparison of the standard redox potentials of NO and NO_2 among other biologically relevant nitrogen species is given in Table 5. A comparison of the atmospheric abundance, life time, and major sources and sinks is given in Table 1. The latter data are compiled from SCOPE reports (Söderlund and Svensson, 1976; Crutzen, 1983). Compared to other atmospheric nitrogen compounds the reactivity of NO_x is quite large and thus, relatively large fluxes are required to maintain even small atmospheric mixing ratios. On the other hand, even small variations in fluxes result in large variations in the atmospheric mixing ratios. Since atmospheric NO_x plays a key role in the chemistry of the atmosphere, the knowledge of the temporal and spatial distribution of sources and sinks of NO_x are extremely important for atmospheric models. The role of soils and of microbial denitrification for NO_x exchange between terrestrial ecosystems and the atmosphere is presently very uncertain and thus of special interest.

ROLE OF NO_x IN ATMOSPHERIC CHEMISTRY

Total reactive nitrogen (NO_y) is the sum of NO, NO_2, HNO_3, HO_2NO_2, PANs (peroxyacetylnitrate and related compounds), HNO_2, N_2O_5, NO_3, and the gaseous nitrates (Singh, 1987). NO_x is defined as the sum of NO and NO_2 which are in photochemical equilibrium (Logan, 1983; Singh, 1987). In the following reaction schemes," hv", symbolizes light; in the first scheme with a wave-length of less than 420 nm.

$$NO + O_3 \rightarrow NO_2 + O_2$$
$$NO_2 + hv \rightarrow NO + O \quad (\lambda < 420\ nm)$$
$$O + O_2 + M \rightarrow M + O_3 \quad (M = N_2 \text{ or } O_2)$$

Although most of the emissions of NO_x occur in the form of NO, it reacts within minutes with atmospheric O_3. NO_2 in turn photolyzes

Denitrification in Soil and Sediment, Edited by N. P. Revsbech and J. Sørensen, Plenum Press, New York, 1990

Table 1. Abundance, lifetime, sources, and sinks of atmospheric nitrogen compounds[a]

	N_2	N_2O	NH_3	NO_x
Mixing ratio in clean air	78 %	300 ppbv	0-3 ppbv	1-100 pptv
Atmospheric abundance (Tg N)	3.9×10^9	1300	1	1
Atmospheric lifetime	15×10^6 yr	100-200 yr	< 9 d	1.5 d
Source/sink flux (Tg N yr^{-1})	224-324	6-11	109-131	25-85
Main Scurces	Denitrification	Denitrification, Nitrification	Volatilization	Anthropogenic, Soils
Main sinks	N_2 Fixation	Oxidation in Stratosphere	Deposition	Oxidation in Troposphere, Deposition

[a] Data are compiled from Söderlund and Svensson (1976) and Crutzen (1983).

rapidly, eventually reforming O_3. NO_x usually makes up the majority of NO_y, but complex interconversion sequences exist. NO_x is finally transformed into HNO_3,

$$NO_2 + OH + M \rightarrow M + HNO_3$$

which is deposited to the ground. HNO_3 is an important component of acid rain (Logan, 1983).

The photochemical equilibrium of NO_x is influenced by a large number of reactions, and in turn influences other reactions. One of the most important reactions is the conversion of NO with hydrogen peroxyl or organic peroxy radicals (RO_2):

$$HO_2 + NO \rightarrow NO_2 + OH$$
$$RO_2 + NO \rightarrow NO_2 + RO$$

These reactions play an important role in the net production of tropospheric O_3 during the decomposition of atmospheric CO, CH_4, and nonmethane hydrocarbons (Crutzen, 1979). As a simple example, two chemical pathways of CO destruction in presence and absence of a sufficiently high NO mixing ratio are given (Crutzen, 1979):

NO < 5 pptv:

$$CO + OH \rightarrow CO_2 + H$$
$$H + O_2 + M \rightarrow M + HO_2 \quad (M = N_2 \text{ or } O_2)$$
$$HO_2 + O_3 \rightarrow O_2 + OH$$

$$\text{net:} \quad CO + O_3 \rightarrow CO_2 + O_2$$

NO > 5 pptv:

$$\begin{array}{lll} CO + OH & \rightarrow & CO_2 + H \\ H + O_2 + M & \rightarrow & M + HO_2 \quad (M = N_2 \text{ or } O_2) \\ HO_2 + NO & \rightarrow & OH + NO_2 \\ NO_2 + h\nu & \rightarrow & NO + O \quad (\lambda < 420 \text{ nm}) \\ O + O_2 + M & \rightarrow & M + O_3 \end{array}$$

$$\text{net: } CO + 2\,O_2 \rightarrow CO_2 + O_3$$

The destruction of CO may result in the net destruction or the net production of O_3, the result being dependent on the presence of sufficient NO in the atmosphere. Similar conclusions are also valid for the chemical decomposition of atmospheric CH_4 or NMHCs, where NO_x also is the rate-limiting compound in O_3 production.

Presently, there seems to be a tendency of increasing tropospheric O_3 mixing ratios (Logan, 1985; Penkett, 1988). The increase of tropospheric O_3 contributes to the greenhouse effect and increases the oxidizing capacity of the troposphere (Ramanathan et al., 1987; Isaksen, 1988). This has not only effects on climate and the chemistry of the atmosphere, but probably also on the health of life, e.g. photochemical smog or modern forest decline (Hinrichsen, 1985).

The increase of O_3 in the troposphere is in contrast to its decrease in the stratosphere (Crutzen 1979; Cicerone 1987). This decrease is partially due to the consumption of O_3 in a catalytical cycle,

$$\begin{array}{lll} NO + O_3 & \rightarrow & O_2 + NO_2 \\ O_3 + h\nu & \rightarrow & O_2 + O \quad (\lambda < 1140 \text{ nm}) \\ NO_2 + O & \rightarrow & O_2 + NO \end{array}$$

$$\text{net: } 2\,O_3 + h\nu \rightarrow 3\,O_2$$

where NO_x is the catalyst. However, since NO_x is extremely reactive and has only a short lifetime within the troposphere, it has no chance to reach the stratosphere. Instead, stratospheric NO_x is produced in-situ from less reactive nitrogen compounds. Especially N_2O which has a long lifetime in the troposphere is decomposed after transport into the stratosphere to yield NO. In summary, N_2O emission affects stratospheric O_3, NO_x emission affects tropospheric O_3.

SOURCES AND SINKS OF NO_x

Global sources of NO_x emission to the troposphere were estimated and compiled by several authors (Ehhalt and Drummond, 1982; Crutzen, 1983; Stedman and Shetter, 1983; Logan, 1983). The results of these estimates vary between 20 and 80 TgN yr-1 (Table 2). They include anthropogenic processes like combustion of fossil fuels and biomass burning, natural abiogenic processes like lightning, oxidation of ammonia, and input from the stratosphere, and microbial processes like emission from soils. The most important sources are localized at the planets surface. The contribution of soil microbial processes to NO_x emission is one of the most uncertain figures.

Table 2. Global sources of NO_x (Tg N yr^{-1})

Source	Ehhalt and Drummond (1982)	Crutzen (1983)	Stedman and Shetter (1983)	Logan (1983)
Fossil fuel combustion	8-19	12-20	18-22	14-28
Biomass burning	6-16	10-40	2-15	4-24
Lightning	2- 8	1-10	2- 6	2-20
NH_3 oxidation	1- 5		1- 2	1-10
Stratospheric input	1		1- 2	< 1
Soils	1-10	1-15	5-20	4-16
Total	19-59	24-85	27-67	26-99

NO_x is lost from the atmosphere by wet and dry deposition which on a global basis were estimated to 15-42 and 7-22 TgN yr-1, respectively (Ehhalt and Drummond, 1982; Logan, 1983). Wet deposition occurs by conversion to HNO_3 and rainout. Dry deposition occurs by direct interception and conversion of NO_x on surfaces such as leaves, soil, stone. Whereas emission of NO_x seems to be dominated by NO, dry deposition of NO_x seems to be dominated by NO_2 (Johansson, 1989).

Because of the inhomogenous distribution of sources and sinks and of the high reactivity and short lifetime of NO_x in the troposphere measurements of NO_x mixing ratios range over 4 orders of magnitude (Logan, 1983; Warneck, 1988). Values in the range of 100 ppbv are only found in urban areas where significant sources of NO exist. Here, the ratio of NO/NO_2 is relatively high. Outside the source areas NO reacts rapidly with O_3 and NO_2 becomes the dominant species. In rural areas, NO_x mixing ratios are in a range of <20 ppbv in valleys and flatlands, and of <100 pptv at mountain sites. Still lower values (<30 pptv) are reached in the free troposphere or in remote marine areas.

In remote, unpolluted areas soils are the most important potential source for NO_x in the continental atmospheric boundary layer. Hence, NO_x exchange rates between terrestrial ecosystems and the atmosphere are of key interest for the tropospheric chemistry over continents (Crutzen et al., 1985; Liu et al., 1987).

FLUX OF NO_x BETWEEN SOIL AND ATMOSPHERE

The flux of gases between soil and atmosphere can be measured by different methods, including measurements of vertical gradients, eddy correlation, accumulation within the planetary boundary layer, or flux chamber systems (Harriss, 1989). For interpretation of the resulting data it is important to realize, that some methods (especially the chamber systems) may disturb in-situ conditions, but others

Table 3. NO_x emission between unfertilized soils and atmosphere

Site	NO flux (ng N m-2s-1)	NO_2 flux (% of NO)	Reference
Grasslands			
Australia	0.6 - 7.3	< 3	Galbally and Roy, 1978
Germany	-6 - 14	9 - 95	Slemr and Seiler, 1984
Colorado	-9 - 28	measured as NO_x	Delany et al., 1986
Colorado	0.03 - 65	10	Williams et al., 1987
England	-12 - 26	measured as NO_x	Colbourn et al., 1987
Virginia	0.003 - 12.5		Anderson & Levine, 1987
Croplands			
Sweden	0.3 - 17	6	Johansson & Granat,1984
Pennsylvania	0.2 - 3.8		Williams et al., 1988
Colorado	0.001 - 16		Williams et al., 1988
Fallow			
Spain	-2 - 250	57 - 95	Slemr and Seiler, 1984
Colorado	ca. 1.4		Williams et al., 1987
Others			
Coniferous forest, Sweden	0.1 - 0.8		Johansson, 1984
Mixed forest Pennsylvania	0.2 - 4.1		Williams et al., 1988
Chaparral California	0 - 35		Anderson et al., 1988
Savanna Venezuela	3 - 15	0	Johansson et al., 1988
Cloud forest Venezuela	0.1 - 2		Johansson et al., 1988
Rainforest Brazil	9.2 - 16		Kaplan et al., 1988

(especially the micrometeorological methods) integrate over all reactions that occur within the atmospheric boundary layer including reactions in soil, plants and atmosphere. These reactions are different for NO and NO_2.

NO

Soils apparently are significant sources for atmospheric NO. The presently available field data have recently been compiled by Johansson et al. (1988) and Johansson (1989). An overview is shown in

Table 3 which emphasizes fluxes from unfertilized soils; fertilization results in increased NO emission rates (see below). It must be emphasized that the studies almost exclusively were conducted on soils covered with vegetation and thus integrate over reactions taking place in soil as well as in plants. Experiments have shown that NO and NO_2 is taken up by grass and other vegetation (Slemr and Seiler, 1984; Johansson and Granat, 1984; Williams et al., 1987; Johansson, 1987) so that one may expect different rates of NO_x emission from vegetated soil compared to bare soil. Burning of vegetation results in a significant and long-lasting stimulation of NO emission (but also of N_2O emission) for not yet clarified reasons (Anderson et al., 1988; Johansson et al., 1988).

Some field studies have shown that soils sometimes may act as a sink rather than a source for atmospheric NO (Table 4). This behaviour was also observed in absence of a vegetation cover (Slemr and Seiler, 1984). Even in those studies where soils acted as a source, the investigators observed that NO mixing ratios in the chamber reached a steady state level (Galbally and Roy, 1978; Slemr and Seiler, 1984; Kaplan et al., 1988). This level is interpreted as a compensation mixing ratio where the rate of NO release equals the rate of NO uptake:

$$J = P - \underline{k}\ m$$

and

$$P = \underline{k}\ mc$$

where J = net flux; P = release rate; $\underline{k}$ = uptake rate constant (first order reaction); m = NO mixing ratio; mc = NO compensation level.

This model has been confirmed for NO by laboratory studies on soil samples (Johansson and Galbally, 1984; Remde et al., 1989). The model implies that the direction of flux between soil and atmosphere depends on the magnitude of the compensation level and of the NO mixing ratio in the ambient atmosphere. At mixing ratios higher than the compensation level, soils will act as a sink for atmospheric NO, at mixing ratios lower than the compensation level as a source. Soils with a given compensation level may act as sources under unpolluted conditions and as sinks during pollution events (Slemr and Seiler, 1984). To model this behaviour of soils, it is crucial to know the compensation level and by which factors it is influenced. A simlar model was developed to relate laboratory measurements with field measurements by Galbally and Johansson (1989). The model was recently extented by replacing the first order uptake term with Michaelis-Menten uptake kinetics to accomodate measurements at higher NO mixing ratios (ppmv range) (Remde and Conrad, in prep.).

The model has also proved useful for studying the flux of H_2 and CO between soil and atmosphere (Conrad and Seiler, 1980; 1985). It may also be useful for the flux of N_2O, which obviously is produced and consumed in soil and exhibits a compensation mixing ratio (Seiler and Conrad, 1981). In contrast to NO, however, the N_2O compensation mixing ratio seems always to be higher than the N_2O mixing ratio of the ambient atmosphere, so that soils act as net source for N_2O except in a few occasions (Ryden, 1981; Slemr et al., 1984). This is apparently different for NO: one reason is the obviously high variability of ambient NO, the other may be a large variability of NO compensation mixing ratios due to variabilities in microbial NO production and consumption (see below).

The data base of NO compensation levels in soils is small.

Table 4. Compensation mixing ratios of NO in soil

System	NO compensation mixing ratio (ppbv)	Reference
Field data		
Grassland, Australia	16	Galbally and Roy, 1978
Grassland, Germany	20	Slemr and Seiler, 1984
Forest, Sweden	0.2 - 2	Johansson, 1984
dto. fertilized	<170	
Cropland, Sweden (partially fertilized)	2 - 75	Johansson and Granat, 1984
Paddy field, Australia (fertilized)	2	Galbally et al., 1987
Laboratory data (aerobic incubation)		
Loamy Clay (pH 7.8)	56 - 76	Remde et al., 1989
Loam (pH 6.3)	250 -400	Johansson and Galbally, 1984
Sandy loam (pH 4.7)	565 -605	Remde et al., 1989

Laboratory measurements on soil samples indicate values of 60-600 ppbv NO under aerobic incubation conditions (Johansson and Galbally, 1984; Remde et al., 1989). Field data range between 0.2 and 20 ppbv NO for unfertilized soil, and reach values of 75-170 ppbv NO in fertilized soil (Table 4). The data base is too small yet to compare and interprete field and laboratory values of NO compensation mixing ratios. However, discrepancies between the two data sets may be due to effects of the vegetation and the effects of NO diffusion within the soil profile (Galbally, 1989). The present data indicate, however, that soils may change between acting as source and as sink for atmospheric NO, when ambient NO mixing ratios change around 20 ppbv as in some rural areas (Warneck, 1988).

NO_2

It is unclear to which extent NO_2 is also emitted from soils. Only some of the field studies reported emission of NO_2 (Galbally and Roy, 1978; Slemr and Seiler, 1984; Johansson and Granat, 1984; Williams et al., 1987). While most of the observations showed NO_2 emission rates ranging between 3 and 10% of the rates of NO emission, Slemr and Seiler (1984) obtained values of up to 95% on vegetation-less soil. Vegetation may be a sink for NO_x, especially for NO_2 (Wesley et al., 1982; Duyzer et al., 1983; Delany et al., 1986). Studies on Scots Pine actually showed that tree branches are only a small sink for NO_x, but uptake of NO_2 was much higher than uptake of NO (Johansson, 1987). It is not unlikely that NO_2 emission is only significant on bare soils and is otherwise intercepted by the plants.

Since the diurnal rhythm of NO_2 emission from bare soil reached a maximum at noon about two hours earlier than NO emission, Slemr and Seiler (1984) suggested different mechanisms for NO_2 production compared to NO production and suggested that NO_2 was produced at the very soil surface, while NO was produced in somewhat deeper soil layers. Laboratory measurements indicate that NO_2 may be chemically produced when dry soil is heated to temperatures of > 60°C (Baumgärtner and Conrad, in prep.). This observation is consistent with the interpretation by Slemr and Seiler (1984), since temperatures regularly reached > 40°C even in a depth of 1 cm. However, more studies are needed to generalize NO_2 production and emission from soils and to understand the production mechanisms of NO_2.

MECHANISMS OF NO PRODUCTION

NO production in soil may be caused by different processes: chemical reactions, denitrification, nitrification, and other microbial processes.

Chemical Reactions

The most important chemical reaction is the decomposition of dissolved NO_2^- under acidic conditions (Chemodenitrification) which is basically a dismutation. The reaction is endergonic under standard conditions (compare Table 5) and only becomes exergonic at low pH:

$$2\ H^+ + 3\ NO_2^- \rightarrow 2\ NO + NO_3^- + H_2O$$
$$2\ H^+ + 2\ NO_2^- \rightarrow NO + NO_2 + H_2O$$

The second reaction producing NO_2 is negligible except under extremely acid conditions so that NO is the major gaseous end product of chemodenitrification (Van Cleemput and Baert, 1976; Chalk and Smith, 1983; Galbally, 1989). NO may also be produced by the reduction with transition metals such as Fe, Cu, or Mn in the reduced state (Chalk and Smith, 1983), e.g.

$$Fe^{2+} + NO_2^- + 2\ H^+ \rightarrow Fe^{3+} + NO + H_2O$$

Other chemical reactions may involve volatilization of HNO_2 or the production of methyl nitrite from HNO_2 and organic matter that both would be detected as an emission of NO_x by most analytical systems (Chalk and Smith, 1983; Galbally, 1989).

Denitrification

Respiratory denitrification is the dissimilatory reduction of NO_3^- or NO_2^- to gaseous products (predominantly N_2) that is coupled to electron transport phosphorylation (Tiedje, 1988). The sequential reduction of NO_3^- can be formulated in several ways depending on the role of NO as intermediate (Henry and Bessieres, 1984). Essentially, two contrasting pathways are distinguished, one in which NO is an obligatory intermediate (Zumft et al., 1988), and another in which NO equilibrates with an intermediate (Averill and Tiedje, 1982; Garber and Hollocher, 1982):

$$NO_3^- \rightarrow NO_2^- \rightarrow NO \rightarrow N_2O \rightarrow N_2$$

$$NO_3^- \rightarrow NO_2^- \rightarrow \underset{\displaystyle (X)}{\overset{\displaystyle NO}{\downarrow\uparrow}} \rightarrow N_2O \rightarrow N_2$$

Despite this controversy, it is clear that NO is a potential product of denitrifiers and can be released into the soil environment. However, production of NO has been explicitely shown and quantified for only a few species of denitrifiers. Among them are the genera Pseudomonas, Alcaligenes, and Rhizobium (Firestone et al., 1979; Betlach and Tiedje, 1981; Carlson and Ingraham, 1983; Anderson and Levine, 1986; Zafiriou et al., 1989) to which the most ubiquitous soil denitrifiers seem to belong (Tiedje, 1988).

Nitrification

Chemoautotrophic nitrification is the oxidation of NH_4^+ to NO_2^- and of NO_2^- to NO_3^- that is coupled to electron transport phosphorylation and CO_2 fixation. Nitrosomonas is able to produce significant amounts of N_2O and NO during oxidation of NH_4^+ (Ritchie and Nicholas, 1972; Lippschultz et al., 1981; Anderson and Levine, 1986). Nitrosospira, Nitrosolobus, and Nitrosococcus produce N_2O as well (Blackmer et al., 1980; Goreau et al., 1980) and therefore, presumably also produce NO. N_2O is produced by reduction of NO_2^- which is used as an alternative electron acceptor to O_2 (Poth and Focht, 1985). N_2 is another product of this reaction (Poth, 1986) which is called nitrifier denitrification.

From a thermodynamic point of view, the reduction of NO_2^- to NO and N_2O with NH_2OH/NO_2^- as reductant is exergonic (Table 5). Recently, we observed that NO and N_2O was produced not only during oxidation of NH_4^+, but also by reduction of NO_2^- with N_2H_4 (hydrazine) as electron donor (Remde, 1989) indicating that both NO and N_2O were produced during denitrification of NO_2^-. A likely reaction sequence of NH_4^+ oxidation and NO and N_2O production is given in Fig. 1. The figure also provides a conceptual explanation why NO and N_2O production by nitrifiers is stimulated at reduced O_2 supply (Lipschultz et al., 1981; Anderson and Levine, 1986), i.e.: reduction of NO_2^- instead of O_2 saves the small amounts of O_2 that are required for activation of NH_4^+ by ammonium monoxygenase. Nitrifier denitrification virtually is an aerobic denitrification.

Recently, Kuenen and Robertson (1988) reviewed aerobic denitrification and the simultaneous occurrence of heterotrophic nitrification plus denitrification in multitalented microorganisms.

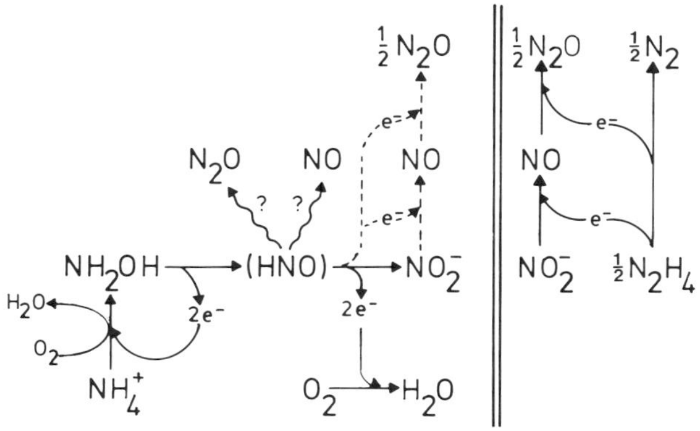

Fig. 1. Proposed pathway for production of NO and N_2O during nitrifier denitrification

Table 5. Standard redoxpotentials of nitrogen compounds[a)]

Redox couple	e- transferred	E_0'(mV)
H^+ / H_2	2	-414
N_2 / NH_4^+	6	-276
NO_2^- / NH_2OH	4	57
NO_2^- / NH_4^+	6	340
NO_2^- / NO	1	345
NO_3^- / NO_2^-	2	432
NO_2 / NO	2	638
NO_2^- / N_2O	4	759
O_2 / H_2O	4	818
NH_2OH / NH_4^+	2	906
NO_2 / NO_2^-	1	931
NO / N_2O	2	1173
N_2O / N_2	2	1355

a) The standard redox potentials were calculated relative to E_0' of H_2 or O_2 by using the standard Gibbs free energies of reduction or oxidation reactions (Thauer et al., 1977).

These include the newly isolated Thiosphaera pantotropha (Robertson and Kuenen, 1984; 1986; Robertson et al., 1988), but also the well-known denitrifiers Alcaligenes faecalis or Pseudomonas fluorescens which can also nitrify heterotrophically (Castignetti and Hollocher, 1984). A. faecalis produces NO and N_2O both when reducing NO_3^- (Anderson and Levine, 1986) and when oxidizing NH_4^+ (Papen et al., 1989). If it can be proven that other ubiquitous soil microorganisms are also able to nitrify, denitrify, and produce NO under aerobic conditions, the produced NO could either originate from the reduction of NO_3^- (or NO_2^-) or from the oxidation of NH_4^+ (or reduced organic N, if NH_4^+ is not used by the heterotrophic nitrifier).

It still has to be shown, however, how widespread NO production really is among heterotrophic nitrifiers. Methanotrophic bacteria are potential nitrifiers (Dalton, 1979) and N_2O is produced by Methylosinus trichosporium, but at lower rates than in Nitrosomonas (Yoshinari, 1985). N_2O was also produced in Methanomonas agilis (Krämer and Conrad, in prep.). However, NO production could neither be observed in this bacterium nor in M. trichosporium, although both methanotrophs were able to consume NO.

Other Microbial Processes

It seems possible that NO production might occur in all microorganisms which produce N_2O. N_2O production was shown in microorganisms with assimilatory NO_3^- reduction, with respiratory NO_3^- reduction to NO_2^-, and with dissimilatory NO_3^- reduction to NH_4^+. All these metabolic groups typically produce N_2O but not N_2, and they do not gain energy by producing N_2O. They have thus been named nonrespiratory N_2O producers in contrast to respiratory N_2O-producing denitrifiers (Tiedje, 1988). As reviewed by Tiedje (1988) and Cole (1988) N_2O seems to be produced by nitrate reductase acting on NO_2^-. However, this evidence is at the moment restricted to Enterobacteriaceae, in particular Escherichia coli (Newman and Cole, 1978; Smith, 1983), and to bean leaves (Klepper, 1987;

Dean an Harper, 1986). These organisms are so far the only examples among nonrespiratory nitrate reducers, where NO production has been demonstrated in addition to N_2O production (Klepper, 1987; Dean an Harper, 1986; Ji and Hollocher, 1988a,b). In these organisms, production of NO actually was more than one order of magnitude higher than production of N_2O. It is possible that NO production serves nitrosation reactions (Ji and Hollocher, 1988a).

By looking on the various ways of microbial NO production it appears that NO is generally produced by reduction of nitrite. This seems to be true for the nitrifiers as well as for the various metabolic groups of nitrate reducers. However, it seems speculative to assume that NO is necessarily produced whenever N_2O is produced and vice versa. Ratios of NO/N_2O production measured in pure bacterial cultures vary tremendously. Variation of NO/N_2O is high even within one species. This variation seems to be influenced by culture conditions, e.g. O_2 partial pressure (Lipschultz et al., 1981; Anderson and Levine, 1986).

MECHANISMS OF NO CONSUMPTION

NO consumption in soil can basically be caused by oxidative or by reductive processes. The chemical oxidation of NO with O_2 to NO_2 is exergonic (ΔG^{C} = -34.7 kJ mol-1) as the redox potential of NO_2/NO is more negative than that of O_2/H_2O (Table 5). However, the homogeneous gas phase reaction is too slow to be of importance, if NO concentrations are low (Finlayson-Pitts and Pitts, 1986). At high mixing ratios of > 100 ppmv NO the half life for the reaction would be one hour, whereas at atmospheric mixing ratios of < 10 ppbv NO, it would be on the order of 10,000 h (Galbally and Roy, 1978). Therefore, chemical oxidation of NO in soil can be ruled out, unless O_3 is available as an oxidant. In fact, chemical oxidation of NO was not observed in autoclaved soil (Johansson and Galbally, 1984; Remde at al, 1989). However, this does not exclude the existence of biological activities which are able to oxidize NO. So far, such an oxidative activity has not been described for NO, but seems to exist for N_2O involving the reaction of soil catalase (Vedenina and Zavarzin, 1978; Vedenina et al., 1980).

Reductive consumption of NO is the most plausible mechanism when considering the positive redox potentials of NO/N_2O and N_2O/N_2 (Table 5). Chemical reduction of NO in soil is unlikely, since the NO consumption activity is destroyed by autoclaving the soil (see above). However, it is not completely ruled out since autoclaving could also destroy the chemical reductant (e.g. H_2S in sediments) of NO. Reductive consumption of NO is also plausible when considering that NO seems to be an product in the microbial pathways of NO_2^- reduction to N_2O or N_2 (see above). However, consumption of NO has so far only occasionally been studied in pure bacterial cultures.

NO consumption has so far not been observed in nitrifiers. Relevant studies only showed a production of NO, but failed to demonstrate whether the accumulated NO would be reutilized or not (e.g. Anderson and Levine, 1986). The same is the case for heterotrophic nitrifying <u>Alkaligenes</u> (Papen et al., 1989). However, heterotrophically nitrifying methanotrophic bacteria were shown to consume NO (Krämer and Conrad, in prep.). Like production, consumption of NO has hardly been studied in nonrespiratory N_2O producers (Ji and Hollocher, 1988).

However, NO consumption was observed in Wolinella succinogenes (Payne et al., 1982) which has a dissimilatory NH_4^+-producing nitrite reductase (Liu et al., 1983), but appears to also possess a nitrite reductase which is coupled to electron-transport phosphorylation (Schröder et al., 1985). W. succinogenes is able to reduce N_2O (Yoshinari, 1980) and contains a N_2O reductase that is different from those found in respiratory denitrifiers (Teraguchi and Hollocher, 1989).

NO consumption has been demonstrated in respiratory denitrifiers. NO reduction has been shown in intact cell suspensions of Pseudomonas aureofaciens (Firestone et al., 1979), P. perfectomarina (Zafiriou et al., 1989), P. aeruginosa, P. fluorescens, P. stutzeri, Paracoccus denitrificans, Azospirillum brasilense, and A. lipoferum (Remde, 1989). Growth on NO as sole electron acceptor was claimed for Thiobacillus denitrificans (Ishaque and Aleem, 1973) and for Bacillus strains (Pichinoty et al., 1978). NO-dependent proton translocation was demonstrated for Paracoccus denitrificans, Rhodobacter sphaeroides, Alcaligenes (Achromobacter) cycloclastes, and Bradyrhizobium japonicum (Shapleigh and Payne, 1985). Hence, utilization of NO as an electron acceptor seems to be coupled to electron transport phosphorylation. The biochemistry of nitrite reduction and the possible occurrence of a NO reductase has recently been reviewed by Hochstein and Tomlinson (1988) and Zumft et al. (1988). It has recently been shown that NO consumption is not affected by mutational loss of nitrite reductase (Zumft et al., 1988) and that a N_2O-forming NO reductase can be solubilized from membranes of P. stutzeri (Heiss et al., 1989).

CONTRIBUTION OF DIFFERENT PROCESSES TO NO TURNOVER IN SOIL

To understand the environmental regulation of NO turnover, it is necessary to know how much the various processes contribute to it. For NO, there are only a few relevant studies and generalizations are hardly possible. More data exist for N_2O production. Here it seems that nitrification, denitrification, but also other microbial processes may contribute significantly to N_2O production. Since the contribution of the individual processes seems to vary with soil type and environmental situation, it is certainly a long way until we understand the production of gaseous nitrogen oxides under a given situation.

Chemical reactions seem only to be important for the production of NO, but not of N_2O (Bremner et al., 1980; Nelson, 1982). It is principally possible to differentiate between biological and chemical mechanisms by performing a killed control, e.g. by autoclaving, gamma irradiation, metabolic poisons. However, killed controls can be biassed. Autoclaving for example not only kills the bacteria, but also destroys NO_2^-, so that chemical NO_2^- decomposition will be underestimated because of lack of NO_2^- (McKenney et al., 1982; Blackmer and Cerrato, 1986; Remde et al., 1989). In general, chemodenitrification will exhaust NO_2^- which must be supplied by biological reactions, either nitrification or denitrification. Under steady state conditions, chemical NO production thus depends on biological processes that produce labile NO_2^- (or labile NH_2OH in case of chemical N_2O production; Bremner et al., 1980). NO production by chemodenitrification is only significant under acidic conditions (Van Cleemput and Baert, 1984). However, microbial acid production together with microbial NO_2^- formation may stimulate chemical NO production, perhaps within microniches (Firestone and Davidson, 1989). Therefore, it is questionable if killed controls really can quantitatively determine the contribution of chemodenitrification to NO production. On the other hand, the question may also be irrelevant unless a soil contains so high amounts of NO_2^- that resupply by biological reactions is negligible.

Differentiation between NO production by nitrification and denitrification would be straigthforward, if one of the processes were strictly aerobic and the other strictly anaerobic. However, this is not the case (see below). Moreover, it is almost impossible to exclude the existence of anaerobic microniches within a soil even when it seems to be well aerated (Parkin, 1987). Differentiations thus depend on a combination of experiments, i.e. aerobic versus anaerobic incubation, stimulation by addition of NH_4^+ versus NO_3^-, use of 15N-labelled nitrogen compounds, and application of metabolic inhibitors, such as acetylene, nitrapyrin, chlorate (Bremner and Blackmer, 1980; Knowles, 1981; 1985). Using this technique of differentiation, we recently observed that in one soil NO was predominantly produced by denitrification, and in another one predominantly by nitrification (Remde, 1989).

Recently, an acetylene inhibition technique was developed that uses increasing acetylene concentrations to inhibit the ammonium monooxygenase (2.1 Pa) of nitrifiers and the N_2O reductase (10 kPa) of denitrifiers (Klemedtsson et al., 1988a). This method avoids stimulation of N-metabolic processes by addition of N-compounds (or 15N-compounds) and thus can be used to differentiate the source of N_2O or NO under unfertilized conditions. However, the results can be biassed if denitrification is limited by NO_2^- or NO_3^- which is supplied by nitrification of NH_4^+. Hence, the acetylene inhibition technique does not differentiate between nitrification and denitrification but between the pools of soil NH_4^+ or NO_3^- as original source for N_2O or NO. So far, this technique has only been applied to N_2O (e.g., Klemedtsson et al., 1988b; Robertson and Tiedje, 1987; Davidson and Swank, 1986). As a result it was found that nitrification, denitrification, but sometimes also other microbial processes (e.g. heterotrophic nitrification) may significantly contribute to N_2O production. Further research, especially with respect to NO is necessary to show the usefulness of this acetylene inhibition technique.

CONTROL OF NO METABOLISM IN SOIL

Respiratory denitrification and autotrophic nitrification-denitrification may be the quantitatively most important microbial processes involved in the production of NO in soil. These processes then would ultimately control the NO production rate. The process regulation of NO and N_2O metabolism during nitrification and denitrification has recently been reviewed by Firestone and Davidson (1989). These authors described a simplistic model (the "hole-in-the-pipe model"; Fig. 2) which nicely emphasizes the tremendous difficulties to quantitatively model the metabolism of NO and N_2O in soil on basis of the control of nitrification and denitrification by environmental factors. The problem is that there are at least to levels of regulation: (1) the regulation of the rate of the process (nitrification, denitrification), and (2) the regulation of the partitioning between NO, N_2O and the main product. A third level of regulation may concern the consumption of extracellular NO and N_2O. As consequence, a change in the soil environment (e.g., temperature decrease) can reduce the rate of product formation, but it can simultaneously increase the proportion of NO or N_2O in the metabolic products and can affect the rates of consumption by other microbial populations.

The reviews of environmental regulation of NO production (Firestone and Davidson, 1989; Galbally, 1989; Johansson, 1989) clearly show the tremendous lack of quantitative data that would allow to model NO metabolism in soil on the basis of process regulation. The same is also still true for N_2O despite the scientific progress and the accumulation of data over the last 10 years (Firestone and Davidson, 1989; Knowles, 1985). The situation is mainly due to our ignorance of the regulation of partitioning factors for NO and N_2O. However, the regulation of rates of nitrification and denitrification also are not yet understood satisfactorily.

Rates of Product Formation

The influence of environmental parameters on rates of nitrification and denitrification have been reviewed by several authors (Garcia, 1975; Focht and Verstraete, 1977; Payne, 1981; Schmidt, 1982; Firestone, 1982; Knowles, 1982; Kuenen and Robertson, 1988; Tiedje, 1988).

Tiedje (1988) presents a conceptual model for the environmental regulation of denitrification that is based on the hierarchy of three important regulators of denitrification: O_2 > NO_3^- > organic carbon (Fig. 3). The model has been further elaborated by Robertson (1989) (reviewed by Firestone and Davidson, 1989). According to this model, O_2 would be the most important proximate regulator of denitrification rates, but would itself be regulated by more distal regulators such as respiration, soil texture, and soil moisture. The next important proximate regulators would be the availability of NO_3^- followed by the availabilty of organic carbon as electron donor in denitrifiaction.

So far, denitrification is believed to be highly sensitive towards O_2, and the concept of O_2 being the most important regulator of denitrification is supported by most observations (Tiedje, 1988). However, examples are accumulating that demonstrate not only the persistence of denitrification under aerobic conditions, but also the existence of bacteria which require at least some O_2 for optimum denitrification. These examples include Aquaspirillum magnetotacticum (Bazylinski and Blakemore, 1983), Thiosphaera

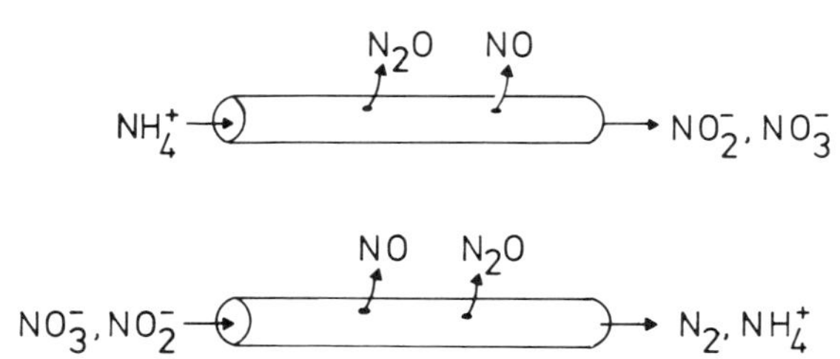

Fig. 2. "Hole-in-the-pipe" model to explain the two levels of regulation of NO and N_2O production via nitrification and denitrification: flow of N through the "process pipes" and holes in the pipes through which NO and N_2O "leak" (adapted from Firestone and Davidson, 1989)

panthotropha (Robertson and Kuenen, 1984), and two strains of *Pseudomonas aeruginosa* (Lloyd et al., 1987). Further research will have to show, whether these or similar denitrifiers are of any importance for nitrogen turnover in soils.

A similar conceptual model as for denitrification has recently been presented for nitrification (Robertson, 1989; reviewed by Firestone and Davidson, 1989). The hierarchy of proximate regulators seems to be that the availability of NH_4^+ largely dominates the rate of nitrification as long as at least some O_2 is available ($NH_4^+ \gg O_2$).

To a great extent, these models describe the controls of nitrification and denitrification rates very well. However, soil temperature and pH were only included as distal regulators affecting almost every soil biochemical process. They probably must also be considered as important proximate regulators of nitrification and denitrification rates. They also control the partitioning of the products (see below). In addition, control of the reaction rate in soil is not necessarily a direct effect on one particular population of nitrifiers or denitrifiers, but can also be based on a shift of substrate flow towards a better adapted population, e.g. heterotrophic nitrification by acid-tolerant fungi (Kreitinger et al., 1985; Lang and Jagnow, 1986).

Partitioning of Products

For the partitioning of N_2O to N_2 in dentrification, there is at least a qualitative understanding (Firestone and Davidson, 1989): N_2O/N_2 seems to increase with increasing NO_3^- and O_2, and with decreasing organic carbon, temperature, and pH. A kinetic model has been proposed to explain the partitioning (Betlach and Tiedje, 1981), but has only occasionally been explored further (McKenney et al., 1982; 1984; Zafiriou, 1989). The partitioning of N_2O to NO_2^- in nitrification is less well understood (Firestone and Davidson, 1989), but N_2O/NO_2^- seems to increase with decreasing O_2 and pH.

The regulation of partitioning factors for NO is presently unknown. However, there are a few examples of simultaneous measurements of NO and N_2O. These data clearly show that partitioning of NO and N_2O can be completely different both in pure cultures and soil (Anderson and Levine, 1986; 1987). Hence, it appears that partitioning of NO is differently regulated than that of N_2O. Temperature for example is one effector which controls the partitioning during denitrification in soil. This control seems to be due to different activation energies of consecutive enzymatic reduction steps (McKenney et al., 1984). The mechanistic basis for the partitioning effect of O_2 (Lipschultz et al., 1981; Anderson and Levine, 1986) is unknown, but differences in synthesis of consecutive enzymes may be important (Tiedje, 1988; Körner and Zumft, 1989). Presently it also cannot be excluded that partitioning effects are due to metabolite control of the activity of key enzymes. Regulation of bacterial metabolism of NO is certainly a field for future research.

Here, it must be emphasized that soil is not simply an enzymatic reactor but contains many microbial populations which have the potential to adapt to changing situations by growth. We know amazingly little about how much environmental changes translate into a change of the composition of the microbial community or just of the activity of the existing microbial community. Parkin et al. (1985) reported on the apparent adaptation of a denitrifier population to a soil with low pH values. This population had a optimum pH of 3.9, whereas populations in neutral soil usually have higher pH optima. Similar adaptations may occur for other environmental parameters, temperature, nutrients etc. It would be interesting to know whether an adapted population shows the same response to environmental change as a non-adapted population, not only with respect to total activity but also to partitioning of the products.

Rates of NO Consumption

Together with the NO production rate (P), the NO consumption rate constant (k) defines the NO compensation mixing ratio (mc = P/k), and consequently also the direction and magnitude of NO flux between soil and atmosphere. At a given NO production rate, increase of NO consumption would result in decreased NO compensation mixing ratios. Very little is known about the regulation of NO consumption in soil. NO consumption is less stimulated by anaerobic conditions than is NO production (Remde et al., 1989). Even under NO-saturating conditions, V_{max} of NO consumption was smaller than rates

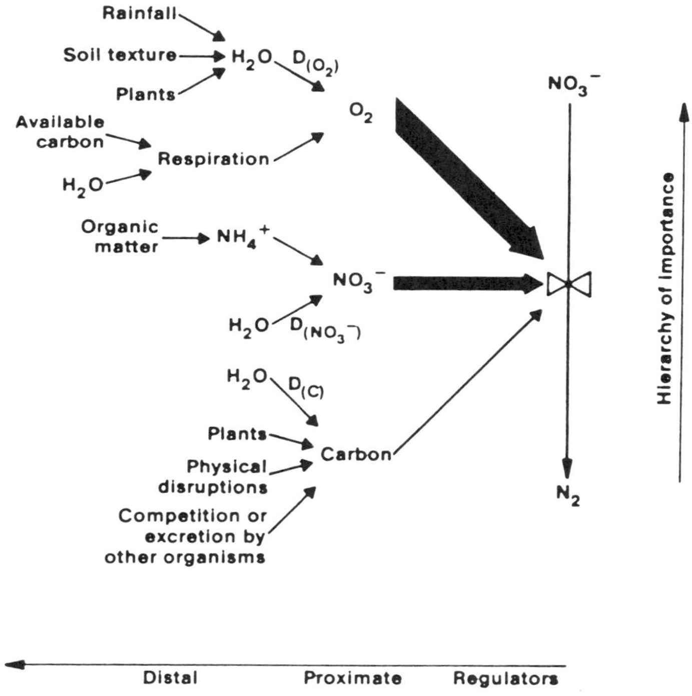

Fig. 3. Conceptual model of environmental denitrification. The vertical dimension shows the hierarchy of importance of the three major regulators of denitrification. The horizontal dimension illustrates the proximity of the regulatory factors or features to the enzymatic process. D refers to an effect on diffusion rates of the indicated regulator (taken from Tiedje, 1988)

of NO production. As a consequence, a compensation level of NO mixing ratios was never reached under anaerobic conditions (Remde et al., 1989; Remde, 1989). Recent observations indicate that NO is consumed by denitrifiers and that NO consumption rates are stimulated when denitrifcation is stimulated by glucose, NO_3^-, or NO_2^- (Baumgärtner and Conrad, in prep.). The relative regulation of NO production versus NO consumption in soil is largely unknown at the moment.

MAN'S IMPACT ON NO_x FLUX BETWEEN SOIL AND ATMOSPHERE

The direction and rate of NOx exchange between soil and atmosphere can be influenced (1) by changing the ambient atmospheric mixing ratio of NO_x, or (2) by changing the NO_x metabolism in soil. Atmospheric mixing ratios can be increased by pollution, e.g. burning of fossil fuels or biomass burning. If they reach a value higher than soil's compensation mixing ratio, there will be a net flux of NO_x into the soil. Because of the photochemical conversion of NO to NO_2, there can be a net flux of NO out of the soil and a net flux of NO_2 into the soil. This situation actually seems to prevail in unpolluted areas, although it occasionally can also be the other way round, especially if NO_2 is produced in the soil (Slemr and Seiler, 1984). If soils are inefficient NO_2 producers, the NO flux out of the soil seems to be higher than the NO_2 flux into the soil, so that the soil acts as a net source for atmospheric NO_x (Johansson and Granat, 1984; Johansson et al., 1988). In polluted areas, however, the soil may act as a net sink for both NO and NO_2.

A chronical net input of NO_x may change the soil ecosystem, its nitrogen metabolism, and possibly also its capacity for NO_x turnover. Recently it was shown by Melillo et al. (1989) that the chronical input of atmospheric nitrogen by deposition of NH_x (mainly NH_3) or NO_x (mainly NO_x and HNO_3) seems to affect the behaviour of soils with respect to N_2O emission (increase), CH_4 deposition (decrease), and emission of COS + CS_2 (increase). Recently, we observed that the nitrifier flora which is causing the corrosion of building stones by oxidizing deposited NH_4^+ produces NO in similar quantities as aerobic soils (Baumgärtner et al., 1990). Hence, feedbacks by deposition of atmospheric NH_x or NO_y seem possible, but information is scarce.

Soil metabolism can also be changed by other human activities. Most obvious is the direct application of nitrogen fertilizers for agricultural purpose. Fertilization results in increased emission rates of NO and in some soils also of NO_2 (Johansson, 1984; Johansson and Granat, 1984; Slemr and Seiler, 1984; Anderson and Levine, 1987). The percentage loss of fertilizer-N as NO_x is in a range of 0.3 to 5.4% and thus, is in a similar range as the loss of N_2O-N (Conrad et al., 1983). However, there is obviously no linear relationship between NO and N_2O production in soils (Anderson and Levine, 1987).

There is a marked difference between the fate of fertilizer lost as N_2O or NO_x. In contrast to N_2O which lasts in the atmosphere for a hundred years and is then destructed in the stratosphere (Crutzen, 1983), NO_x is soon deposited again after it has been emitted from the soil. Fertilizer loss as NO_x thus means nitrogen loss from a particular field plot, but also means unintended fertilization of other continental ecosystems. This disturbance of the natural nutrient cycle may have unsuspected impacts on the regional biosphere.

SUMMARY

NO_x is a trace constituent in the clean troposphere, but nevertheless plays a central role in its chemistry, e.g. generation of O_3. Beside combustion of fossil fuel or biomass, soils are a major source in the budget of atmospheric NO_x. NO_x seems to be released from soils mainly as NO. The emission of NO is a net flux, the direction and the magnitude of which depend on the atmospheric NO concentration and on the activities of NO production and consumption in the soil. NO is possibly produced by chemical reactions, by denitrification, by nitrification, and by other microbial processes. The same or similar processes seem to be involved in the consumption of NO. Since NO obviously is an intermediate rather than an end product of microbial metabolism, the control of NO turnover in soil by environmental parameters is very complicated. Human activities may influence NO metabolism by acute or chronic fertilization with nitrogen. The percentage loss of fertilizer-N as NO apparently is in a similar range as that of N_2O.

ACKNOWLEDGEMENTS

This work profited from stimulating discussions with M. Baumgärtner, M. Krämer, H. Papen, A. Remde, and F. Slemr and from financial support by the Bundesministerium für Forschung und Technologie (07EU719) and the Fonds der Chemischen Industrie.

REFERENCES

Anderson, I.C., and Levine, J.S., 1986, Relative rates of nitric oxide and nitrous oxide production by nitrifiers, denitrifiers, and nitrate respirers, Appl. Environ. Microbiol., 51:938-945.

Anderson, I.C., and Levine, J.S., 1987, Simultaneous field measurements of biogenic emissions of nitric oxide and nitrous oxide, J. Geophs. Res., 92:965-976.

Anderson, I.C., Levine, J.S., Poth, M.A., and Riggan, P.J., 1988, Enhanced biogenic emissions of nitric oxide and nitrous oxide following surface biomass burning, J. Geophys. Res., 93:3893-3898.

Averill, B.A., and Tieje, J.M., 1982, The chemical mechanism of microbial denitrification, FEBS Lett., 138:8-12.

Baumgärtner, M., Remde, A., Bock, E., and Conrad, R., 1990, Release of nitric oxide from building stones into the atmosphere, Atmos. Environ. 24B:87-92.

Bazylinski, D.A., and Blakemore, R.P., 1983, Denitrification and assimilatory nitrate reduction in Aquaspirillum magnetotacticum, Appl. Environ. Microbiol., 46:1118-1124.

Betlach, M.R., and Tiedje, J.M., 1981, Kinetic explanation for accumulation of nitrite, nitric oxide, and nitrous oxide during bacterial denitrification, Appl. Environ. Microbiol., 42:1074-1084.

Blackmer, A.M., Bremner, J.M., and Schmidt, E.L., 1980, Production of nitrous oxide by ammonia-oxidizing chemoautotrophic microorganisms in soil, Appl. Environ. Microbiol., 40:1060-1066.

Blackmer, J.M., and Cerrato, M.E., 1986, Soil properties affecting formation of nitric oxide by chemical reaction of nitrite, Soil Sci. Soc. Am. J., 50:1215-1218.

Bremner, J.M., Blackmer, A.M., and Waring, S.A., 1980, Formation of nitrous oxide and dinitrogen by chemical decomposition of hydroxylamine in soils, Soil Biol. Biochem., 12:263-269.

Bremner, J.M., and Blackmer, A.M., 1980, Mechanisms of nitrous oxide production in soils, in: "Biogeochemistry of ancient and modern environments", P.A. Trudinger, M.R. Walter and B.J. Ralph, eds., p. 279-291, Springer, Berlin.

Carlson, C.A., and Ingraham, J.L., 1983, Comparison of denitrification by Pseudomonas stutzeri, Pseudomonas aeruginosa, and Paracoccus denitrificans, Appl. Environ. Microbiol., 45:1247-1253.

Castignetti, D., and Hollocher, T.C., 1984, Heterotrophic nitrification among denitrifiers, Appl. Environ. Microbiol., 47:620-623.

Chalk, P.M., and Smith, C.J., 1983, Chemodenitrification. Dev. Plant Soil Sci. 9:65-89.

Cicerone, R.J., 1987, Changes in stratospheric ozone, Science, 237:35-42.

Colbourn, P., Ryden, J.C., and Dollard, G.J., 1987, Emission of NO_x from urine-treated pasture, Environ. pollut., 46:253-261.

Cole, J.A., 1988, Assimilatory and dissimilatory reduction of nitrate to ammonia, in: "The nitrogen an sulphur cycles", J.A. Cole and S. Ferguson, eds., pp. 281-329. Cambridge University Press, Cambridge.

Conrad, R., and Seiler, W., 1980, Contribution of hydrogen production by biological nitrogen fixation to the global hydrogen budget, J. Geophys. Res., 85:5493-5498.

Conrad, R., and Seiler, W., 1985, Destruction and production rates of carbon monoxide in arid soils under field conditions, in: "Planetary ecology", D.E. Caldwell, J.A. Brierley, and C.L. Brierley, eds., p. 112-119. Van Nostrand Reinhold, New York.

Conrad, R., Seiler, W., and Bunse, G., 1983, Factors influencing the loss of fertilizer nitrogen into the atmosphere as N_2O, J. Geophys. Res., 88:6709-6718.

Crutzen, P.J., 1979, The role of NO and NO_2 in the chemistry of the troposphere and stratosphere, Ann. Rev. Earth Planet. Sci., 7:443-472.

Crutzen, P.J., 1983, Atmospheric interactions - Homogeneous gas reactions of C, N, and S containing compounds, in: "The major biogeochemical cycles and their interactions", B. Bolin and R.B. Cook, eds., SCOPE 21, p. 67-114. Wiley, Chichester.

Crutzen, P.J., Delany, A.C., Greenberg, J., Haagenson, P., Heidt, L., Lueb, R., Pollock, W., Seiler, W., Wartburg, A., and Zimmerman, P., 1985, Tropospheric chemical composition measurements in Brazil during the dry season, J. Atmos. Chem., 2:233-256.

Dalton, H., 1977, Ammonia oxidation by the methane oxidising bacterium Methylococcus capsulatus strain Bath, Arch. Microbiol., 114:273-279.

Davidson, E.A., and Swank, W.T., 1986, Environmental parameters regulating gaseous nitrogen losses from two forested ecosystems via nitrification and denitrification, Appl. Environ. Microbiol., 52:1287-1292.

Dean, J.V., and Harper, J.E., 1986, Nitric oxide and nitrous oxide production by Soybean and Winged Bean during the in vivo nitrate reductase assay, Plant Physiol., 82:718-723.

Delany, A.C., Fitzjarrald, D.R., Lenschow, D.H., Pearson Jr., R., Wendel, G.J., and Woodruff, B., 1986, Direct measurements of nitrogen oxides and ozone fluxes over grassland, J. Atmos. Chem., 4:429-444.

Duyzer, J.H., Meyer, G.M., and van Aalst, R.M., 1983, Measurement of dry deposition velocities of NO, NO_2 and O_3 and the influence of chemical reactions, Atmos. Environ., 17:2117-2120.

Enhalt, D.M., and Drummond, J.W., 1982, The tropospheric cycle of NO_x, in: "Chemistry of the unpolluted and polluted troposphere", H.W. Georgii, and W. Jaeschke, eds., p. 219-251. Reidel, Dordrecht.

Finlayson-Pitts, B.J., and Pitts Jr., J.N., 1986, "Atmospheric chemistry: fundamentals and experimental techniques", Wiley, New York.

Firestone, M.K., Firestone, R.B., and Tiedje, J.M., 1979, Nitric oxide as an intermediate in denitrification: evidence from nitrogen-13 isotope exchange, Biochem. Biophys. Res. Comm., 91:10-16.

Firestone, M.K., and Davidson, E.A., 1989, Microbiological basis of NO and N_2O production and consumption, in: "Exchange of trace gases between terrestrial ecosystems and the atmosphere", M.O. Andreae and D.S. Schimel, eds., Dahlem Konferenzen, p. 7-21, Wiley, Chichester.

Focht, D.D., and Verstraete, W., 1977, Biochemical ecology of nitrification and denitrification, Adv. Microb. Ecol., 1:135-214.

Galbally, I.E., 1989, Factors controlling NOx emission from soils. in: "Exchange of trace gases between terrestrial ecosystems and the atmosphere", M.O. Andreae and D.S. Schimel, eds., Dahlem Konferenzen, p. 23-27, Wiley, Chichester.

Galbally, I.E., and Roy, C.R., 1978, Loss of fixed nitrogen from soils by nitric oxide exhalation, Nature, 275:734-735.

Galbally, I.E., and Johansson, C., 1989, A model relating laboratory measurements of rates of nitric oxide production and field measurements of nitric oxide emission from soils, J. Geophys. Res., 94:6473-6480.

Galbally, I.E., Freney, J.R., Muirhead, W.A., Simpson, J.R., Trevitt, A.C.F., and Chalk, P.M., 1987, Emission of nitrogen oxides (NO_x) from a flooded soil fertilized with urea: relation to other nitrogen loss processes, J. Atmos. Chem., 5:343-365.

Garber, E.A.E., and Hollocher, T.C., 1982, Nitrogen-15, oxygen-18 tracer studies on the activation of nitrite by denitrifying bacteria. Nitrite/water-oxygen exchange and nitrosation reactions as indicators of electrophilic catalysis, J. Biol. Chem., 257:8091-8097.

Garcia, J.L., 1975, La dénitrification dans les sols, Bull. Inst. Pasteur, 73:167-193.

Goreau, T.J., Kaplan, W.A., Wofsy, S.C., McElroy, M.B., Valois, F.W., and Watson, S.W., 1980, Production of NO_2^- and N_2O by nitrifying bacteria at reduced concentrations of oxygen, Appl. Environ. Microbiol., 40:526-532.

Harriss, R.C., 1989, Experimental design for geophysiological research, in: "Exchange of trace gases between terrestrial ecosystems and the atmosphere", M.O. Andreae and D.S. Schimel, eds., Dahlem Konferenzen. Wiley, Chichester. (in press)

Heiss, B., Frunzke, K., and Zumft, W.G., 1989, Formation of the N-N bond from nitric oxide by membrane-bound cytochrome bc complex of nitrate-respiring (denitrifying) Pseudomonas stutzeri, J. Bacteriol., 171:3288-3297.

Henry, Y., and Bessieres, P., 1984, Denitrification and nitrite reduction: Pseudomonas aeruginosa nitrite-reductase. Biochimie 66:259-289.

Hinrichsen, D., 1985, Multiple pollutants and forest decline. Ambio 15:258-265.

Hochstein, L.I., and Tomlinson, G.A., 1988, The enzymes associated with denitrification, Ann. Rev. Microbiol., 42:231-262.

Isaksen, I.S.A., 1988, "Is the oxidizing capacity of the atmosphere changing?" in: "The changing atmosphere", F.S. Rowland and I.S.A.

Isaksen, eds., Dahlem Konferenzen. p. 141-157. Wiley, Chichester.

Ishaque, M., and Aleem, M.I.H., 1973, Intermediates of denitrification in the chemoautotrophic Thiobacillus denitrificans, Arch. Mikrobicl., 94:269-282.

Ji, X.B., and Hollocher, T.C., 1988a, Mechanism for nitrosation of 2,3-Diaminonaphtalene by Escherichia coli: Enzymatic production of NO followed by O2-dependent chemical nitrosation, Appl. Environ. Microbiol., 54:1791-1794.

Ji, X.B., and Hollocher, T.C., 1988b, Reduction of nitrite to nitric oxide by enteric bacteria, Biochem. Biophys. Res. Comm., 157:106-

Johansson, C., 1984, Field measurements of emission of nitric oxide from fertilized and unfertilized forest soils in Sweden, J. Atmos. Chem., 1:429-442.

Johansson, C., 1987, Pine forest: a negligible sink for atmospheric NO_x in rural Sweden. Tellus 39B:426-438.

Johansson, C., 1989, Fluxes of NO_x above soil and vegetation, in: "Exchange of trace gases between terrestrial ecosystems and the atmosphere", M.O. Andreae and D.S. Schimel, eds., Dahlem Konferenzen, p. 229-246, Wiley, Chichester.

Johansson, C., and Galbally, I.E., 1984, Production of nitric oxide in loam under aerobic and anaerobic conditions, Appl. Environ. Microbiol., 47:1284-1289.

Johansson, C., and Granat, L., 1984, Emission of nitric oxide from arable land. Tellus 36B:25-37.

Johansson, C., Rhode, H., and Sanhueza, E., 1988, Emission of NO in a tropical savanna and a cloud forest during the dry season, J. Geophys. Res., 93:7180-7192.

Kaplan, W.A., Wofsy, S.C., Keller, M., and DaCosta, J.M., 1988, Emission of NO and deposition of O_3 in a tropical forest system, J. Geophys. Res., 93:1389-1395.

Klemedtsson, L., Svensson, B.H., and Rosswall, T., 1988a, A method of selective inhibition to distinguish between nitrification and denitrification as sources of nitrous oxide in soil, Biol. Fertil. Soils, 6:112-119.

Klemedtsson, L., Svensson, B.H., and Rosswall, T., 1988b, Relationship between soil moisture content and nitrous oxide production during nitrification and denitrification, Biol. Fertil. Soils, 6:106-111.

Klepper, L.A., 1987, Nitric oxide emissions from Soybean leaves during in vivo nitrate reductase assays, Plant Physiol., 85:96-99.

Knowles, R., 1981, Denitrification, in: "Soil biochemistry", vol.5, E.A. Paul and J.N. Ladd, eds., p.323-369. Marcel Dekker, New York.

Knowles, R., 1982, Denitrification, Microbiol. Rev., 46:43-70.

Knowles, R., 1985, Microbial transformations as sources and sinks for nitrous oxides, in: "Planetary ecology", D.E. Caldwell, J.A. Brierley and C.L. Brierley, eds., p.411-426. Van Nostrand Reinhold, New York.

Kreitinger, J.P., Klein, T.M., Novick, N.J., and Alexander, M., 1985, Nitrification and characteristics of nitrifying microorganisms in an acid forest soil, Soil Sci. Soc. Am. J., 49:1407-1410.

Kuenen, J.G., and Robertson, L.A., 1988, Ecology of nitrification and denitrification, in: "The nitrogen and sulphur cycles", J.A. Cole and S. Ferguson, eds., p.162-218. Cambridge University Press, Cambridge.

Lang, E., and Jagnow, G., 1986, Fungi of a forest soil nitrifying at low pH values, FEMS Microbiol. Ecol., 38:257-265.

Lipschultz, F., Zafiriou, O.C., Wofsy, S.C., McElroy, M.B., Valois, F.W., and Watson, S.W., 1981, Production of NO and N_2O by soil nitrifying bacteria, Nature, 294:641-643.

Liu, M.C., Liu, M.Y., Payne, W.J., Peck Jr., H.D., and LeGall, J., 1983, Wolinella succinogenes nitrite reductase: purification and properties, FEMS Microbiol. Lett., 19:201-206.

Liu, S.C., Trainer, M., Fehsenfeld, F.C., Parrish, D.D., Williams, E.J., Fahey, D.W., Hübler, G., and Murphy, P.C., 1987, Ozone production in the rural troposphere and the implications for regional and global ozone distribution, J. Geophys. Res., 92:4191-4207.

Lloyd, D., Boddy, L., and Davies, K.J.P., 1987, Persistence of bacterial denitrification capacity under aerobic conditions: the rule rather than the exception, FEMS Microbiol. Ecol., 45:185-190.

Logan, J.A., 1983, Nitrogen oxides in the troposphere: global and regional budgets, J. Geophys. Res., 88:10785-10807.

Logan, J.A. 1985. Tropospheric ozone: seasonal behavior, trends, and anthropogenic influence, J. Geophys. Res., 90:10,463-10,482.

McKenney, D.J., Shuttleworth, K.F., Vriesacker, J.R., and Findlay, W.I., 1982, Production and loss of nitric oxide from denitrification in anaerobic Brookston clay, Appl. Environ. Microbiol., 43:534-541.

McKenney, D.J., Johnson, G.P., and Findlay, W.I., 1984, Effect of temperature on consecutive denitrification reactions in Brookston clay and Fox sandy loam, Appl. Environ. Microbiol., 47:919-926.

Melillo, J.M., Steudler, P.A., Aber, J.D., and Bowden, R.D., 1989, Atmospheric deposition and nutrient cycling, in: "Exchange of trace gases between terrestrial ecosystems and the atmosphere", M.O. Andreae and D.S. Schimel, eds., Dahlem Konferenzen, p. 263-280, Wiley, Chichester.

Nelson, D.W., 1982, Gaseous losses of nitrogen other than through denitrification, Agronomy, 22:327-364.

Newman, B.M., and Cole, J.A., 1978, The chromosomal location and pleiotropic effects of mutations of the $nirA^+$ gene of Escherichia coli K12: the essential role of $nirA^+$ in nitrite reduction and other anaerobic reactions, J. Gen. Microbiol., 10:1-12.

Papen, H., Von Berg, R., Hinkel, I., Thoene, B., and Rennenberg, H., 1989, Heterotrophic nitrification by Alcaligenes faecalis: NO_2^-, NO_3^-, N_2O, and NO production in exponentially growing cultures, Appl. Environ. Microbiol., 55:2068-2072.

Parkin, T.B., 1987, Soil microsites as a source of denitrification variability, Soil. Sci. Soc. Am. J., 51:1194-1199.

Parkin, T.B., Sexstone, A.J., and Tiedje, J.M., 1985, Adaptation of denitrifying populations to low soil pH, Appl. Environ. Microbiol., 49:1053-1056.

Payne, W.J., 1981, "Denitrification", Wiley, New York.

Payne, W.J., Grant, M.A., Shapleigh, J., and Hoffman, P., 1982, Nitrogen oxide reduction in Wolinella succinogenes and Campylobacter species, J. Bacteriol., 152:915-918.

Penkett, S.A., 1988, Indications and causes of ozone increase in the troposphere, in: "The changing atmosphere", F.S. Rowland and I.S.A. Isaksen, eds., Dahlem Konferenzen. p. 91-103. Wiley, Chichester.

Pichinoty, F., Garcia, J.L., Job, C., and Durand, C., 1978, Isolement de bacteries utilisant en anaerobiose l'oxyde nitrique comme acepteur d'electrons respiratoire. C.R., Acad. Sci. Ser. D., 286:1403-1405.

Poth, M., 1986, Dinitrogen production from nitrite by a Nitrosomonas isolate, Appl. Environ. Microbiol., 52:957-959.

Poth, M., and Focht, D.D., 1985, 15N kinetic analysis of N_2O production by Nirosomonas europaea: an examination of nitrifier denitrification, Appl. Environ. Microbiol., 49:1134-1141.

Ramanathan, V., Callis, L., Cess, R., Hansen, J., Isaksen, I., Kuhn, W.,

Lacis, A., Luther, F., Mahlman, J., Reck, R., and Schlesinger, M., 1987, Climate-chemical interactions and effects of changing atmospheric trace gases, Rev. Geophys., 25:1441-1482.

Remde, A., 1989, "Umsetzung von NO_x in Böden und Bodenmikroorganismen", PhD thesis, University of Konstanz, Konstanz, F.R.G.

Remde, A., Slemr, F., and Conrad, R., 1989, Microbial production and uptake of nitric oxide in soil, FEMS Microbiol. Ecol., 62:221-230.

Ritchie, G.A.F., and Nicholas, D.J.D., 1972, Identification of the sources of nitrous oxide produced by oxidative and reductive processes in Nitrosomonas europaea, Biochem. J., 126:1181-1191.

Robertson, G.P., 1989, Nitrification and denitrification in humid tropical ecosystems: potential controls on nitrogen retention, in: "Mineral nutrients in tropical forest and savanna ecosystems", J. Procter, ed., Blackwell Scientific, Oxford.

Robertson, G.P., and Tiedje, J.M., 1987, Nitrous oxide sources in aerobic soils: nitrification, denitrification and other biological processes, Soil Biol. Biochem., 19:187-193.

Robertson, L.A., and Kuenen, J.G., 1984, Aerobic denitrification: a controversy revived, Arch. Microbiol., 139:351-354.

Robertson, L.A., and Kuenen, J.G., 1986, Heterotrophic nitrification in Thiosphaera pantotropha: oxygen uptake and enzyme studies, J. Gen. Microbiol., 134:857-863.

Robertson, L.A., Van Niel, E.W.J., Torremans, R.A.M., and Kuenen, J.G., 1988, Simultaneous nitrification and denitrification in aerobic chemostat cultures of Thiosphaera panthotropha, Appl. Environ. Microbiol., 54:2812-2818.

Ryden, J.C., 1981, N_2O exchange between a grassland soil and the atmosphere. Nature 292:235-237.

Schmidt, E.L., 1982. Nitrification in soil. Agronomy 22:253-288.

Schröder, I., Robertson, A.M., Bokranz, M., Unden, G., Bocher, R., and Kröger, A., 1985, The membraneous nitrite reductase involved in the electron transport of Wolinella succinogenes, Arch. Microbiol., 140:380-386.

Seiler, W., and Conrad, R., 1981, Field measurements of natural and fertilizer induced N_2O release rates from soils, J. Air Poll. Contr. Assoc., 31:767-772.

Shapleigh, J.P., and Payne, W.J., 1985, Nitric oxide-dependent proton translocation in various denitrifiers, J. Bacteriol., 163:837-840.

Singh, H.B., 1987, Reactive nitrogen in the troposphere, Environ. Sci. Technol., 21:320-327.

Slemr, F., and Seiler, W., 1984, Field measurements of NO and NO_2 emissions from fertilized and unfertilized soils, J. Atmos. Chem., 2:1-24.

Slemr, F., Conrad, R., and Seiler, W., 1984, Nitrous oxide emissions from fertilized and unfertilized soils in a subtropical region (Andalusia, Spain), J. Atmos. Chem., 1:159-169.

Smith, M.S., 1983, Nitrous oxide production by Escherichia coli is correlated with nitrate reductase activity, Appl. Environ. Microbiol., 45:1545-1547.

Söderlund, R., and Svensson, B.H., 1976, The global nitrogen cycle, in: "Nitrogen, phosphorus and sulphur-global cycles", B.H. Svensson and R. Söderlund, eds., SCOPE 7, p. 23-73. Stockholm.

Stedman, D.M., and Shetter, R.E., 1983, The global budget of atmospheric nitrogen species, in: "Trace atmospheric constituents", S.E. Schwartz, ed., p. 411-454. Wiley, New York.

Teraguchi, S., and Hollocher, T.C., 1989, Purification and some characteristics of a cytochrome c-containing nitrous oxide reductase from Wolinella succinogenes, J. Biol. Chem., 264:1972-1979.

Thauer, R.K., Jungermann, K., and Decker, K., 1977, Energy conservation in chemotrophic anaerobic bacteria, Bact. Rev., 41:100-180.

Tiedje, J.M., 1988. Ecology of denitrification and dissimilatory nitrate reduction to ammonia, in: "Biology of anaerobic microorganisms", A.J.B. Zehnder, ed., p.179-244. Wiley, New York.

Van Cleemput, O., and Baert, L., 1976, Theoretical considerations on nitrite self-decomposition reactions in soils, Soil Sci. Soc. Am. J., 40:322-324.

Van Cleemput, O., and Baert, L., 1984, Nitrite: a key compound in N loss processes under acid conditions, Plant Soil, 76:233-241.

Vedenina, I.Y., and Zavarzin, G.A., 1977, Biological removal of nitrous oxide under oxidizing conditions, Mikrobiologiya, 46:898-903.

Vedenina, I.Y., Miller, Y.M., Kapustin, O.A, and Zavarzin, G.A., 1980, Oxidation of nitrous oxide during decomposition of hydrogenperoxide by catalase, Mikrobiologiya, 49:5-8.

Warneck, P., 1988, "Chemistry of the natural atmosphere", Academic Press, London.

Wesely, M.L., Eastman, J.A., Stedman, D.H., and Yalvac, E.D., 1982, An eddy-correlation measurement of NO_2 flux to vegetation and comparison to O_3 flux, Atmos. Environ., 16:815-820.

Williams, E.J., Parrish, D.D., and Fehsenfeld, F.C., 1987, Determination of nitrogen oxide emissions from soils: results from a grassland site in Colorado, United States, J. Geophys. Res. 92:2173-2179.

Williams, E.J., Parrish, D.D., Buhr, M.P., Fehsenfeld, F.C., and Fall, R., 1988, Measurement of soil NO_x emissions in central Pennsylvania, J. Geophys. Res., 93:9539-9546.

Yoshinari, T., 1980, N_2O reduction by Vibrio succinogenes, Appl. Environ. Microbiol., 39:81-84.

Yoshinari, T., 1985, Nitrite and nitrous oxide production by Methylosinus trichosporium, Can. J. Microbiol., 31:139-144.

Zafiriou, O.C., Hanley, Q.S., and Snyder, G., 1989, Nitric oxide and nitrous oxide production and cycling during dissimilatory nitrite reduction by Pseudomonas perfectomarina, J. Biol. Chem., 264:5694-5699.

Zumft, W.G., Döhler, K., Körner, H., Löchelt, S., Viebrock, A., and Frunzke, K., 1988, Defects in cytochrome cd1-dependent nitrite respiration of transposon Tn5-induced mutants from Pseudomonas stutzeri, Arch. Microbiol., 149:492-498.

Zumft, W.G., Viebrock, A., and Körner, H., 1988. Biochemical and physiological aspects of denitrification, in: "The nitrogen and sulphur cycles", J.A. Cole and S. Ferguson, eds., p.245-279. Cambridge University Press, Cambridge.

EMISSIONS OF N_2O FROM VARIOUS ENVIRONMENTS - THE USE OF STABLE ISOTOPE COMPOSITION OF N_2O AS TRACER FOR THE STUDIES OF N_2O BIOGEOCHEMICAL CYCLING

Tadashi Yoshinari

Wadsworth Center for Laboratories and Research
New York State Department of Health, and
School of Public Health
State University of New York at Albany
Albany, N.Y. 12201-0509
U.S.A.

INTRODUCTION

There is a considerable interest in the emissions of N_2O from various environments to the atmosphere. This is because the gas entering the atmosphere is (1) a natural source of stratospheric NO, which is an important element in the ozone chemistry in the stratosphere (Crutzen, 1970; McElroy and McConnel, 1971; Johnston, 1972; Cicerone 1987), and (2) a greenhouse gas, like CO_2 and CH_4, contributing to the increase of average surface temperature of the Earth by absorption of infrared radiation (Lacis et al. 1981; Dickinson and Cicerone 1986). The consequence of the destruction of the ozone layer is the increase of health hazards by the penetration of more intense UV irradiation. Tropospheric N_2O concentration, currently about 300 ppb, has been increasing at the rate of approximately 0.2-0.4% annually (Weiss 1981; Khalil and Rasmussen 1983; Rasmussen and Khalil 1986).

The atmospheric N_2O concentration before the industrial revolution has been estimated to be 280-290 ppb (Pearman et al. 1986). The increase by about 8% since then and the current increase are attributed to mainly anthropogenic processes (Weiss 1981; Hao et al. 1987).

Nitrous oxide is one of the key compounds in the nitrogen cycle on the Earth since it is an intermediate of biological denitrification as well as a byproduct of biological nitrification. There is considerable interest in the relative significance of nitrification and denitrification in N_2O emissions from soils, sediments and water columns. The use of the stable isotope signature of N_2O appears promising for providing information about N_2O sources difficult to obtain by conventional approaches.

The purpose of this manuscript is first to review (1) available data on the N_2O emissions from various environments and the global budget of

Denitrification in Soil and Sediment, Edited by N. P. Revsbech and
J. Sørensen, Plenum Press, New York, 1990

atmospheric N_2O, (2) different mechanisms of N_2O production and decomposition, and (3) the production of N_2O in the soil, sediment and aquatic systems by nitrification and denitrification. Then, an outline of a new method is described following review of isotope fractionation by various N transformation processes and available data for $\delta^{15}N$ and $\delta^{18}O$ in N_2O.

ATMOSPHERIC SOURCES AND SINKS OF N_2O

There are considerable uncertainties concerning the budget of atmospheric N_2O and individual source strength. This is because the emission rates from major sources cannot be assessed accurately. The estimated sources and a sink of atmospheric N_2O are summarized in Table 1.

The global budget of N_2O emissions from various sources to the atmosphere should be 14-17 Tg y^{-1}, which is the sum of 3-6 Tg y^{-1} [the annual increase of N_2O emissions (0.2-0.4% y^{-1})] and 11 Tg y^{-1} [the amount of N_2O removed in the stratosphere].

Above all the soil systems are the major sources of atmospheric N_2O (Table 1). Especially, tropical forest floor seems to be important with respect to total N_2O emissions from the pedosphere (Keller et al. 1983; 1986; 1988; Livingston et al. 1988; McElroy and Wofsy 1986). However, the relative importance of nitrification and denitrification for overall N_2O fluxes from the surface of soils remains uncertain.

Contrary to earlier estimates (Elkins et al.1978; Cohen and Gordon 1979), the oceans seem to be a weak source for atmospheric N_2O (Table 1). In fact in large areas of the ocean the N_2O concentrations are nearly in equilibrium with the atmosphere (Weiss 1981; Cline et al. 1987).

The atmospheric fluxes of N_2O from coastal wetlands (Smith et al. 1983), coastal marine sediments (Nishio et al. 1983; Jensen et al. 1984; Seitzinger et al. 1983; Butler et al. 1987) and freshwater systems (Wissmar et al. 1987; Seitzinger 1988; Hemond and Duran 1989) are estimated to be of little importance in terms of the global budget of N_2O emissions to the atmosphere.

There are some more possible atmospheric N_2O sources that were not considered in the above estimates (Bowden and Bormann 1986; Ronen et al. 1988). When these are added to current estimates, the total source strengths of N_2O will become even larger. Stratospheric reaction of N_2O with $O(^1D)$ and photolysis is the only well established sink of atmospheric N_2O. There are no known sinks of N_2O on the ground level. Unless there is an unknown sink for atmospheric N_2O on the surface of the Earth, as was suggested by Elkins et al. (1978) and Ryden (1981), many of the source strengths must have been overestimated.

Recently Hao et al.(1987) have confirmed earlier results (Weiss and Craig 1976; Pierotti and Rasmussen 1976) that N_2O from the combustion of fossil fuel is a significant source of atmospheric N_2O, and estimated a total atmospheric N_2O budget of 12 Tg y^{-1}, in which 8.8 Tg y^{-1} is from the biosphere and 3.2 Tg y^{-1} from fossil fuel combustion. However, there are some uncertainties as to their estimates, since Muzio and Kramlich (1988) and Muzio et al. (1989) have found that N_2O concentrations in the effluent from fossil fuel combustion increase to extremely high levels during the storage of the moist combustion products containing SO_2 and

Table 1. Sources and sink of N_2O

Sources	Type	Production rate (10^{12}g N yr^{-1})	Reference
Soils	Nitrification	?	
	Denitrification	?	
	Cultivated soils	1-3	Crutzen 1983
	Fertilized soils	<3	Crutzen 1983
		0.02-8	Conrad et al.1983
	Unfertilized soils	4.5	Slemr et al. 1984
	Tropical forest	1.5-5.8	Banin et al. 1984
		6.1	Keller et al.1986
	Temperate forest	0.3-6	Banin et al. 1984
Ocean		1-2	Crutzen 1983
		1	Codispoti and Christensen 1985
		2.8	Cline et al. 1987
Fossil fuel combustion		1.8	Crutzen 1983
		3.2	Hao et al. 1987
Biomass burning		1-2	Crutzen 1983
Total		17-20	Crutzen 1983
		29	Khalil and Rasmussen 1983
		15	Cicerone 1987
Sink			
Stratosphere	react. with $O(^1D)$	6-11	Crutzen 1983
	and Photolysis	11	Cicerone 1987

NO. Removing SO_2 prior to storage in the sample container appears to eliminate the formation of excess N_2O. Muzio et al. (1989) proposed that the mechanism of formation of N_2O formed from NO involves the formation and subsequent degradation of nitroxyl (HNO). A quantitative evaluation of the artifact may be required for assessing the validity of the present data base on N_2O emissions from combustion sources and the role of combustion for the generation of N_2O in terms of the atmospheric N_2O budget.

FACTORS AFFECTING N_2O PRODUCTION AND REDUCTION BY DENITRIFICATION AND NITRIFICATION

N_2O is an obligate intermediate in denitrification (Payne 1981; Knowles 1982). Since it can be further reduced to N_2, denitrification in the natural environments can act either as a source or sink of atmospheric N_2O. In case of ammonium oxidation by chemoautotrophic nitrifiers, such as *Nitrosomonas europaea*, NO_2^- is normally the end product but a small amount of N_2O is also formed.

The mole fraction of N_2O, $N_2O/(N_2O + N_2)$, has been used as a relative index for comparing the effects of different environmental factors that change the composition of gaseous nitrogen. On the other hand, N_2O yield, (N_2O/NO_2^-), is commonly used to estimate the efficiency of N_2O produced by nitrification. In natural environments N_2O yield may be estimated from N_2O/NO_3^-, assuming that NO_2^- produced by ammonium oxidizing bacteria has been oxidized quantitatively to NO_3^- by NO_2^- oxidizing bacteria. Oxygen, NO_2^-, NO_3^-, pH, temperature, organic carbon and other environmental factors affect both the mole fraction of N_2O and rate of denitrification. Likewise, some of these parameters are known to control the rate of N_2O production by nitrification. The effects of O_2, temperature and pH on the production of N_2O by nitrification and denitrification are summarized below.

1. Oxygen

Oxygen is generally considered to be inhibitory for denitrifying enzymes (Payne 1981; Knowles 1982). The mole fraction of N_2O increases but denitrification activity decreases as the O_2 concentration increases (Firestone et al. 1980; Betlach and Tiedje 1981). The results suggested that for some species of denitrifiers O_2 is a nonspecific inhibitor of each step in denitrification and increase the mole fraction of N_2O, whereas for others O_2 may specifically inhibit N_2O reductase and increase the mole fraction of N_2O. Although the critical levels of O_2 varied among different species of denitrifying bacteria, Seda and Ottow (1985) demonstrated that N_2O was not produced significantly when both O_2 and organic carbon concentrations were high.

The N_2O yield during nitrification activity is inversely correlated with the concentration of dissolved oxygen (Goreau et al. 1980; Lipschultz et al. 1981; Hynes and Knowles 1984; Anderson and Levine 1986). The yield increased from 0.001 to 0.1 as the concentration of O_2 decreased below 10 μM (Goreau et al. 1980). However, below 8.8 μM O_2 the rates of N_2O and NO_2^- production started to decline (Hynes and Knowles 1984). On the other hand, the rate of NH_4^+ uptake started to decline below 29.4 μM O_2, and at 8.8 μM O_2 the rate became approximately one third of maximum rate of NH_4^+ oxidation.

From the patterns of N_2O production by N. europaea in the presence of various combinations of inorganic N and O_2 concentrations, Hynes and Knowles (1984) have suggested that there are two different mechanisms of N_2O production by N. europaea: Under aerobic conditions N_2O may be formed from the hypothetical intermediate nitroxyl (NOH). Another pathway for N_2O production is by the involvement of NO_2^- reductase; this may function only under low O_2 or anaerobic conditions. The production of N_2O by the latter mechanism under O_2 limiting conditions was independently proposed (Poth and Focht 1985). Analytical data of isotope composition of N_2O derived from added ^{15}N-labelled NO_2^- and ^{14}N-ammonium to washed cell suspensions of N. europaea supported the kinetic model based on denitrification as the sole mechanism of N_2O production. Thus, it was concluded that N. europaea is a denitrifier under conditions of oxygen stress, using NO_2^- as a terminal electron acceptor to produce N_2O. The observation that N_2O produced by N. europaea was highly depleted in ^{15}N (Yoshida 1988) agrees with the conclusion that N_2O is produced at the NO_2^- reduction step (Hynes and Knowles (1984) and Poth and Focht (1985)). However, unlike the first mechanism of N_2O production under aerobic conditions (Hynes and Knowles, 1985), Yoshida's results demonstrate that the extent of the depletion of ^{15}N in N_2O was independent of the dissolved O_2 concentrations (10-190 μM).

2. Temperature

Since the values of the temperature quotient (Q_{10}) for both nitrifiers and denitrifiers appear to be almost (Focht and Cheng 1975), both nitrification and denitrification are expected to respond similarly to seasonal changes of temperature in natural environments. In a soil system, it was shown that the mole fraction of N_2O was inversely correlated with the temperature up to 37°C (Keeney et al. 1979). However, the effect of temperature on the rate of N_2O production by nitrification is not known.

3. pH

It is known that nitrification in acid soils is very slow (Alexander 1965). Laboratory studies indicated that the optimal pH for the production of N_2O and NO_2^- by N. europaea was 8.5, with a sharp decrease in these activities below 6.5 (Hynes and Knowles 1984). This suggests that N_2O production by nitrification by N. europaea is insignificant in the environment where pH is below 6.

Denitrification seems to be active in a much wider range of pH than chemoautotrophic nitrification, although the optimal pH is about 7. The mole fraction of N_2O was observed to increase as the pH was lowered from 6.5 to 4.9 (Firestone et al. 1980). Since there was little indication of N_2O produced by nitrification in the northern forest soil from the C_2H_2 blockage technique (Melillo et al. 1983) and in the Brazilian forest soil from enrichment studies (Livingston et al. 1988), it was concluded that the N_2O emission from these forest floors was derived from denitrification. The same conclusion could have been drawn from the evidence that the soil pH of the northern forest and Brazilian forest was 3.7-4.0 and 3.5- 3.9, respectively, since the production of N_2O by chemoautotrophic ammonium oxidizing bacteria is probably precluded under such low pH conditions. Based on the results of enrichment experiments, Keller et al. (1988) have reached the same conclusion that most of N_2O emitted from the Brazilian forest floor is associated with denitrification of soil nitrate. Although pH of the soil was not shown in their report, it could well be in a similar range of pH as that found by Livingston et al. (1988).

PRODUCTION AND DECOMPOSITION OF N_2O BY OTHER MECHANISMS

There are many reports which indicate that N_2O is produced by various other types of microorganisms, such as non-denitrifying NO_3^- reducers (Smith and Zimmerman 1981; Smith 1982; Smith 1983; Kaspar 1981; Anderson and Levine 1986), fungi (Bollag and Tung 1972; Bleakley and Tiedje 1982), cyanobacteria (Weathers and Niedzielski 1986), green algae (Weathers 1984), and a methane oxidizing bacterium (Yoshinari 1985). The presence of NO_2^- appears to be prerequisite for the formation of N_2O by these organisms. The ability of N_2O production by these organisms can be clearly differentiated from that of denitrifiers: 1. The production of N_2O is not affected by the presence or absence of C_2H_2 (Kaspar 1982; Weathers and Niedzielski 1986). 2. The reduction of NO_2^- is not coupled with energy yielding reactions (Smith 1984; Kaspar 1982; Anderson and Levine 1986; Bleakley and Tiedje 1982). 3. The rate of N_2O production is much less than by denitrification (Kaspar 1982). 4. The apparent K_m for NO_2^- reduction to N_2O is an order of magnitude higher than that of K_m for NO_2^- reduction (Smith 1982). The mechanisms for the production of

N_2O by these non-denitrifiers are not known in detail. Bollag and Tung (1972) suggested a chemical reaction, such as the oxidation of hydroxylamine, to explain their finding that N_2O was simultaneously released during the reduction of NO_2^- by Fusarium species in growing culture and in resting cell experiments at low oxygen tensions. Smith (1982) considered the formation of NH_2OH by Citrobacter sp. to explain the formation of N_2O. However, this scheme was rejected on the ground that the presence of NH_2OH as high as 0.1 mM was required to account for a rate comparable to that observed with cell suspensions. Two enteric bacteria, Escherichia coli and Serratia marcescens, produced N_2O only after reaching the stationary phase (Bleaky and Tiedje 1982). The possibility of its formation through chemical decomposition of NH_2OH was rejected since NH_2OH could not be detected in the supernatant of the bacterial cultures. In view of the fact that these bacteria cannot reduce NO_2^-, their results clearly indicate that N_2O is produced only in the presence of high concentration of NO_2^-. In the case of Propionibacteria, NO_2^- was converted to N_2O quantitatively (Kaspar 1982). Both the respiratory nitrate reductase system and NO_2^- seem to be involved in the production of N_2O by Escherichia coli (Smith 1983). A methane oxidizing bacterium, Methylosinus trichosporium OB3b, apparently produces N_2O by the reduction of NO_2^- under conditions of oxygen stress (Yoshinari, 1985). The organism is morphologically and physiologically similar to chemoautotrophic nitrifiers and is capable of oxidizing NH_4^+ to NO_2^- by methane monooxidase. In the presence of NO_2^-, green algae (Weathers 1984) and cyanobacteria (Weathers and Niedzielski 1986) also produce N_2O, but the mechanism of its formation by these organisms is not known.

The formation of N_2O by the microorganisms considered above is reminiscent of "chemodenitrification" (Smith and Chalk 1980a; 1980b), which also requires the presence of NO_2^- for N_2O production. It is possible that some labile organic compounds produced during heterotrophic metabolism first react with NO_2^- per se or in the presence of nitrate reductase, and then produce N_2O from the following decomposition of the organic compounds.

The formation of N_2O by chemical decomposition of NH_2OH (Bremner et al. 1980), chemical reduction of NO_2^- by Fe^{2+} at pH 5 (Chalamet and Bardin 1977), and in the pH range from 6 to 8 (Moraghan and Buresh 1977) may become significant in certain soil conditions. The formation by chemical reaction of NO_2^- and hydroxylamine (Bothner-By and Friedman 1952) does not seem to be important since there was no significant increase of the rate of N_2O production by the addition of NO_2^- to NH_2OH in soils (Bremner et al. 1980).

Earlier, it was thought that production of N_2O by these organisms might be important for emission to the atmosphere because organisms of some types are numerous in soils and because the rates were high under certain conditions (Smith and Zimmerman, 1981). However, a reliable assessment of the environmental significance of N_2O production by these organisms has yet to be established, ideally by using a method that can differentiate the production of N_2O by this mechanism from that by other mechanisms. Until further information becomes available, it may be reasonable to assume that the production of N_2O in soil and other ecosystems by the mechanism considered here is of minor importance as a source for atmospheric N_2O since the reaction becomes significant only in the presence of relatively high NO_2^- concentration (>1 mM), which is not commonly found in natural environments.

It is generally believed that N_2O is stable under aerobic conditions. Accordingly, little is known about the oxidative removal of N_2O by biogenic or chemical reactions. Vedenina and Zavarzin (1977) first reported that N_2O may be oxidized under aerobic conditions. Subsequently, Vedenina et al. (1980) proposed the oxidation of N_2O through the interaction of catalase. The rapid reduction of N_2O by $Fe(OH)_2$ in the presence of Cu^{2+} ions has been reported (Moraghan and Buresh 1977). While environmental significance has yet to be assessed, these chemical and biochemical processes of the removal of N_2O under aerobic conditions may deserve more attention as a possible sink for atmospheric N_2O.

N_2O EMISSIONS FROM SOIL, SEDIMENT AND AQUATIC SYSTEMS

1. Soil system

The soil system generally acts as a source of atmospheric N_2O. Studies on various environmental factors that affect the emission of N_2O from the soil environments have been reviewed (Rolston 1981; Knowles 1982; Mosier et al. 1983; Bowden 1986).

A large data base on N_2O emission from nitrification in soils has been established (Blackmer et al. 1980; Breitenbeck et al. 1980; Bremner and Blackmer 1978; Bremner et al. 1981a; Bremner et al. 1981b; Ryden 1981; Seiler and Conrad 1981). However, considerable indirect evidence suggests that N_2O in the soil arises from both processes of nitrification and denitrification (McKenney et al. 1980; Conrad and Seiler 1980; Mosier and Hutchinson 1981; Seiler and Conrad 1981; Duxbury and McConnaughey 1986). The relative contribution of the oxidative and reductive pathways to the production of N_2O must be determined by a complex combination of various environmental parameters. On top of the parameters considered in preceding section, moisture content is an important factor because the amount of water leads to restricted aeration and the development of anaerobic microsites (Focht and Verstraete 1977; Sexstone et al. 1985). Below 60% of water saturation, denitrification proceeds very slowly (Freney et al. 1979; Smith and Tiedje 1979; Breitenbeck et al. 1980; Aulakh et al. 1984). Therefore, most of N_2O formed in soil with water content below 60% may be considered as the product from nitrification.

The relative importance of nitrification and denitrification for the production of N_2O was estimated by incubation of the soil with and without N-serve [2-chloro-6-(trichloromethyl)-pyridine] or C_2H_2 (Aulakh et al. 1984; Duxbury and McConnaughey 1986). Duxbury and McConnaughey (1986) estimated that N loss as N_2O from the zero- and NO_3^--treatments were 20% and 12.5% of (N_2 + N_2O), respectively. In contrast N loss from the urea treatment was equally divided between N_2 and N_2O, attributing the observed N_2O formation to nitrification and that of N_2 to denitrification.

While the loss rates of fertilizer-N as N_2O are generally in the range of 0.01-3% [see Table 2 of Keller et al. (1988) for the range reported by other investigators], Bremner et al. (1981b) found the loss rates as much as 4-7% when the fields were fertilized with anhydrous ammonia.

Most of N_2O emitted from soils to the atmosphere seems to originate from the uppermost soil layer or soil surface (Denmead et al. 1979; Seiler and Conrad 1981; Conrad et al. 1983; Slemr et al. 1984). Since there was no direct correlation between the rate of N_2O emission from the soil surface and the existence of a strong vertical profile of N_2O in the soil column, Seiler and Conrad (1981) concluded that N_2O produced at depths must have been consumed in the upper soil layer during the upward transport by a diffusive process. This process of N_2O reduction to N_2 during diffusion would be enhanced if the soil is sufficiently wet, since the diffusion coefficient of N_2O is much less than that of N_2 (Letey et al. 1980a).

Spatial and temporal variations appear to be significant sources of error in estimates of N_2O emission. The spatial variability may partly be due to heterogeneity of the soil system. Since the majority of the N_2O flux occurs in local "hot spots", intensive sampling is necessary to develop reasonable areal flux estimates (Livingston et al. 1988). Temporal variability has been reported to be much greater than spatial variability (Goodroad and Keeney 1984; Livingston et al. 1988). This phenomenon must be related to the observation that the rate of N_2O reduction increases rapidly following longer period of incubation of soil in the presence of N_2O (Garcia 1975) or anaerobic preincubation (Firestone and Tiedje 1979; Letey et al. 1980b and 1980c). Since the presence of N_2O or NO_3^- during preceding anaerobic incubation was required for enhanced activity of N_2O reduction for both soil slurries and _Flavobacterium_ sp., Firestone and Tiedje (1979) concluded that the increase in N_2O-reducing activity resulted from _de novo_ enzyme synthesis. Letey et al.(1980c) found that anoxic preincubation resulted in almost no N_2O production during the incubation period even though high NO_3^- concentrations were added.

2. _Sediment system_

The sediment system acts as a sink or source of N_2O, depending on the environmental conditions (Wissmar et al. 1987; Seitzinger 1988; Hemond and Duran 1989). The flux must be the sum of all three pathways of $NH_4^+ \rightarrow N_2O$, $NO_3^- \rightarrow N_2O$, and $N_2O \rightarrow N_2$ (Seitzinger et al. 1983).

Using a suspension of coastal surface sediment Jørgensen et al. (1984) have demonstrated that the O_2 concentration is the primary factor that determines the relative proportion of N_2O produced by nitrification and denitrification. Their results indicate that denitrification is the main source of N_2O at 0-10 kPa O_2, and that N_2O produced by nitrification can be significant only in a very narrow range of O_2 concentration. Since the pattern of denitrification (Andersen et al. 1984) and N_2O emission (Jensen et al. 1984) was coincidental, Jensen et al. (1984) reached to the same conclusion that denitrification was the major source of observed N_2O flux in the coastal sediment system.

Recent information (Christensen et al. 1989) suggests that N_2O is produced just above the oxic/anoxic interface where the O_2 concentration is 0-20 µM. Although N_2O may be derived mostly from denitrification, a large proportion of nitrate consumed by denitrification seems to be produced by nitrification at the same interface (Henriksen 1981; Hansen et al. 1981; Nishio et al. 1983; Jenkins and Kemp 1984; Seitzinger et al. 1984). Therefore, the magnitude of N_2O flux must be positively correlated with the extent of the nitrification-denitrification coupling. The findings that the production of N_2O and the rate at which the nutrient is transported in coastal marine sediments (Seitzinger et al. 1983) and in large microcosms (Seitzinger an Nixon 1985) are correlated may be a case in point. The low rate of N_2O emission or negative flux of N_2O (Jensen et al., 1984; Hemond and Duran 1989) and diminished rate of denitrification (Jenkins and Kemp 1984) in sediments during summer are followed by a sharp decline in nitrification (Hansen et al. 1981; Jenkins and Kemp 1984). This may be explained as the consequence of the O_2 depletion at the surface layer of sediment. Hemond and Duran (1989) have demonstrated that the benthic N_2O flux during summer can be predicted from a diffusion model for oxygen penetration. Temperature may be one of the main variables for benthic N_2O production and consumption, because it is directly related to respiratory O_2 demand in the sediment.

The production of N_2O by nitrification in the stream sediment system (Wissmer et al. 1988) and coastal marine sediments (Nishio et al. 1983) appears to be much less than that by denitrification. While nitrification rates were in the same order of magnitude in the sediments of Tama estuary and Odawa Bay, the former was 16 times higher in denitrification and 26 times in N_2O flux than in Odawa Bay (Nishio et al. 1983). These results agree with the conclusion (Jensen et al. 1984; Jørgensen et al. 1984) that the contribution of nitrification to the production of N_2O is much less than that of denitrification.

3. Aquatic system

Nitrous oxide in the open ocean has been agreed to originate mostly from nitrification. This is based on the evidence that there was a linear correlation between ΔN_2O (N_2O[observed] - N_2O[saturated]) and AOU (Apparent Oxygen Utilization) (Yoshinari 1976; Cohen and Gordon, 1979, Elkins et al., 1978, Cline et al., 1987), and that production of NO_3^- is correlated with AOU (Redfield 1934). However, this hypothesis seems to demand rigorous scrutiny for the following reasons. Pierotti and Rasmussen (1980) have pointed out that, since the oxidation of ammonia or organic amines accounts for an extremely small part of the total biological oxygen demand, the ΔN_2O:AOU plots actually demonstrate a relationship between N_2O production and total non-photosynthetic biological activity in the water since it left the surface. From a laboratory observation that N_2O produced by bacterial nitrification was highly depleted in ^{15}N, it was suggested that the role of nitrification in the production of N_2O in the ocean could have been overestimated (Yoshida 1988). It is possible that N_2O primarily produced by nitrification at the oxygen minimum layers in the open ocean is modified to some extent by the reductive processes of $NO_3^- \rightarrow N_2O$ and $N_2O \rightarrow N_2$.

The eastern tropical South Pacific can be a source or sink for atmospheric N_2O, depending on the exact nature of the dissolved O_2 distribution in the area (Cohen and Gordon 1978; Elkins et al. 1978; Codispoti and Christensen 1985). The role of denitrification for net production of N_2O in the area is to be clarified by future investigation.

There are reports on the accumulation or production of N_2O in freshwater lakes (Knowles et al. 1981; Lemon and Lemon 1981; Vincent et al. 1981; Yoh et al. 1983), streams (McElroy et al. 1978; Wofsy et al. 1981; Wissnar et al. 1987; Hemond and Duran 1989), and coastal water columns (Jensen et al. 1984; McCarthy et al. 1984; Butler et al. 1987). Earlier studies have been reviewed by Knowles (1982) and Kaplan and Wofsy (1985). The rates of N_2O production in oxygenated water columns are always much smaller than those in the sediments (Jensen et al. 1984; Butler et al. 1987; Hemond and Duran 1989).

Recently Downes (1988) reported that, prior to hypolimnetic anoxia, a large amount of NO_2^- and N_2O accumulated in the water column of Lake Rotoiti, New Zealand. Circumstantial evidence indicated that both NO_2^- and most of N_2O at low O_2 (6.3 μM <$[O_2]$< 31.2 μM) were mainly derived from nitrification. The accumulation of NO_2^- was attributed to a greater affinity for O_2 by NH_4^+ oxidizers than by NO_2^- oxidizers under the low O_2 concentrations. At O_2 concentrations below 6.3 μM, the process, $NO_2^- \rightarrow N_2O \rightarrow N_2$, leads to complete denitrification.

ISOTOPE FRACTIONATION IN THE NITROGEN CYCLE

The abundance of stable isotopes (δ) is expressed in terms of per mil notation.

$$\delta^{15}N \text{ or } \delta^{18}O = [(R[sample])/(R[standard]) -1] \times 1000$$

where R is either $^{15}N/^{14}N$ or $^{18}O/^{16}O$, and R[standard] for nitrogen and oxygen is atmospheric nitrogen and SMOW (standard mean ocean water), respectively.

To express the changes in the isotope composition, fractionation factor (α) and enrichment factor (ε) are commonly used. The former is defined as follows: α = R(substrate)/R(product), where R(product) is usually smaller than R(substrate) and α therefore larger than 1.0. This is because organisms preferentially utilize lighter elements. The ε value is defined by Mariotti et al. (1981): $\varepsilon = 1000 \times (1 - \alpha)$.

In the process of biological fixation of atmospheric nitrogen, the discrimination of ^{15}N over ^{14}N is nearly zero. An estimated average of the α value for this step is 1.002 (Wada and Hattori 1989).

In the process of nitrification ($NH_4^+ \rightarrow NO_3^-$), and denitrification ($NO_3^- \rightarrow N_2O + N_2$), ammonium nitrogen and nitrate nitrogen are enriched with heavy nitrogen, because the organisms preferentially utilize light nitrogen. The reported α values for both processes are about 1.01-1.03. It is obvious from this that, when the nitrogen substrates are present in excess, all the processes, except the nitrogen fixation, tend to leave behind more of heavy nitrogen.

Experimental results for the biological discrimination of heavy nitrogen or oxygen during the metabolism of inorganic nitrogen compounds are summarized in Table 2, in which data for the extent of isotope discrimination during assimilation, nitrification and denitrification are expressed in terms of the enrichment factor (ε). Since α is generally greater than 1.0, ε becomes negative. ε_N and ε_O in these columns are the enrichment factors for nitrogen and oxygen, respectively.

1. Assimilation

While *Azotobacter vinelandii* hardly discriminates during N_2 assimilation the organism discriminates during N_2O reduction followed by the assimilation, of the nitrogen in N_2 (Yamazaki et al. 1987). However, the order of the discrimination in N_2O assimilation by *A. vinelandii* has about the same magnitude as for of algal ammonium uptake (Sharp, personal communication).

2. Nitrification

For the process of ammonium oxidation to nitrite ε_N is -35. When it comes to the formation of N_2O from ammonia oxidation, the value becomes much less. Yoshida (1988) attributes this to the two step reaction. When N_2O is considered as the reduction product from nitrite, there are two processes of isotope discrimination: $NH_4^+ \rightarrow NO_2^-$ and $NO_2^- \rightarrow N_2O$. When the isotope fractionation from nitrite to N_2O is calculated with this assumption, the value became of the same order as the $NO_2^- \rightarrow N_2O$ reaction in denitrification ($\alpha = 1.030$).

3. Denitrification

The ε_N value for the reduction of NO_3^- to N_2O by soil denitrifiers is -29 and -25 at 20^0C and 30^0C, respectively (Mariotti, et al. 1981). The ε_0 value for the reduction of N_2O to N_2 by Pseudomonas aeruginosa is -42 and -37 at 10^0C and 26^0C, respectively (Wahlen and Yoshinari 1985). It may be noticed that both ε_N and ε_0 are negatively correlated with the temperature.

Table 2. Enrichment factor (ε) for the transformation of various inorganic nitrogen during assimilation, nitrification and denitrification

Type of reaction	Organism resp. for reaction	ε_N	ε_0	Reference
Assimilation				
$N_2 \rightarrow$ cell	A. vinelandii	-2		Yamazaki et al. 1987
$N_2O \rightarrow$ cell	A. vinelandii	-34 to -17		Yamazaki et al. 1987
$NH_4^+ \rightarrow$ cell	S. costatum	-20 to -14		Sharp (pers. comm.)
Nitrification				
$NH_4^+ \rightarrow NO_2^-$	N. europaea	-35		Mariotti et al. 1981
$NH_4^+ \rightarrow N_2O$	N. europaea	-60		Yoshida 1988
Denitrification				
$NO_3^- \rightarrow N_2O$	Soil denitrif.	-29 (20^0C)		Mariotti et al. 1981
	- -	-25 (30^0C)		- - - 1981
$N_2O \rightarrow N_2$	P. aeruginosa		-42 (10^0C) -37 (26^0C)	Wahlen & Yoshinari 1985

STABLE ISOTOPE COMPOSITION OF N_2O IN THE ENVIRONMENT

The $\delta^{15}N$ data for atmospheric and oceanic N_2O obtained by Yoshida and Matsuo (1983) and Yoshida et al. (1984) indicate that the average $\delta^{15}N$ value of N_2O in the air over the Pacific Ocean (7.2 o/oo) is higher than that of dissolved N_2O in the subsurface waters above the oxygen minimum layer in eastern tropical Pacific (5.2 o/oo), but slightly lower than that of land surface air (8.1 o/oo). These results suggest that the N_2O produced in the subsurface and/or the oxygen minimum layer is depleted in ^{15}N-N_2O and consequently N_2O in the maritime air is lighter than that in the land air. The $\delta^{15}N$ value of N_2O from the waters with an extremely low concentration of dissolved oxygen in the eastern tropical Pacific was very high (12.6 o/oo), suggesting the occurrence of isotopic enrichment of N in N_2O during denitrification.

An average of $\delta^{18}O$ in N_2O from nine air samples in Albany, New York, was 45 ± 0.8 o/oo. The samples from other environments show a considerable range in $\delta^{18}O$ with both lighter and heavier values than that in the atmosphere (Wahlen and Yoshinari 1985). The lowest values of $\delta^{18}O$ (21 o/oo) were obtained in N_2O from the air in an enclosed housing of a waste treatment facility where nitrification was actively taking place on biological contractors. A quantitative oxidation of ammonium to nitrate via nitrite was confirmed from the budget of nitrogen species of NH_4^+, NO_2^-, NO_3^- and organic nitrogen between the inlet and outlet of the wastewater tank (Yoshinari and Wahlen 1985). The highest value in N_2O (93 o/oo) was obtained in the water collected from the N_2O maximum layer at Green Lake, New York, and it was attributed to the enrichment of $\delta^{18}O$ by the process of denitrification (Wahlen and Yoshinari 1985). The values of $\delta^{18}O$ in N_2O from denitrification are expected to be variable, and to correlate with the extent of the removal of N_2O by denitrification in the N_2O pool.

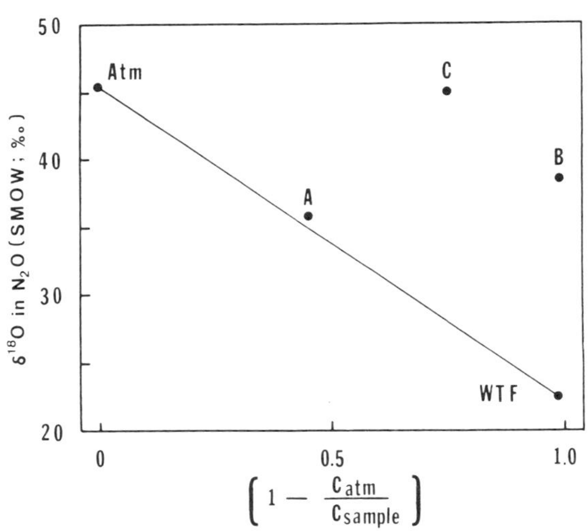

Figure 1. $\delta^{18}O$ in N_2O from three different soil environments. A: N_2O from soil heavily fertilized with cow manure, B: N_2O from the soil with chemical fertilizer, C: N_2O from unfertilized soil. Atmospheric N_2O (Atm) and N_2O collected from nitrification plant (WTF) are included in this figure. C [atm] and C [sample] are the concentration of atmospheric and sample N_2O, respectively.

These results seem to indicate that a significant amount of N_2O produced by denitrification and nitrification is reduced to N_2 before escaping to the atmosphere. Another possibility is that production of N_2O by denitrification is a stronger source than that by nitrification for atmospheric N_2O.

The $\delta^{18}O$ values of N_2O in soil air samples which were collected from three different locations (Wahlen and Yoshinari, unpublished results) with those of N_2O in the atmosphere and nitrification plant (Wahlen and Yoshinari 1985; Yoshinari and Wahlen 1985) are shown in Fig. 1.

These results indicate that (1) the $\delta^{18}O$ value of N_2O derived from the soil fertilized heavily with cow manure (A) was lower than that from chemically fertilized soil (B), and (2) the $\delta^{18}O$ value of N_2O derived from unfertilized soil (C) was about the same as that from the atmosphere. It is possible that N_2O from soil with cow manure (A) was mostly derived from the nitrification. The value from fertilized soil (B) was much higher than that from nitrification plant, although the rate of N_2O production was about the same. It is possible that N_2O in B was derived either from denitrification alone ($NO_3^- \rightarrow N_2O$ and $N_2O \rightarrow N_2$) or combined reactions of nitrification and denitrification ($NH_4^+ \rightarrow N_2O$, $NO_3^- \rightarrow N_2O$ and $N_2O \rightarrow N_2$). In both cases, isotopic enrichment for N_2O in the soil environment must have been carried out by the reaction of $N_2O \rightarrow N_2$. The fact that the $\delta^{18}O$ value (40 o/oo) of N_2O in B was still lower than that of atmosphere (45 o/oo) seems to deserve further attention. However, more work is required to clarify whether this is a general pattern for $\delta^{18}O$ in N_2O emitted from the surface of various types of soil.

METHODS FOR STABLE ISOTOPE DETERMINATION IN N_2O

The results shown above for oxygen and nitrogen compositions of N_2O were determined by laser infrared spectrometry and mass spectrometry, respectively. Each method can determine the stable isotope composition of either the nitrogen or oxygen in N_2O. The method for the measurement of $\delta^{18}O$ (Wahlen and Yoshinari, 1985) is by high resolution IR spectroscopy with a tunable diode as detector, which requires about 20 μmoles (0.5 ml) of N_2O. This corresponds to the requirement of a volume of 500 l to 3000 l seawater sample per analysis since the dissolved N_2O concentration in seawater is in the range of 0.15 and 1 μl l^{-1}. For the measurement of $\delta^{15}N$ in N_2O it requires installing another appropriate diode detector in the housing, and the volume of sample for the analysis is yet to be determined. The method for $\delta^{15}N$ developed by Yoshida and Matsuo (1985) is based on measurement of N_2 by ratio mass spectrometry, which requires only about 1-2 μmoles of N_2O. The method can analyze only the isotopic composition of N in N_2O since the oxygen in N_2O is removed by Cu during its combustion.

To overcome the shortcomings of these methods, a new method that can determine the isotopic compositions of both N and O in N_2O has been developed in this laboratory. This method, as is described below, requires less than 2 μmoles of N_2O. While the requirement was reduced to 1/10 of that for laser infrared spectrometry, it still requires as much as 300 l of seawater sample per analysis. An additional advantage of the novel technique is the capability for analyzing both N and O isotope compositions of N_2O by converting into N_2 and CO_2 over a heated graphite rod, and subsequently analyzing by mass spectrometry. The use of a combined set of both N and O stable isotope ratios in N_2O samples will be undoubtedly more powerful than that of a single isotope ratio of N or O for the studies of the origin and mass balance of N_2O in the ocean and atmosphere.

In the new method, which can determine both $\delta^{15}N$ and $\delta^{18}O$ in N_2O, N_2O must first be converted to N_2 and CO_2 in the presence of graphitic carbon. Then, the $\delta^{15}N$ in N_2 and $\delta^{18}O$ in CO_2 are measured by a ratio mass spectrometer. The reaction is analogous to the measurement of $\delta^{18}O$ in O_2, which was developed by Kroopnick (1971), but it was found to be more complex than that with O_2.

There were two major problems in obtaining quantitative conversion of N_2O. Earlier results indicated that traces of two different byproducts (CO and NO) were formed during the reaction. While there was no CO formation for the conversion of O_2 to CO_2 with graphitic carbon when heated at 900^0C (Kroopnick, 1971), trace amounts of both CO and NO were formed during the reaction of N_2O with graphitic carbon at 750^0C.

The formation of these byproducts may be explained partly by the known mechanism of thermal decomposition of N_2O. The generally accepted reaction mechanism for thermal decomposition of N_2O (Loirat et al. 1985) is as follows:

$$N_2O + M \rightarrow N_2 + O + M \qquad (1)$$
$$N_2O + O \rightarrow N_2 + O_2 \qquad (2)$$
$$N_2O + O \rightarrow 2NO \qquad (3)$$
$$O + NO + M \rightarrow NO_2 + M \qquad (4)$$
$$NO_2 + O \rightarrow NO + O_2 \qquad (5)$$
$$NO + N_2O \rightarrow NO_2 + N_2 \qquad (6)$$
$$O + \text{wall} \rightarrow 1/2\ O_2 + \text{wall} \qquad (7)$$

The main reaction (1) is the breakage of the N-O bond in the presence of a third body. The oxygen atom thus formed can react in two ways, and form ($N_2 + O_2$) and 2 NO [(2) and (3)]. The first reaction is rate limiting. The reactions (2) and (3) proceed with about the same rate. Because of the reactions (4) and (5), NO is a minute component of the final products. However, the proportion of NO as final product increases greatly as the reaction temperature is increased (Johnston, 1951). Most of the NO appears to be formed in the initial phase of reaction (Loirat et al., 1985).

To achieve quantitative conversion of N_2O to N_2 and CO_2, it was essential to suppress the side reaction that form NO and CO to an absolutely minimum. Although it is common practice to convert contaminating NO to N_2 with heated copper oxide (Hachey et al., 1987), it was not desirable to lose part of the N_2O-O as CuO, since that would affect the determination of $\delta^{18}O$. The lowest possible temperature was therefore set for the conversion of N_2O to CO_2 to minimize the formation of NO and CO. Fig. 2 shows a schematic diagram of the apparatus to prepare N_2 and CO_2 from N_2O.

Approximately 50 μl (~2 μmoles) of pure N_2O fractionated by gas chromatography is first introduced into the U-tube (B) of known volume, previously chilled with liquid nitrogen. The volume of N_2O gas in B is estimated from reading the internal pressure with the pressure transducer (c) attached to the end of the U-tube after the gas is expanded to room temperature. Then, it is transferred to A, and valve V_1

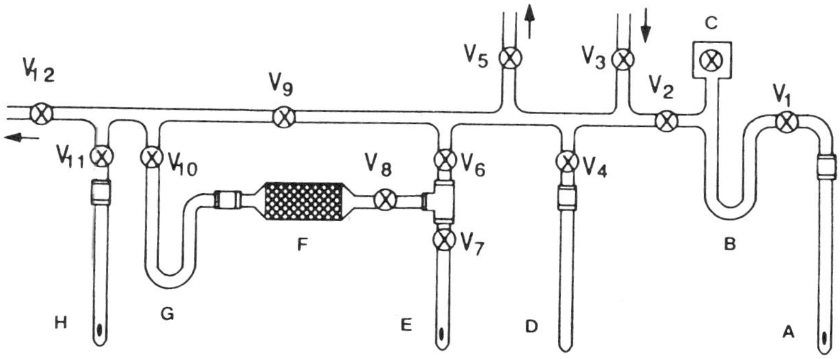

Figure 2. Schematic diagram of vacuum line to transfer N_2O to the combustion chamber, and to recover the CO_2 and N_2 fractions after the combustion of N_2O in the presence of graphitic rod. A: Reaction tube with a small rod of graphitic carbon, B: U-tube for collection of N_2O and CO_2, C: Pressure transducer sensor connected to a digital transducer indicator, D: Pyrex tube for the collection of CO_2, E: Cold finger with a MS 5A pellet to trap N_2, F: Catalytic tube filled with CuO and Cu, G: U-tube to trap CO_2 formed from oxidation of trace CO, and H: Pyrex tube with a MS 5A pellet for collection of N_2. For gas transfer, high vacuum seal glass valves (V_1 - V_{12}) are opened or closed as required. The end of V_3 is the inlet for N_2O samples purified by gas chromatography. The ends of V_5 and V_{12} are connected to a mercury diffusion pump.

is closed. After the reaction at 615^0C for 90 min., the condensible gas (CO_2) is trapped in the same U-tube with liquid N_2, and then transferred to the glass tube (A) after measuring the internal pressure of CO_2 in B at the room temperature. The remaining non-condensible gas (N_2) is transferred to E and V_7 is closed. Then, after desorbing the gas from the MS 5A pellet, it is passed through a catalytic column filled with CuO and Cu (F). This is necessary to convert trace CO to CO_2, which is removed in the U-tube (G). The N_2 thus purified is then transferred to the glass tube with a pellet of MS 5A (H). The ends of tubings (D and H) are sealed with a flame.

The samples thus prepared are later analyzed by ratio mass spectrometry. From the pressure readings, the recovery % of CO_2 can be calculated. Under the conditions described above, the recovery is generally quantitative (100-102 %).

The precision of the method is currently 0.2 and 0.5 o/oo [S.E.] for $\delta^{15}N$ and $\delta^{18}O$ with tank N_2O that has δ values of 2.5 and 37.0 o/oo for $\delta^{15}N$ and $\delta^{18}O$, respectively.

This method has been developed in collaboration with Dr. Brian Fry at the Ecosystems Center, Woods Hole, Massachusetts, who has done all the isotope measurements. At the early stage for the development of this method, Dr. Noriyuki Tanaka helped in designing and constructing the analytical system.

SUMMARY

Our present understanding of the emission of N_2O to the atmosphere and biogeochemical cycling of N_2O in natural environments may be summarized as follows. (1) N_2O emitted from the surface of soils and sediments is the product of both nitrification and denitrification, (2) N_2O production by both nitrification and denitrification is primarily determined by the oxygen concentration of the soil, sediment and water environments, where either one or both of these processes take place, (3) N_2O emission from natural ecosystems is the product of complex interactions between transport, biological production and biological consumption, (4) the soil surface may be the major source for atmospheric N_2O, and (5) the measurements of stable isotope composition of N_2O at the natural abundance level may be of use to help clarify the relative significance of N_2O emission by nitrification and denitrification.

REFERENCES

Alexander, M., 1965, Nitrification, Agronomy 10:307-343, American Society of Agronomy, Madison, WI.

Andersen, T.K., Jensen M.H., and Sørensen, J., 1984, Diurnal variation of nitrogen cycling in coastal, marine sediments, I. Denitrification. Mar. Biol. 83:171-176.

Anderson, I.C. and Levine, J.S., 1986, Relative rates of nitric oxide and nitrous oxide production by nitrifiers, denitrifiers, and nitrate respirers, Appl. Environ. Microbiol. 51:938-945.

Aulakh, M.S., Rennie, D.A., Paul, E.A., 1984, Acetylene and N-Serve effects up upon N_2O emissions from NH_4^+ and NO_3^- treated soils under aerobic and anaerobic conditions, Soil Biol. Biochem., 16:351-356.

Banin, A., Lawless, J.G., and Whitten, R.C., 1984, Global N_2O cycles-terrestrial emissions, atmospheric accumulation and biospheric effects, Adv. Space Res., 4:207-216.

Betlack, M., and Tiedje, J.M., 1980, Kinetic explanation for accumulation of nitrite, nitric oxide and nitrous oxide during denitrification, Appl. Environ. Microbiol., 46:1118-1124.

Blackmer, A.M., Bremner,J.M., and Schmidt, E.L., 1980, Production of nitrous oxide by ammonia oxidizing chemoautotrophic microorganisms in soil, Appl. Environ. Microbiol., 40:1060-1066.

Bleakley, B.H., and Tiedje, J.M., 1982, Nitrous oxide production by organisms other than nitrifiers or denitrifiers, Appl. Environ. Microbiol. 44:1342-1348.

Bollag, J.M., and Tung, G., 1972, Nitrous oxide release by soil fungi, Soil Biol. Biochem., 4:271-276.

Bothner-By, A.A., and Friedman, L., 1952, The reaction of nitrous acid with hydroxylamine, J. Chem. Phys., 20:459-462.

Bowden, W.B., 1986. Gaseous nitrogen emissions from undisturbed terrestrial ecosystems: An assessment of their impacts on local and global nitrogen budgets, Biogeochemistry, 2:249-279.

Bowden, W.B., and Borman, F.H., 1986, Transport and loss of nitrous oxide in soil water after forest clear-cutting, Science 233:867-869.

Breitenbeck, G.A., Blackmer, A.M., and Bremner, J.M., 1980, Effects of different nitrogen fertilizers on emission of nitrous oxide from soil, Geophys. Res. Lett, 7:85-88.

Bremner, J.M., and Blackmer, A.M., 1978, Nitrous oxide emission from soils during nitrification of fertilizer nitrogen, Science 199:295-296.

Bremner, J.M., Blackmer, A.M., and Waring, S.A., 1980, Formation of nitrous oxide and dinitrogen by chemical decomposition of hydroxylamine in soils, Soil Biol. Biochem., 12:263-269.

Bremner, J.M., Breitenbeck, G.A., and Blackmer, A.M., 1981, Effect of nitrapyrin on emission of nitrous oxide from soil fertilized with anhydrous ammonia, Geophys. Res. Lett., 8:353-356.

Bremner, J.M., Breitenbeck, G.A., and Blackmer, A.M., 1981, Effect of anhydrous ammonia fertilization on emission of nitrous oxide from soils, J. Environ. Qual., 10:77-80.

Butler, J.H., Jones, R.D., Garber, J.H., and Gordon, L.I., 1987, Seasonal distributions and turnover of reduced trace gases and hydroxylamine in Yaquina Bay, Oregon, Geochem. Cosmochim. Acta 51:697-706.

Chalamet, A., and Bardin, R., 1977, Action des ferreux sur la reduction de l'acide nitreux dans les sols hydromorphes, Soil Biol. Biochem. 9:281-285.

Christensen, P.B., Nielsen, L.P., Revsbech, N.P., and Sørensen, J., 1988, Microzonation of denitrification activity in stream sediments as studied with a combined oxygen and nitrous oxide micro sensor, Appl. Environ. Microbiol., 55:1234-1241.

Cicerone, R.J., 1987, Changes in stratospheric ozone. Science, 237:35-42.

Cline, J.D., Wisegarver, D.P., and Kelly-Hansen, K., 1987, Nitrous oxide and vertical mixing in the equatorial Pacific during the 1982-1983 El Nino, Deep-Sea Res., 34:857-873.

Codispoti, L.A., and Christensen, J.P., 1985, Nitrification, denitrification and nitrous oxide cycling in the eastern tropical South Pacific Ocean, Mar. Chem., 16:277-300.

Cohen, Y. and Gordon, L.I., 1978, Nitrous oxide in the oxygen minimum of the eastern tropical North Pacific: Evidence for its consumption during denitrification and possible mechanisms for its production, Deep-Sea Res., 25:509-524.

Cohen, Y. and Gordon, L.I., 1979, Nitrous oxide production in the ocean, J. Geophys. Res., 84:347-353.

Conrad, R., Seiler, W., and Bunse, G., 1983, Factors influencing the loss of fertilizer nitrogen into the atmosphere as N_2O, J. Geophys. Res., 88:6709-6718.

Conrad, R., and Seiler, W., 1980, Field measurements of the loss of fertilizer nitrogen into the atmosphere as nitrous oxide, Atmos. Environ., 14:555-558.

Crutzen, P.J., 1970, The influence of nitrogen oxides on the atmospheric ozone content, Quart. J. Royal, Met. Soc. 96:320-325.

Crutzen, P.J., 1983. Atmospheric interactions - homogeneous gas reactions of C, N and S containing compounds, In: "The Major Biogeochemical Cycles and Their Interactions," (Eds. B. Bolin and R.B. Cook), pp 67-114. SCOPE.

Denmead, O.T., Freney, J.R., Simpson, J.R., 1979, Nitrous oxide emission during denitrification in a flooded field, Soil Sci. Soc. Am. J., 43:726-728.

Dickinson, R.E., and Cicerone, R.J., 1986, Future global warming from atmospheric trace gases, Nature, 319:109-115.

Downes, M.T. 1988. Aquatic nitrogen transformations at low oxygen concentrations, Appl. Environ. Microbiol., 54:172-175.

Duxbury, J.M., and McConnaughey, P.R., 1986, Effect of fertilizer source on denitrification and nitrous oxide emissions in a maize-field, Soil Sci. Soc. Am. J. 50:644-648.

Elkins J.W., Wofsy, S.C., McElroy, M.B., Kolb, C.E., and Kaplan, W.A., 1978, Aquatic sources and sinks for nitrous oxide, Nature 275:602-606.

Firestone, M.K., Firestone, R.T., and Tiedje, J.M., 1980, Nitrous oxide from soil denitrification: Factors controlling its biological production, Science, 208:749-751.

Firestone, M.K., Smith, M.S., R.B., Firestone, and Tiedje, J.M., 1979, The influence of nitrate, nitrite, and oxygen on the composition of the gaseous products of denitrification in soil, Soil Sci. Soc. Am. J. 43:1140-1144.

Firestone, M.K., and Tiedje, J.M., 1976, Temporal changes in nitrogen oxide and dinitrogen from denitrification following onset of anaerobiosis, Appl. Environ. Microbiol., 38:673-679.

Focht, D.D., and Chang, A.C., 1975, Nitrification and denitrification processes related to waste water treatment, Adv. Appl. Microbiol. 19:153-186.

Focht, D.D., and Verstraete, W., 1977, Biochemical ecology of nitrification and denitrification, Adv. Microbial Ecol., 1:135-214.

Freney, J.R., Denmead, O.T., and Simpson, J.R., 1979. Nitrous oxide emission from soils at low moisture contents, Soil Biol. Biochem. 11:167-173.

Garcia, J.-L., 1975, Evaluation de la denitrification dans les rizieres par la methode de reduction de N_2O, Soil Biol. Biochem., 7:251-256.

Goodroad, L.L., and Keeney, D.R., 1984, Nitrous oxide emissions from forest, marsh, and prairie ecosystems, J. Environ. Qual. 13:448-452.

Goodroad, L.L., Keeney, D.R., and Peterson, L.A., 1984, Nitrous oxide emissions from agricultural soils in Wisconsin, J. Environ. Qual. 13:557-561.

Goreau, T.J., Kaplan, W.A., Wofsy, S.C., McElroy, M.B., Valois, F.W., and Watson, S.W., 1980, Production of NO_2^- and N_2O by nitrifying bacteria at reduced concentrations of oxygen, Appl. Environ. Microbiol., 40:526-532.

Hachey, D.L., Wong, W.W., Boutton, T.W., and Klein, P.D., 1987, Isotope ratio measurements in nutrition and biomedical research, Mass Spectrometry Rev. 6:289-3248.

Hansen, J.I., Henriksen, K., and Blackburn, T.H., 1981, Seasonal distribution of nitrifying bacteria and rates of nitrification in coastal marine sediments. Microbial Ecol. 7:297-304.

Hao, W.M., Wofsy, S.C., McElroy, M.B., Beer, J.M., and Toquan, M.A.,1987, Sources of atmospheric nitrous oxide from combustion, J. Geophys. Res. 92:3098-3104.

Hemond, H.F., and Duran, A.P., 1989, Fluxes of N_2O at the sediment-water and water-atmosphere boundaries of a nitrogen-rich river, Water Resources Res. 25:839-846.

Henriksen, K., Hansen, J.I., and Blackburn, T.H., 1981, Rates of nitrification, distribution of nitrifying bacteria, and nitrate fluxes in different types of sediment from Danish waters, Mar. Biol. 61:299-304.

Hynes, R.K., and Knowles, R., 1984, Production of nitrous oxide by Nitrosomonas europaea: effects of acetylene, pH and oxygen, Can. J. Microbiol. 30:1397-1404.

Jenkins, M.C., and Kemp, W.M., 1984, The coupling of nitrification and denitrification in two estuarine sediments, Limnol. Oceanogr. 29:609-619.

Jensen, H.B., Jørgensen, K.S. and Sørensen, J., 1984, Diurnal variation of nitrogen cycling in coastal, marine sediments. II. Nitrous oxide emission, Mar. Biol., 83:177-183.

Johnston, H.J., 1951, Interpretation of the data on the thermal decomposition of nitrous oxide, J. Chem. Phys., 19:663-667.

Johnston, H., 1972, Newly recognized vital nitrogen cycle, Proc. Natl. Acad. Sci. U.S.A., 69:2369-2372.

Jorgensen, K.S., Jensen, H.B. and Sørensen, J., 1984, Nitrous oxide production from nitrification and denitrification in marine sediment at low oxygen concentrations, Can. J. Microbiol., 30:1073-1078.

Kaplan, W., and Wofsy, S.C., 1985, The biochemistry of nitrous oxide: A review. In: "Advances in Aquatic Microbiology," vol 3., H.W. Jannasch and P. J-B. Williams, eds., pp. 181-206. Academic Press, Florida.

Kaspar, H.F., 1982, Nitrite reduction to nitrous oxide by propionibacteria: Oxidation mechanism, Arch. Microbiol. 133:126-130.

Keller, M.T., Goreau, T.J., Wofsy, S.C., 1986, Emissions of N_2O, CH_4, and CO_2 from tropical forest soils, J. Geophys. Res. 91:11791-11802.

Keller, M., Kaplan, W.A., Wofsy, S.C. and DaCosta, J.M., 1988, Emissions of N_2O from tropical forest soils: Response to fertilization with NH_4^+, NO_3^- and PO_4^{3-}, J. Geophys. Res., 93:1600-1604.

Keeney, D.R., Fillery, I.R. and Marx, G.P., 1979, Effect of temperature on the gaseous nitrogen products of denitrification in a silt loam soil, Soil Sci. Soc. Am. J., 43:1124-1128.

Khalil, M.A.K. and Rasmussen, R.A., 1983, Increase and seasonal cycles of nitrous oxide in the earth's atmosphere, Tellus 35B:161-169.

Knowles, R., Lean, D.R.S. and Chan, Y.K., 1981, Nitrous oxide concentrations in lakes: variations with depth and time, Limnol. Oceanogr. 26:855-866.

Knowles, R., 1982, Denitrification, Microbiol. Rev. 46:43-70.

Kroopnick, P.M., 1971, Oxygen and carbon in the oceans and atmosphere: Stable isotope as tracers for consumption, production and circulation models, Ph.D. Thesis, University of California, San Diego.

Lacis, A., Hansen, J., Lee, P., Mitchell,T., and Lebedeff, S., 1981, Greenhouse effect of trace gases, 1970-1980, Geophys. Res. Lett. 8:1035-1038.

Lemon, E., and Lemon, D., 1981, Nitrous oxide in fresh waters of the Great Lakes basin., Limnol. Oceanogr., 26:867-879.

Letey, J., Hadas, A., Valoras, N., and Focht, D.D., 1980a, Effect of preincubation treatments on the ratio of N_2O/N_2 evolution, J. Environ. Qual., 9:232-235.

Letey, J., Jury, W.A., Hadas, A., and Valoras, N., 1980b, Gas diffusion as a factor in laboratory incubation studies on denitrification, J. Environ. Qual. 9:223-227.

Letey, J., Valoras, N., Hadas, A., and Focht, D.D., 1980c, Effect of air-filled porosity, nitrate concentration and time on the ratio of N_2O/N_2 evolution during denitrification, J. Environ. Qual. 9:227-231.

Lipschultz, F., Zafiriou, O.C., Wofsy, S.C., McElroy, M.B., Valois, F.W., and Watson, S.W., 1981, Production of NO and N_2O by soil nitrifying bacteria, Nature. 294:641-643.

Livingston, G.P., Vitousek, P.M., and Matson, P.M., 1988, Nitrous oxide flux and nitrogen transformations across a landscape gradient in Amazonia, J. Geophys. Res. 93:1593-1599.

Loirat, H., Caralap, F., Forst, W., and Schoenberger, C., 1985, Thermal unimolecular decomposition of nitrous oxide at low pressures, J. Phys. Chem. 89:4586-4591.

Mariotti, A., Germon, J.C., Hubert, P., Kaiser, P., Letolle, R., Tradieux, A., and Tradieus, P., 1981. Experimental determination of nitrogen kinetic isotope fractionations: Some principles; illustration for the denitrification and nitrification processes, Plant Soil 62:413-430.

Mariotti, A., Germon, J.C., and Leclerc, A., 1982, Nitrogen isotope fractionation associated with the $NO_2^- \rightarrow N_2O$ step of denitrification in soils, Can. J. Soil Sci., 62:227-241.

McCarthy, J.J., Kaplan, W.A., and Nevins, J.L., 1984, Chesapeake Bay nutrient and plankton dynamics. 2. Sources and sinks of nitrite, Limnol. Oceanogr. 29:84-98.

McElroy, M.B., and McConnel, J.C., 1971, Nitrous oxide: A natural source of stratospheric NO, J. Atmos. Sci. 28:1095-1098.

McElroy, M.B., Elkins, J.W., Wofsy, S.C., Kolb, C.E., Duran, A.P., And Kaplan, W.A., 1978. Production and release of N_2O from the Potomac Estuary, Limnol. Oceanogr., 23:1168-1182.

McElroy, M.B., and Wofsy, S.C., 1986, Tropical forests: Interactions with the atmosphere, In: "Tropical Rain Forests and World Atmosphere," G.T. Prance, ed., pp. 33-61, Advancement of Science, Selected Symposium 1001", Washington D.C.

McKenney, D.J., Shuttleworth, K.F., and Findlay, W.I., 1980,Nitrous oxide evolution rates from fertilized soils: Effects of applied nitrogen, Can. J. Soil Sci., 60:429-438.

Melillo, J.M., Aber, J.D., Steudler, P.A., and Schimel, J.P., 1983, Denitrification potentials in a successional sequence of northern hardwood forest stands, In: "Environmental Biogeochemistry", Ecol. Bull., 35," R. Hallberg, ed., pp 217-228. SCOPE.

Moraghan, J.T., and Buresh, R.J., 1977, Chemical reduction of nitrite and nitrous oxide by ferrous ion, Soil Sci. Soc. Am. J. 41:47-50.

Mosier, A.R., and Hutchinson, G.L., 1981, Nitrous oxide emission from cropped fields, J. Environ. Qual. 10:169-173.

Mosier, A.R., Stillwell, M., Parton, W.J., and Woodmansee, R.G., 1981, Nitrous oxide emissions from a native shortgrass prairie, Soil Sci. Soc. Am. J. 45:617-627.

Mosier, A.R., Parton, W.J., and Hutchinson, G.L., 1983, Modelling nitrous oxide evolution from cropped and native soils, In: "Environmental Biogeochemistry", Ecol. Bull. (Stockholm) 35, R. Hallberg, ed., pp 229-241.

Muzio, L.J., Kramlich, J.C., 1988, An artifact in the measurement of N_2O from combustion sources, Geophys. Res. Lett. 15:1369-1372.

Muzio, L.J., Teague, M.E., Kramlich, J.C., Cole, J.A., McCarthy, J.M., and Lyon, R.K., 1989, Errors in grab sample measurements of N_2O from combustion sources, J. Air Poll. Cont. Assoc. 39:287-293.

Nishio, T., Koike, I., Hattori, A., 1983, Estimates of denitrification and denitrification in coastal and estuarine sediments, Appl. Environ. Microbiol. 45:444-450.

Payne, W.J., 1981, Denitrification, 214 pp. John Wiley, New York.

Pearman, G.I., Etheridge, D., de Silva, F., and Fraser, P.J., 1986, Evidence of changing concentrations of atmospheric CO_2, N_2O and CH_4 from air bubbles in Antarctic ice, Nature, 320:248-250.

Pierotti, D., and Rasmussen, R.A., 1976, Combustion as a source of nitrous oxide, Geophys. Res. Lett. 3:265-267.

Pierotti, D. and Rasmussen, R.A., 1980, Nitrous oxide measurements in the eastern tropical Pacific Ocean, Tellus 32:56-72.

Poth, M., and Focht, D.D., 1985, ^{15}N kinetic analysis of N_2O production by Nitrosomonas europaea: an examination of nitrifier denitrification, Appl. Environ. Microbiol, 49:1134-1141.

Rasmussen, R.A., and Khalil, M.A.K., 1986, Atmospheric trace gases: Trends and distributions over the last decade, Science 232: 1623-1624.

Redfield, A.C., 1934, On the proportions of organic derivatives in sea water and their relation to the composition of plankton. pp.176-192. James Johnston Memorial Volume, Liverpool.

Rolston, D.E., 1981, Nitrous oxide and nitrogen gas production in fertilizer loss, In: "Denitrification, nitrification, and atmospheric nitrous oxide", C.C. Delwiche, ed, pp. 127-149, John Wiley and Sons.

Ronen, D., Magaritz, M., and Almon, E., 1988, Contaminated aquifers are a forgotten component of the global N_2O budget, Nature 335:57-59.

Ryden, J.C., 1981, N_2O exchange between a grassland and the atmosphere, Nature 292:235-237.

Seada, M.N.I.A., and Ottow, J.C.G., 1985, Effect of increasing oxygen concentration on total denitrification and nitrous oxide release from soil by different bacteria, Biol. Fert. Soils 1:31-38.

Seiler, W., and Conrad, R., 1981, Field measurements of natural and fertilizer-induced N_2O release rates from soils, J. Air Pollut. Cont. Assoc., 31:767-772.

Seitzinger, S.P., 1988, Denitrification in freshwater and coastal marine ecosystems: Ecological and geochemical significance, Limnol. Oceanogr. 33:702-724.

Seitzinger, S.P., and Nixon, S.W., 1985, Eutrophication and the rate of denitrification and N_2O production in coastal marine sediments. Limnol. Oceanogr. 30:1332-1339.

Seitzinger, S.P., Nixon, S.W., and Pilson, M.E.Q., 1984, Denitrification and nitrous oxide production in a coastal marine ecosystem, Limnol. Oceanogr. 29:73-83.

Seitzinger, S.P., Pilson, M.E.Q., and Nixon, S.W., 1983, Nitrous oxide production in nearshore marine sediments, Science 222:1244-1245.

Sexstone, A.J., Revsbech, N.P., Parkin, T.B., and Tiedje, J.M., 1985, Direct measurement of oxygen profiles and denitrification rates in soil aggregates, Soil Sci. Soc. Am. J. 49:645-651.

Slemr, F., Conrad, R., Seiler, W., 1984, Nitrous oxide emissions from fertilized and unfertilized soils in a subtropical region (Andalusia, Spain), J. Atm. Chem. 1:159-169.

Smith, C.J. and Chalk, P.M., 1980, Fixation and loss of nitrogen during transformations of nitrite in soils, Soil Sci. Soc. Am. J. 44:288-291.

Smith, C.J. and Chalk, P.M., 1980, Gaseous nitrogen evolution during nitrification of ammonia fertilizer and nitrite transformations in soils, Soil Sci. Soc. Am. J. 44:277-282.

Smith, C.J., DeLaune, R.D., and Patrick, W.H., 1983, Nitrous oxide emission from Gulf Coast wetlands, Geochem. Cosmochim. Acta 47:1805-1814.

Smith, M.S., and Tiedje, J.M., 1979, Phases of denitrification following oxygen depletion in soil, Soil Biol. Biochem. 11:261-267.

Smith, M.S., 1982, Dissimilatory reduction of NO_2^- to NH_4^+ and N_2O by soil Citrobacter sp. applied and Environ. Microbiol. 43: 854-860.

Smith, M.S., 1983, Nitrous oxide production by Escherichia coli is correlated with nitrate reductase activity, Appl. Environ. Microbiol. 45:1545-1547.

Smith, M.S., and Zimmerman, K., 1981, Nitrous oxide production by nondenitrifying soil nitrate reducers, Soil Sci. Soc. Am. J. 45:865-871.

Vedenina, I.Ya., and Zavarzin, G.A., 1977, Biological removal of nitrous oxide under oxidizing conditions, Mikrobiologiya 46:898-903.

Vedenina, I.Ya., Miller, Yu.M., Kapustin, O.A., and Zavarzin, G.A., 1980, Oxidation of nitrous oxide during decomposition of hydrogen peroxide by catalase, Mikrobiologiya 49:5-8.

Vincent, W.F., Downes, M.T., and Vincent, C.L., 1981, Nitrous oxide cycling in Lake Vanda, Antarctica, Nature 292:618-620.

Wada, E., and Hattori, A., 1989, Nitrogen in the sea; Forms, abundance and rate process, CRC press, in press.

Wahlen, M. and Yoshinari, T., 1985, Oxygen isotope ratios in N_2O from different environments, Nature 313:780-782.

Weathers, P.J., 1984, N_2O evolution by green algae, Appl. Environ. Microbiol. 48:1251-1253.

Weathers, P.J. and Niedzielski, J.J., 1986, Nitrous oxide production by cyanobacteria, Arch. Microbiol. 146:204-206.

Weiss, R.F., and Craig, H., 1976, Production of atmospheric nitrous oxide by combustion, Geophys. Res. Lett. 3:751-753.

Weiss, R.F, 1981, The temporal and spacial distribution of tropospheric nitrous oxide, J. Geophys. Res. 86:7185-7195.

Wissmar, R.C., Lilley, M.D., and deAngelis, M., 1987, Nitrous oxide release from aerobic riverine deposits, J. Freshwater Ecol. 4:209-218.

Wofsy, S.C., McElroy, M.B., and Elkins, J.W., 1981, Transformations of nitrogen in a polluted estuary: Nonlinearities in the demand for oxygen at low flow, Science 213:754-757.

Yamazaki, T., Yoshida, N., Wada, E., and Matsuo, S., 1987, N_2O reduction by *Azotobacter vineiandii* with emphasis on kinetic nitrogen isotope effects, Plant Cell Physiol. 28:263-271.

Yoh, M., Terai, H., and Saijo, Y., 1983, Accumulation of nitrous oxide in the oxygen deficient layer of freshwater lakes, Nature 301:327-329.

Yoshida, N., 1988, ^{15}N-depleted N_2O as a product of nitrification, Nature 335:528-529.

Yoshida, N., Hattori, A., Saino, T., Matsuo, S., and Wada, E., 1984, $^{15}N/^{14}N$ ratio of dissolved N_2O in the eastern tropical Pacific Ocean, Nature 307:442-444.

Yoshida, N. and Matsuo, S., 1983, Nitrogen isotope ratio of atmospheric N_2O as a key to the global cycle of N_2O, Geochem. J. 17:231-239.

Yoshinani, T., 1976, N_2O in the sea, Mar. Chem. 4:189-202.

Yoshinari, T., 1985, Nitrite and nitrous oxide production by *Methylosinus trichosporium*, Can. J. Microbiol. 31:139-144.

Yoshinari, T., and Wahlen, M., 1985, Oxygen isotope ratios in N_2O from nitrification at a wastewater treatment facility, Nature 317:349-350.

ACETYLENE INHIBITION TECHNIQUE: DEVELOPMENT, ADVANTAGES, AND POTENTIAL PROBLEMS

Roger Knowles

Macdonald College of McGill University
21,111 Lakeshore Road, Ste. Anne de Bellevue
Quebec, H9X 1CO, Canada

INTRODUCTION

The discovery in 1966 of the significant interaction of C_2H_2 with nitrogenase (reviewed in Burns and Hardy, 1975) was rapidly followed by the recognition that many other small double- or triple-bonded molecules such as N_2O could act as substrates (Hardy and Knight, 1966). Interactions between C_2H_2, N_2O, and nitrogenase have since been studied in more detail and N_2O is in fact a poor substrate, with apparent Km values of 24 kPa for Klebsiella pneumoniae (Jensen and Burris, 1986) and 50 kPa for Azotobacter vinelandii (Liang and Burris, 1988) components in vitro.

In the '60s it became clear that complete denitrification involved the reduction of N_2O to N_2 by an N_2O reductase (N_2O-R), a reaction essentially similar to that catalyzed by nitrogenase (Knowles, 1982). It seemed reasonable to ask in what way the gases N_2O and C_2H_2 might interact with N_2O-R and whether this enzyme might reduce C_2H_2. A partial answer was provided in 1973 by a Soviet group developing methods for detecting life with a Mars probe (Fedorova et al., 1973). They wrote "It is recommended to determine the denitrifying property of soils by means of blocking with acetylene the reduction $N_2O \rightarrow N_2$ stage", and they pointed to the possibility of detecting "in one test" both denitrification and N_2 fixation. Their findings were confirmed and extended quite independently in 1976 by our group (Yoshinari and Knowles, 1976) and that of Payne (Balderston et al., 1976). The inhibition of N_2O-R was observed in several different denitrifying bacteria; the other denitrifying reductases were unaffected; C_2H_2 was not reduced; and stoichiometric conversion of NO_3^- to N_2O occurred.

APPLICATIONS

The following year it was shown that the N_2O-R inhibition phenomenon could be usefully adapted for the assay of denitrification in soils (Klemedtsson et al., 1977; Yoshinari et al., 1977) and that nitrogenase activity could be detected in the same system (Yoshinari et al., 1977). The reduction of N_2O was completely inhibited by 0.1 kPa of C_2H_2 and 10 kPa of C_2H_4 had very little

Denitrification in Soil and Sediment, Edited by N. P. Revsbech and J. Sørensen, Plenum Press, New York, 1990

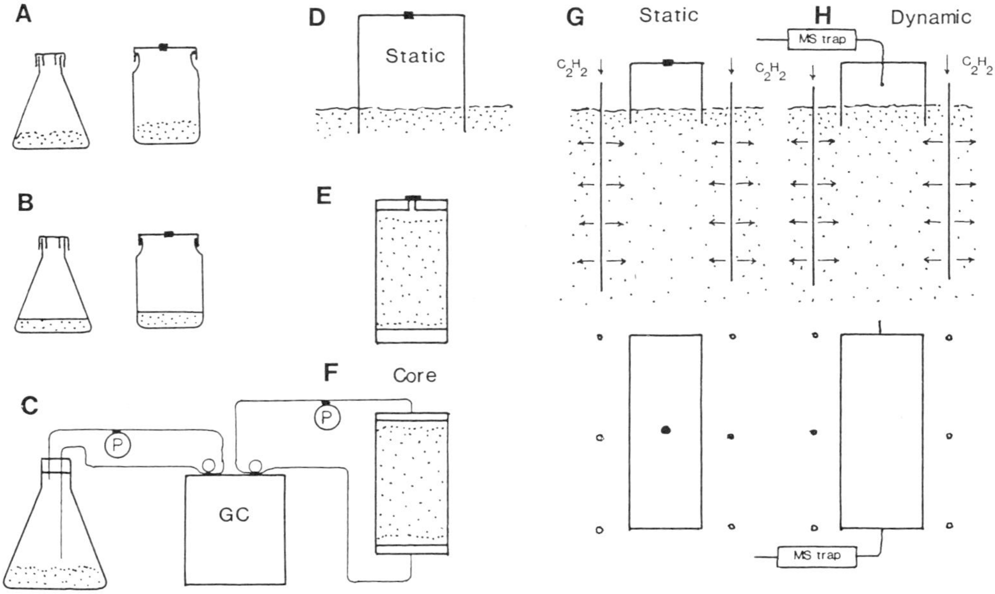

Fig. 1. Examples of soil denitrification assays. A, field moist soil. B, soil/water slurry. C, recirculating with automated analysis. D, static enclosure with C_2H_2 in atmosphere. E, enclosed core. F, core with recirculation. G, static and H, dynamic enclosures with C_2H_2 introduced into the soil.

effect. There was good stoichiometric conversion of NO_2^- or NO_3^- to N_2 (Yoshinari et al., 1977). Further laboratory evaluations of the potential of this method for soil studies employed comparisons with ^{13}N (Smith et al., 1978) and ^{15}N (Ryden et al., 1979) and in most cases the C_2H_2 inhibition method appeared to provide a valid estimate of denitrification rate. The three GC detectors available for N_2O measurements differ greatly in their sensitivity (thermal conductivity: ultrasonic: electron capture = 1: (55-130): 7300; Y.K Chan and R. Knowles, unpublished data). That it was possible to determine low concentrations of soil NO_3^- (<0.5 μg NO_3^- - N g^{-1}) by promoting its denitrification in the presence of C_2H_2 and then measuring the N_2O produced by thermal conductivity detection (Yoshinary et al., 1977) was confirmed and extended to even lower concentrations by using sensitive electron capture detection (Lensi et al., 1985). This approach was modified for any NO_3^- -containing sample by adding, in the absence of C_2H_2, cells of a denitrifying bacterium which lacks N_2O-R (Christensen and Tiedje, 1988), and concentrations of as little as 0.2 μg NO_3^--N L^{-1} (14 nM) were determined.

One of the first applications of the C_2H_2 inhibition technique to aquatic systems was that of Sørensen (1978b) in which intact sediment cores were treated with C_2H_2-saturated water to introduce a relatively uniform C_2H_2 concentration in the pore water. After incubation the core was sectioned and the N_2O extracted and measured. It is beyond the scope of this article to describe the many types of applications in terrestrial and aquatic systems, some of which are covered in other contributions to this book. However, Figs. 1 and 2 summarize diagramatically some of the major variations in laboratory and field techniques which have been employed for the assay of samples from soils and sediments,

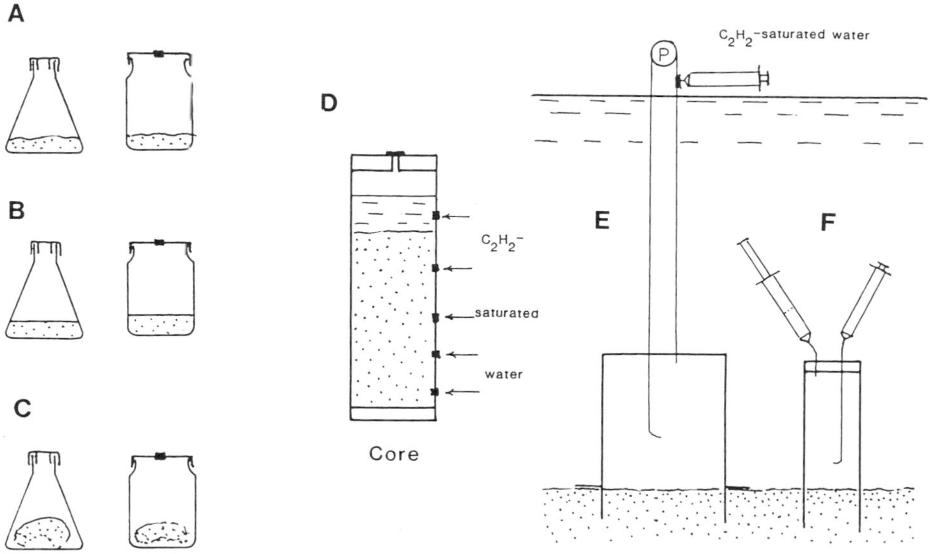

Fig. 2. Examples of sediment denitrification assays. A, sediment alone. B, sediment/water slurry. C, sediment core. D, intact core with C_2H_2 introduced into the sediment. E and F, in situ enclosures with C_2H_2-saturated water introduced.

respectively. It is worth noting that the limit of detection of the denitrification of soil NO_3^- in situ in the field using the dynamic enclosure method (Fig. 1H) is about 1 g N ha^{-1} day^{-1} (Ryden and Dawson, 1982), or about three orders of magnitude more sensitive than the ^{15}N-based technique.

MECHANISM OF INHIBITION

Nitrous oxide reductase has only recently been purified (chapters 1 and 3, this book). It is a dimeric enzyme which contains 3-4 Cu atoms per subunit, perhaps at the active site (Riester et al., 1989; Zumft and Matsubara, 1982). C_2H_2 inhibition of partially purified N_2O-R is non-competitive with a K_i of 28 μM (Kristiansson and Hollocher, 1980). The N_2O-R from Wolinella succinogenes is very much less sensitive to C_2H_2 (Teraguchi and Hollocher, 1989). In view of the many effects of C_2H_2 on metalloenzymes (Table 1 and Hyman and Arp, 1988) it is assumed that a Cu-C_2H_2 interaction may be involved. Whether C_2H_2 acts as a suicide substrate of N_2O-R, as is the case for the Cu-containing ammonia and methane monooxygenases (reviewed in Bédard and Knowles, 1989), must await more detailed studies.

POTENTIAL PROBLEMS

Users of inhibitors for experimental purposes must be aware of possible side effects and other potential problems. C_2H_2 is no exception, and the list of biological effects of C_2H_2 has grown over the years as evidenced by the summary in Table 1 (reviewed in Hyman and Arp, 1988; Payne, 1984). Certain of these effects impinge directly or indirectly on the validity of the C_2H_2 inhibition measurement of denitrification. Although yet to be demonstrated experimentally,

Table 1. Summary of the biological effects of C_2H_2

Effect	References[a]
Inhibition of	
Nitrate reductase in spinach	1
Nitrous oxide reductase in denitrifiers	2, 3, 4
Nitrogen fixation (thus can cause N limitation)	5
Nitrogenase-catalyzed hydrogen evolution	6
Proliferation of clostridia	7
Hydrogenases	
Conventional hydrogenases	8
Some uptake hydrogenases	8, 9, 10
Ammonia mono-oxygenase	11, 12, 13, 14,15
Ammonia oxidation in soil	16, 17
Methanogenesis	18, 19, 20, 21, 22
Methane mono-oxygenase	11, 23
Ethylene cooxidation	24
Growth of sulfate-reducing bacteria (*Desulfovibrio desulfuricans*)	25
Can be utilized by	
Rhodococcus rhodochrous (aerobic)	26, 27, 28, 29
Pelobacter acetylenicus (anaerobic)	30
Aerobic soil and sediment	31, 32, 33, 34, 35
Anaerobic soil and sediment	36, 37

[a] 1, Maldonado et al., 1981; 2, Balderston et al., 1976; 3, Fedorova et al., 1973; 4, Yoshinari and Knowles, 1976; 5, Burns and Hardy, 1975; 6, Hwang et al., 1973; 7, Brouzes and Knowles, 1971; 8, Smith et al., 1976; 9, Chan et al., 1980; 10, Tibelius and Knowles, 1984; 11, Bédard and Knowles, 1989; 12, Hyman and Wood, 1985; 13, Hynes and Knowles, 1978; 14, Hynes and Knowles, 1982; 15, Hynes and Knowles, 1984; 16, Bremner and Blackmer, 1979; 17, Walter et al., 1979; 18, Elleway et al., 1971; 19, Macgregor and Keeney, 1973; 20, Oremland and Taylor, 1975; 21, Raimbault, 1975; 22, Sprott et al., 1982; 23, de Bont and Mulder, 1976; 24, de Bont, 1976; 25, Payne and Grant, 1982; 26, de Bont and Peck, 1980; 27, Germon and Knowles, 1988; 28, Kanner and Barthe, 1979; 29, Kanner and Bartha, 1982; 30, Schink, 1985; 31, Germon, 1980a; 32, Germon, 1980b; 33, Tam et al., 1983; 34, Terry and Duxbury, 1985; 35, Topp and Germon, 1986; 36, Culbertson et al., 1981; 37, Watanabe and de Guzman, 1980.

C_2H_2 could inhibit the activity of (1) any denitrifiers growing at the expense of a hydrogenase system, or (2) any methylotrophic or other denitrifiers which are coupled to organic carbon products of methanotrophs.

The major problems in the application of the C_2H_2 inhibition method to soils and sediments appear to be related to five main groups of phenomena. These are (1) the possible lack of an inhibitory effect at low NO_3^- concentrations, (2) inhibition of nitrification by C_2H_2, (3) phenomena resulting from metabolism of the C_2H_2, (4) phenomena associated with the presence of sulfide in the sample to be studied, and (5) impurities in the C_2H_2. These will be considered in turn.

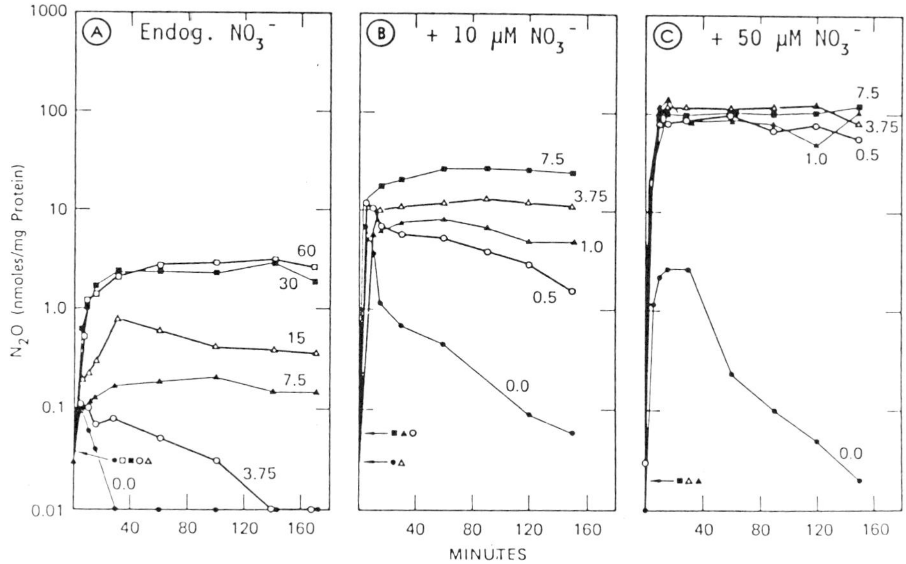

Fig. 3. Reduction of N_2O by washed cells of a marine pseudomonad with the levels of C_2H_2 (kPa) indicated. A, no NO_3^- added. B, 10 μM $NaNO_3$ added. C, 50 μM $NaNO_3$ added. Reproduced with permission from Oremland et al., 1984.

Low NO_3^- Concentrations

When samples to be studied contain very low NO_3^- concentrations, the inhibition achieved by concentrations of C_2H_2 in the range 1 to 5 kPa and even higher is not always complete and some reduction of N_2O to N_2 may occur (Kaspar, 1982; Oremland et al., 1984; Smith et al., 1978; Yeomans and Beauchamp, 1978). It is notable that during denitrification of $^{13}NO_3^-$ in soil in the presence of 5 kPa C_2H_2, the addition of 2 ppm of $^{14}NO_3^-$-N raised the proportion of $^{13}N_2O$ as a percentage of the total ^{13}N-labelled gaseous products from 69 to 96% (Smith et al., 1978). In marine sediments, it is interesting that the pore water NO_3^- concentration below which the inhibition of N_2O reduction by C_2H_2 seems to become incomplete is about 1 to 10 μM, and the relief of the inhibition appears to require the presence of relatively high densities of particulate material or microorganisms (Oremland et al., 1984; Slater and Capone, 1989).

This phenomenon has been observed mainly in marine sediments (Kasper, 1982; Oremland et al., 1984; Slater and Capone, 1989; Van Raalte and Patriquin, 1979) of which the redox potentials are very low. It is not generally reported in pure culture or enrichment culture studies (Slater and Capone, 1989) but partial inhibition occurred at 3.75 kPa C_2H_2 with < 10 μM NO_3^- in cell suspensions of a marine pseudomonad (Fig. 3). Although it cannot be completely excluded, it seems that a direct effect of the presence or absence of NO_3^- on the denitrifiers or the N_2O-R per se is not involved. It is possible that there may be a decrease in redox potential, at least in certain micro-environments, following the virtual disappearance of the NO_3^-/NO_2^- couple. However, the elucidation of the mechanism of this effect must await further study.

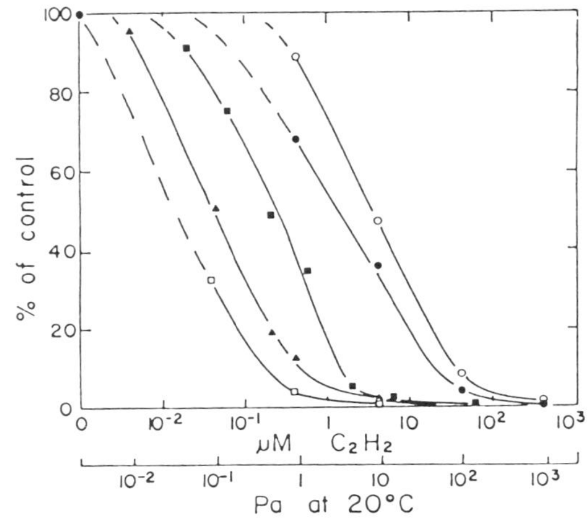

Fig. 4. Inhibition of nitrification (□, ▲, ■) and denitrification (●, ○) by C_2H_2 in pure culture and soil. The nitrification systems used were Nitrosomonas europaea in pure culture (■) and added to sterile soil (□), and soil alone(▲). Reduction of N_2O by soil (●) and by Pseudomonas perfectomarinus (○). Reproduced with permission from Knowles (1985).

Inhibition of Nitrification

In systems in which steady-state nitrification is coupled to denitrification with a resulting small NO_3^- pool size, it is clearly important that the production of NO_3^- should not be perturbed if a valid estimate of denitrification is desired. It was thus disappointing to discover that the nitrification process was also inhibited by C_2H_2 (Hynes and Knowles, 1978, and references in Table 1). The sensitive step is the monooxygenase-catalyzed oxidation of ammonia to hydroxylamine (Hynes and Knowles, 1982), a reaction for which C_2H_2 evidently acts as a suicide substrate (Hyman and Wood, 1985).

Heterotrophic nitrifiers tested so far are apparently not sensitive to C_2H_2 (Hynes and Knowles, 1982), thus suggesting a test (in addition to nitrapyrin) for distinguishing between chemolithotrophic and heterotrophic nitrification. Methanotrophic nitrifiers are sensitive to C_2H_2 by virtue of its effect on the methane monooxygenase (Bédard and Knowles, 1989).

Nitrification is much more sensitive to C_2H_2 than is denitrification (Fig. 4) and this difference has been used to inhibit differentially the two processes and to deduce their relative importance as sources of N_2O (Klemedtsson et al., 1988).

Metabolism of C_2H_2

The metabolism of C_2H_2 in aerobic soils is usually significant only after long incubation periods (of the order of 20 days) but may be appreciable during shorter periods in certain cases (Table 2). Rhodococcus (Nocardia) rhodochrous is sometimes isolated from C_2H_2-utilizing enrichments and it can grow on C_2H_2 as the sole carbon and energy source, producing acetaldehyde, ethanol, acetate, and CO_2 as the major products (de Bont and Peck, 1980; Germon and Knowles, 1988; Kanner and Bartha, 1979; 1982).

Table 2. Metabolism of C_2H_2 in soils and sediments

Atmosphere	System	C_2H_2 metabolized	References[a]
Aerobic	Soil	<2% of 1 kPa in 7 d[a] to 100% of 1 kPa in 29 d_b	1
	Soil	<1% of 1 kPa to CO_2 in 1-6 d	2
	Soil	2% of 10 kPa in 23 d to 100% of 10 kPa in 8 d	3
	Soil	100% of 1 kPa in 8 d	4
	Soil	25-50% of 1 kPa in 20 d	5
	Sediment	100% of 10 kPa in 18-21 d	6
Anaerobic	Soil	100% of 13 kPa in 4-6 d	7
	Sediment	100% of 10 kPa in 3-6 d	8

[a] 1, Terry and Duxbury, 1985; 2, Haider et al., 1983; 3, Germon, 1980b; 4, Topp and Germon, 1986; 5, Yeomans and Beauchamp, 1982a; 6, Tam et al., 1983; 7, Watanabe and de Guzman, 1980; 8, Culbertson et al., 1981.
[b] Roughly equal amounts recovered as $^{14}CO_2$ and soil organic ^{14}C.

In anaerobic soils and sediments, C_2H_2 may be metabolized relatively rapidly and completely (Table 2) or not at all (Van Raalte and Patriquin, 1979). Products reported include formic, acetic, propionic, and butyric acids. NO_3^- (5 μmol g^{-1}) inhibited this metabolism by an unknown mechanism (Watanabe and de Guzman, 1980). The fermentation products produced from C_2H_2 by the obligate

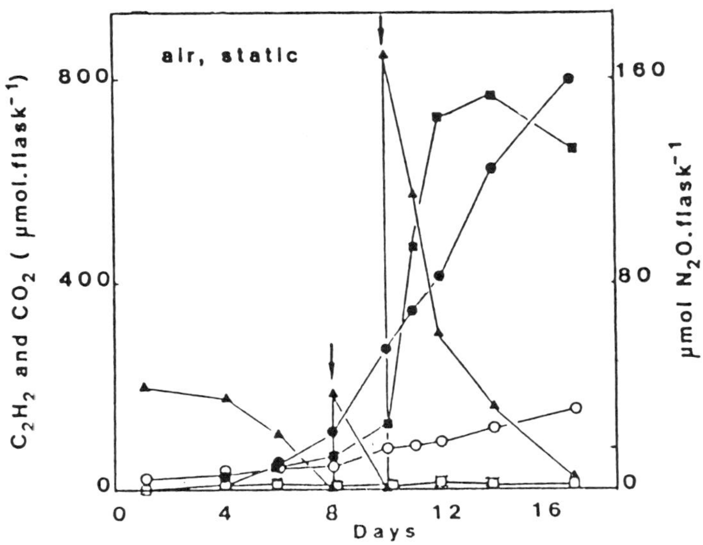

Fig. 5. Metabolism of C_2H_2 (▲) and production of N_2O (□, ■) and CO_2 (○, ●) by soil in the presence (closed symbols) and absence (open symbols) of 1 kPa C_2H_2. The arrows indicate subsequent additions. Reproduced with permission from Topp and Germon (1986).

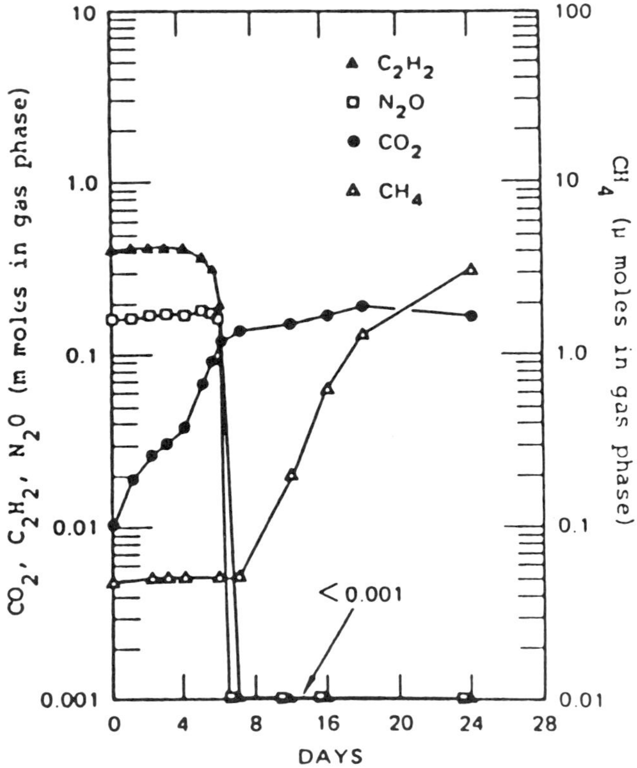

Fig. 6. Metabolism of C_2H_2, reduction of N_2O, and production of CO_2 and CH_4 by estuarine sediment. Reproduced with permission from Culbertson et al. (1981).

anaerobe Pelobacter acetylenicus included acetaldehyde, ethanol, and acetate (Schink, 1985).

Depending on the extent to which C_2H_2 metabolism and disappearance occurs it may stimulate denitrification or relieve the inhibitory effect on N_2O reduction, or it may possibly exert both of these effects.

Stimulation of Denitrification. There are apparently no reports of denitrifying bacteria able to utilize C_2H_2. However, at least some of the products of the aerobic or anaerobic utilization of C_2H_2 by other organisms referred to above can be utilized by denitrifiers. This is probably the basis of the stimulation of NO_3^- reduction and denitrification reported for some soils (Germon, 1980a and b; Haider et al, 1983; and Fig. 5). In one case, the aerobic production of acetaldehyde by an C_2H_2-utilizer and the stimulation of denitrification by the acetaldehyde was implicated (Topp and Germon, 1986). The isolation of the C_2H_2 utilizer showed that it was a Rhodococcus rhodochrous which produced acetaldehyde, ethanol, acetate, CO_2, and biomass in pure culture (Germon and Knowles, 1988). The effects of C_2H_2 on the soil denitrification could be duplicated by the addition of culture supernatant or of acetaldehyde to the soil (Topp and Germon, 1986).

In another case in which C_2H_2 accelerated NO_3^- reduction in soil, less than 1% of the $^{14}C_2H_2$ added was released as $^{14}CO_2$ but up to 28% was retained in

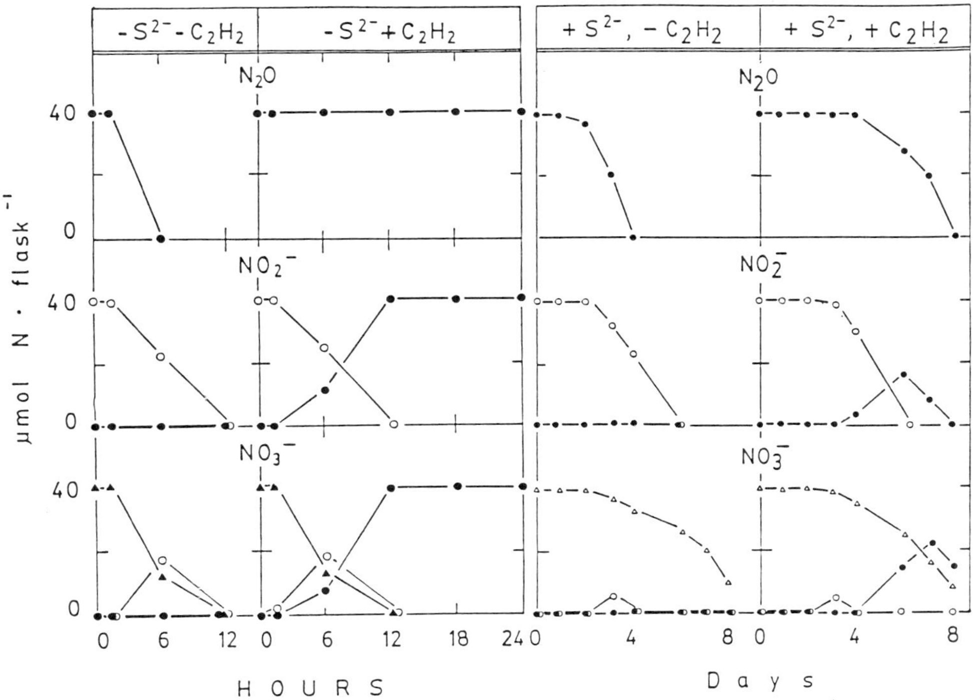

Fig. 7. To the left are shown the effects of C_2H_2 (4 kPa) in the absence of sulfide and to the right the effects of C_2H_2 in the presence of sulfide (0.4 mM) on the reduction of N_2O (●), NO_2^- (○), and NO_3^- (▲), by anaerobic pure cultures of a gliding bacterium in nutrient broth. Reproduced with permission from Adkins and Knowles (1986).

the soil in some form which was subsequently and gradually metabolized to CO_2 in the presence of NO_3^- (Haider et al., 1983). C_2H_2 also increased the rate of metabolism of both soil carbon and added glucose, again providing NO_3^- was non-limiting (Haider et al., 1983). The mechanisms of these effects are unknown, but irreversible binding of products of C_2H_2 metabolism may occur. This is thought to be the basis of the suicidal inhibition observed with ammonia and methane monooxygenases (Bédard and Knowles, 1989; Hyman and Arp, 1988).

Relief of C_2H_2 Inhibition of N_2O Reduction. Whether the metabolism of C_2H_2 will be sufficient to alleviate its inhibitory effect depends on its initial concentration and the percentage decrease in concentration which occurs. Thus, for the aerobic systems listed in Table 2, 1 or 2% metabolism is unlikely to affect the degree of inhibition, whereas 25 to 50% metabolism of 1 kPa C_2H_2 may decrease the C_2H_2 concentration below the inhibitory level, especially in soils with high moisture contents.

Much more rapid metabolism of C_2H_2 was reported for waterlogged soils and estuarine sediment (Table 2) where it was associated with lack of inhibition of N_2O reduction (Fig. 6). However, in a marine sediment in which the C_2H_2 failed to inhibit N_2O-R there was reportedly no disappearance of the C_2H_2 (Van Raalte and Patriquin, 1979). To minimize the risk that C_2H_2 metabolism could stimulate denitrification or relieve its inhibitory effect it is evident that the length

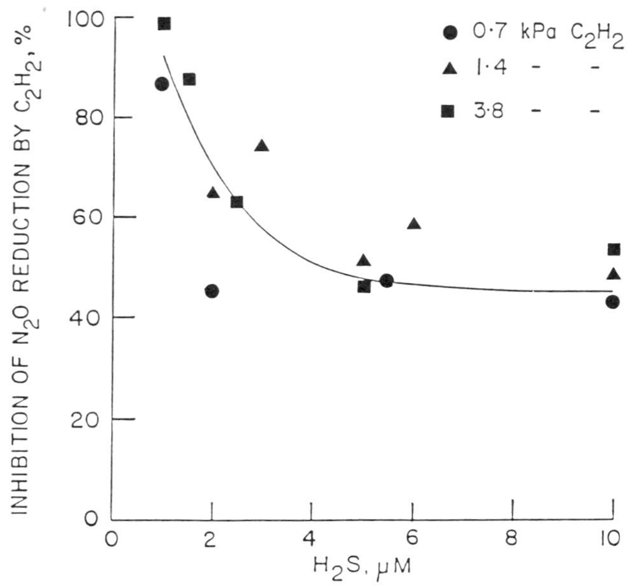

Fig. 8. Inhibition of N_2O reduction by C_2H_2 at different concentrations of H_2S in washed cell suspensions of a marine Pseudomonas fluorescans. The levels of C_2H_2 as indicated. Redrawn with permission from Sørensen et al. (1987).

of exposure of the system to C_2H_2 should be minimized. This must be done by decreasing incubation times and avoiding repeated exposure to C_2H_2 in successive assays of the same site or of the same material.

Interactions Between C_2H_2, N_2O, and Sulfide

In some early applications of the C_2H_2 inhibition technique to temperate and arctic marine sediments (Van Raalte and Patriquin, 1979; T. Yoshinari & R. Knowles, unpublished data) and certain freshwater sediments (Knowles, 1979) there was incomplete inhibition. Reduction of the N_2O produced from NO_3^- eventually occurred even though there was no or negligible disappearance of the C_2H_2. A common feature of many of the sediments was their sulfidic nature and we therefore suspected a possible effect of sulfide. Sørensen had also noted the relatively large concentrations of NO and N_2O in coastal marine sediment (Sørensen, 1978a) and also suspected a sulfide effect.

Inhibition of Reduction of Gaseous N Oxides by Sulfide. The reduction of N_2O is inhibited by Na_2S or H_2S added to soil (Tam and Knowles, 1979), or to pure cultures of Pseudomonas fluorescens (Sørensen et al., 1980) or P. aeruginosa (Tam and Knowles, 1979). Reduction of NO is inhibited in P. fluorescens but to a smaller extent (Sørensen et al., 1980) and the ionic nitrogen oxides are less sensitive (Adkins and Knowles, 1986). The apparent K_i for sulfide inhibition of N_2O reduction in P. fluorescens is about 1 μM (Sørensen et al., 1987). Such effects could be an explanation for the NO and N_2O reported in marine sediment (Sørensen, 1978a).

Sulfide Relieves C_2H_2 Inhibition of N_2O-R. As discussed above, sulfide alone inhibits the reduction of N_2O. However, when it is present along with normally completely inhibitory concentrations of C_2H_2 (1-10 kPa) the inhibition is partially relieved. The phenomenon is observed in soil (Evans et al., 1985; Tam and Knowles, 1979; Yeomans and Beauchamp, 1982b), and in cultures of a marine P. fluorescens isolated from the Baltic Sea (Sørensen et al., 1987). The relief of C_2H_2 inhibition by sulfide is reported also for a gliding bacterium isolated from a soil enrichment with C_2H_2, N_2O and sulfide (Adkins and Knowles, 1984, 1986; Fig. 7) and by another unidentified bacterium from a similarly enriched soil (Trevors and Beauchamp, 1985). The gliding bacterium, which was at first suspected to be a Cytophaga, now appears to be a strain of Flexibacter canadensis (A. Jones, A.M. Adkins & Knowles, unpublished data, and H. Reichenbach, personal communication), an authentic strain of which also shows sulfide relief of C_2H_2 inhibition (A. Jones & R. Knowles, unpublished data).

In soils, the effective sulfide concentrations are in the range 1-4 μmol Na_2S g^{-1} (Tam and Knowles, 1979; Yeomans and Beauchamp, 1982b) but is is not known, of course, what are the dissolved liquid phase concentrations of free sulfide. In cultures of the marine P. fluorescens, partial relief of inhibition occurs with sulfide in the range 2-10 μM in short-term experiments (Fig. 8). In longer-term experiments with Flexibacter canadensis, relief of inhibition occurs in the range 20-400 μM (A. Jones & R. Knowles, unpublished data).

Other sulfur-containing materials which also show an effect in soil similar to that of sulfide are cysteine and alfalfa meal (Yeomans and Beauchamp, 1982b), and thiosulfate (Tam and Knowles, 1979). These compounds require an additional 2 or 3 days before their relieving effect is fully manifest suggesting that their conversion and reduction to sulfide is a necessary prerequisite (Tam and Knowles, 1979; Yeomans and Beauchamp, 1982b). However, the detailed fate of sulfur compounds causing relief of C_2H_2 inhibition has not so far been reported for any system. Nor have binding studies been attempted for either sulfide or C_2H_2. On the other hand it is reported that for the relief of C_2H_2 inhibition to occur, it is necessary for the sulfide to be added before the C_2H_2, at least in short-term experiments (Sørensen et al., 1987).

The mechanism by which sulfide influences the interaction between C_2H_2 and N_2O-R remains elusive. Since the N_2O-R is a copper-containing enzyme in which at least some of the Cu atoms may be associated with the active site, it is assumed, but not proved, that C_2H_2 binds to some of these Cu atoms. Sulfide may interfere with this binding in some way, thus permitting expression of catalytic activity.

It should be added that another possible effect of the presence of sulfide, apart from any effect on N_2O reduction, is a lowering of redox potential which in turn can increase the proportion of NO_3^- which undergoes dissimilatory reduction to NH_4^+. Such an effect was reported in soil supplemented with glucose. The further addition of 19 μmol Na_2S g^{-1} caused a very significant increase in the $^{15}NH_4^+$ produced from $^{15}NO_3^-$ and a similar decrease in the production of $^{15}N_2O$ (de Catanzaro et al., 1987).

Impurities in the C_2H_2

Assays of N_2 fixation activity using the C_2H_2 reduction method are sometimes affected by impurities present in the C_2H_2 used in the assay (e.g., Tough and Crush, 1979). A scrubbing train was suggested which removed acetone, H_2S, NH_3, PH_3, and SO_2 contaminants. More recently, the question of contamination of different sources of C_2H_2 (commercial gases as well as carbide-generated) was re-examined. The list of contaminants (with some maximum concentrations given as % v/v) includes H_2 (0.023), O_2 (0.779), N_2 (3.78), CH_4 (0.073), PH_3 (0.06), acetone (1-10), CO, C_2H_4, C_3H_8, propyne, and ethanol (Hyman and Arp, 1987). A scrubbing and cryogenic trapping system was recommended which could remove virtually all the contaminants, including the fixed gases.

There do not appear to be any reports of C_2H_2 contaminants interfering with an C_2H_2 inhibition assay of denitrification. Nevertheless, it is a potential problem of which all those using such inhibitor tools should be aware (Oremland and Capone, 1988).

SUMMARY

Since the development of the acetylene inhibition technique for the measurement of denitrification, its simplicity, low cost, and sensitivity have led to its frequent use in many natural systems. The mechanism by which C_2H_2 inhibits N_2O reductase, however, remains unknown. It is important to remember potential problems which include (1) the possible lack of an inhibitory effect at low NO_3^- concentrations, (2) the inhibition of nitrification by C_2H_2, (3) phenomena resulting from the metabolism of C_2H_2, (4) phenomena associated with the presence of sulfide in the sample to be studied, and (5) possible effects of impurities in the C_2H_2.

ACKNOWLEDGMENTS

The financial support of the Natural Sciences and Engineering Research Council of Canada, the Inland Waters Directorate of Environment Canada, Agriculture Canada, and FCAC-Quebec, and the important collaboration of students and co-workers is gratefully acknowledged.

REFERENCES

Adkins, A.M., and Knowles, R., 1984, Reduction of nitrous oxide by a soil Cytophaga in the presence of acetylene and sulfide, FEMS Microbiol. Lett. 23:171.

Adkins, A.M., and Knowles, R., 1986, Denitrification by Cytophaga johnsonae strains and by a gliding bacterium able to reduce nitrous oxide in the presence of acetylene and sulfide, Can. J. Microbiol. 32:421.

Balderston, W.L., Sherr, B., and Payne, W.J., 1976, Blockage by acetylene of nitrous oxide reduction in Pseudomonas perfectomarinus, Appl. Environ. Microbiol. 31:504.

Bédard, C., and Knowles, R., 1989, Physiology, biochemistry, and specific inhibitors of CH_4, NH_4^+, and CO oxidation by methanotrophs and nitrifiers, Microbiol. Rev. 53:68.

Bremner, J.M., and Blackmer, A.M., 1979, Effects of acetylene and soil water content on emission of nitrous oxide from soils, Nature 280:380.

Brouzes, R., and Knowles, R., 1971, Inhibition of growth of Clostridium pasteurianum by acetylene: inplication for nitrogen fixation assay, Can. J. Microbiol. 17:1483.

Burns, R.C., and Hardy, R.W.F., 1975, Nitrogen fixation in bacteria and higher plants, Springer-Verlag, Heidelberg.

Chan, Y.K., Nelson, L.M., and Knowles, R. 1980, Hydrogen metabolism of Azospirillum brasilense in nitrogen-free medium, Can. J. Microbiol. 26:1126.

Christensen, S., and Tiedje, J.M., 1988, Sub-parts-per-billion nitrate method: use of an N_2O-producing denitrifier to convert NO_3^- or $^{15}NO_3^-$ to N_2O, Appl. Environ. Microbiol. 54:1409.

Culbertson, C.W., Zehnder, A.J.B., and Oremland, R.S., 1981, Anaerobic oxidation of acetylene by estuarine sediments and enrichment cultures, Appl. Environ. Microbiol. 41:396.

de Bont, J.A.M., 1976, Bacterial degradation of ethylene and the acetylene reduction test, Can. J.Microbiol. 22:1060.

de Bont, J.A.M., and Mulder, E.G., 1976, Invalidity of the acetylene reduction assay in alkane-utilizing, nitrogen-fixing bacteria, Appl. Environ. Microbiol. 31:640.

de Bont, J.A.M., and Peck, M.W., 1980, Metabolism of acetylene by Rhodococcus, Arch. Microbiol. 127:99.

de Catanzaro, J.B., Beauchamp, E.G., and Drury, C.F., 1987, Denitrification versus dissimilatory nitrate rediction in soil with alfalfa, straw, glucose and sulfide treatments, Soil Biol. Biochem. 19:583.

Elleway, R.F., Sabine, J.R., and Nicholas, D.J.D., 1971, Acetylene reduction by rumen microflora, Arch. Mikrobiol. 76:277.

Evans, D.G., Beauchamp, E., and Trevors, J.K., 1985, Sulfide alleviation of the acetylene inhibition of nitrous oxide reduction in soil, Appl. Environ. Microbiol. 49:217.

Fedorova, R.I., Milekhina, E.I., and Il∩yukhina, N.I., 1973, Evaluation of the method of "gas metabolism" for detecting extraterrestrial life. Identification of nitrogen-fixing microorganisms, Izv. Akad. Nauk, SSSR, Ser. Biol. 1973 (6):797.

Germon, J.C., 1980a, Etude quantitatie de la dénitrification biologique dans le sol à l'aide de l'acétyléne. I. Application á différents sols. Ann. Microbiol. (Inst. Pasteur) 131B:69.

Germon, J.C., 1980b, Etude quantitative de la dénitrification biologique dans le sol à l'aide de l'acétyléne. II. Evolution de l'effet inhibiteur de l'acétyléne sur la N_2O-réductase; incidence de l'acétyléne sur la vitesse de dénitrification et sur la réorganisation de l'azote nitrique, Ann. Microbiol. (Inst. Pasteur) 131B:81.

Germon, J.C., and Knowles, R., 1988, Metabolism of acetylene and acetaldehyde by Rhodococcus rhodochrous, Can. J. Microbiol. 34:242.

Haider, K., Mosier, A.R., and Heinemeyer, O., 1983, Side effects of acetylene on the conversion of nitrate in soil, Z. Pflanzenernaehr. Bodenk. 146:623.

Hardy, R.W.F., and Knight, E., Jr., 1966, Reduction of N_2O by biological N_2-fixing systems, Biochem. Biophys. Res. Commun. 23:409.

Hwang, J.C., Chen, C.H., and Burris, R.H., 1973, Inhibition of nitrogenase-catalyzed reductions, Biochim. Biophys. Acta. 292:256.

Hyman, M.R., and Arp, D.J., 1987, Quantification and removal of some contaminating gases from acetylene used to study gas-utilizing enzymes and microorganisms, Appl. Environ. Microbiol. 53:298.

Hyman, M.R., and Arp, D.J., 1988, Acetylene inhibition of metalloenzymes, Analyt. Biochem. 173:207.

Hyman, M.R., and Wood, P.M., 1985, Suicidal labelling and inactivation of ammonia mono-oxygenase by acetylene, Biochem. J. 227:719.

Hynes, R.K., and Knowles, R., 1978, Inhibition by acetylene of ammonia oxidation in Nitrosomonas europaea, FEMS Microbiol. Lett. 4:319.

Hynes, R.K., and Knowles, R., 1982, Effect of acetylene on autotrophic and heterotrophic nitrification, Can. J. Microbiol. 28:334.

Hynes, R.K., and Knowles, R., 1984, Production of nitrous oxide by Nitrosomonas europaea: effects of acetylene, pH, and oxygen, Can. J. Microbiol. 30:1397.

Jensen, B.B., and Burris, R.H., 1986, N_2O as a substrate and as a competitive inhibitor of nitrogenase, Biochemistry 25:1083.

Kanner, D., and Bartha, R., 1979, Growth of Nocardia rhodochrous on acetylene gas, J. Bacteriol. 139:225.

Kanner, D., and Bartha, R., 1982, Metabolism of acetylene by Nocardia rhodochrous, J. Bacteriol. 150:989.

Kaspar, H.F., 1982, Denitrification in marine sediment: measurement of capacity and estimate of in situ rate, Appl. Environ. Microbiol. 43:522.

Klemedtsson, L., Svensson, B.H., Lindberg, T., and Rosswall, T., 1977, The use of acetylene inhibition of nitrous oxide redictase in quantifying denitrification in soils, Swedish J. Agr. Res. 7:179.

Klemedtsson, L., Svensson, B.H., and Rossvall, T., 1988, A method of selective inhibition to destinguish between nitrification and denitrification as sources of nitrous oxide in soil, Biol. Fertil. Soils 6:112.

Knowles, R., 1979, Denitrification, acetylene reduction and methane metabolism in lake sediment exposed to acetylene, Appl. Environ. Microbiol. 38:486.

Knowles, R., 1982, Denitrification, Microbiol. Revs. 46:43.

Knowles, R., 1985, Some effects of inhibitors on nitrogen transformations, p. 363. In K.A. Malik, S.H.M. Naqvi, and M.I.H. Aleem (ed.), Nitrogen and the environment. Nuclear Institute for Agriculture and Biology, Faisalabad, Pakistan.

Kristjansson, J.K., and Hollocher, T.C., 1980, First practical assay for soluble nitrous oxide reductase of denitrifying bacteria and a partial kinetic characterization, J. Biol. Chem. 255:704.

Lensi, R., Gourbiere, F., and Josserand, A., 1985, Measurement of small amounts of nitrate in an acid soil by N_2O production, Soil Biol. Biochem. 17:733.

Liang, J., and Burris, R.H., 1988, Interactions among N_2, N_2O, and C_2H_2 as substrates and inhibitors of nitrogenase from Azotobacter vinelandii, Biochemistry 27:6726.

Macgregor, A.N., and Keeney, D.R., 1973, Methane formation by lake sediments during in vitro incubation, Water Res. Bull. 9:1153-1158.

Maldonado, J.M., Vargas, M.A., Maurino, S.G., and Aparicio, P.J. 1981. Inactivation by acetylene of spinach nitrate redictase, Biochim. Biophys. Acta. 661:112.

Oremland, R.S., and Capone, D.G., 1988, Use of "specific" inhibitors in biogeochemistry and microbial ecology, Adv. Microb. Ecol. 10:285.

Oremland, R.S., and Taylor, B,F., 1975, Inhibition of methanogenesis in marine sediments by acetylene and ethylene: validity of the acetylene reduction assay for anaerobic microcosms, Appl. Microbiol. 30:707.

Oremland, R.S., Umberger, C., Culbertson, C.W., and Smith, R.L., 1984, Denitrification in San Francisco Bay intertidal sediments, Appl. Environ. Microbiol. 47:1106.

Payne, W.J. 1984, Influence of acetylene on microbial and enzymatic assays, J. Microbiol. Methods 2:117.

Payne, W.J., and Grant, M.lA., 1982, Influence of acetylene on growth of sulfate-respiring bacteria, Appl. Environ. Microbiol. 43:727.

Raimbault, M., 1975, Etude de l'influence inhibitrice de l'acétyléne sur la formation biologique du méthane dans un sol de riziére, Ann. Microbiol. (Inst. Pasteur), 126A:247.

Riester, J., Zumft, W.G., and Kroneck, P.M.H., 1989, Nitrous oxide reductase from Pseudomonas stutzeri. Redox properties and spectroscopic characterization of different forms of the multicopper enzyme, Eur. J. Biochem. 178:751.

Ryden, J.C., and Dawson, K.P., 1982, Evaluation of the acetylene-inhibition technique for the measurement of denitrification in grassland soils, J. Sci. Food Agric. 33:1197.

Ryden, J.C., Lund, L.J., and Focht, D.D., 1979, Direct measurement of denitrification loss from soils: I. Laboratory evaluation of acetylene inhivition of nitrous oxide reduction, Soil Sci. Soc. Amer. J. 43:104.

Schink, B., 1985, Fermlentation of acetylene by an obligate anaerobe, Pelobacter acetylenicus sp. nov., Arch. Microbiol. 142:295.

Slater, J.M., and Capone, D.G., 1989, Nitrate requirement for acetylene inhibition of nitrous oxide reduction in marine sediments, Microbial Ecol. 17:143.

Smith, L.A., Hill, S., and Yates, M.G., 1976, Inhibition by acetylene of conventional hydrogenase in nitrogen-fixing bacteria, Nature 262:209.

Smith, M.S., Firestone, M.K., and Tiedje, J.M., 1978, The acetylene inhibition method for short-term measurement of soil denitrification and its evaluation using nitrogen-13, Soil Sci. Soc. Amer. J. 42:611.

Sørensen, J., 1978a, Occurrence of nitric and nitrous oxides in a coastal marine sediment, Appl. Environ. Microbiol. 36:809.

Sørensen, J., 1978b, Denitrification rates in a marine sediment as measured by the acetylene inhibition technique, Appl. Environ. Microbiol. 36:139.

Sørensen, J., Rasmussen, L.K., and Koike, I., 1987, Micromolar sulfide concentrations alleviate acetylene blockage of nitrous oxide reduction by denitrifying Pseudomonas fluorescens, Can. J. Microbiol. 33:1001.

Sørensen, J., Tiedje, J.M., and Firestone, R.B., 1980, inhibition by sulfide of nitric and nitrous oxide reduction by denitrifying Pseudomonas fluorescens, Appl. Environ. Microbiol. 39:105.

Sprott, G.D., Jarrell, K.F., Shaw, K.M., and Knowles, R. 1982, Acetylene as an inhibitor of methanogenic bacteria, J. Gen. Microbiol. 128:2453.

Tam, T.-Y., and Knowles, R., 1979, Effects of sulfide and acetylene on nitrous oxide reduction by soil and by Pseudomonas aeruginosa, Can. J. Microbiol. 25:1133.

Tam, T.-Y., Mayfield, C.I., and Inniss, W.E., 1983 Aerobic acetylene utilization by stream sediment and isolated bacteria, Curr. Microbiol. 8:165.

Teraguchi, S., and Hollocher, T.C., 1989, Purification and some characteristics of a cytochrome c-containing nitrous oxide reductase from Wolinella succinogenes, J. Biol. Chem. 264:1972.

Terry, R.E., and Duxbury, J.M., 1985, Acetylene decomposition in soils, Soil Sci. Soc. Amer. J. 49:90.

Tibelius, K.H., and Knowles, R., 1984, Hydrogenase activity in Azospirillum brasilense is inhibited by nitrite, nitric oxide, carbon monoxide, and acetylene, J. Bacteriol. 160:103.

Topp, E., and Germon, J.-C., 1986, Acetylene metabolism and stimulation of denitrification in an agricultural soil, Appl. Environ. Microbiol. 52:802.

Tough, H.J., and Crush, J.R., 1979, Effect of grade of avetylene on ethylene production by white clover (Trifolium repens L.), during acetylene reduction assays of nitrogen fixation, N.Z. J. Agric. Res. 22:581.

Trevors, J.T., and Beauchamp, E., 1985, Reduction of nitrous oxide (N_2O) by a bacterial isolate in the presence of sulfide and acetylene, J. Microbiol. Methods 4:127.

Van Raalte, C.D., and Patriquin, D.G., 1979, Use of the "acetylene blockage" technique for assaying denitrification in a salt marsh, Marine Biol. (Berlin) 52:315.

Walter, H.M., Keeney, D.R., and Fillery, I.R., 1979, Inhibition of nitrification by acetylene, Soil Sci. Soc. Amer. J. 43:195.

Watanabe, I., and de Guzman, M.R., 1980, Effect of nitrate on acetylene disappearance from anaerobic soil, Soil Biol. Biochem. 12:193.

Yeomans, J.C., Beauchamp, E.G., 1978, Limited inhibition of nitrous oxide reduction in soil in the presence of acetylene, Soil Biol. Biochem. 10:517.

Yeomans, J.C., and Beauchamp, E.G., 1982a, Acetylene as a possible substrate in the denitrification process, Can. J. Soil Sci. 62:139.

Yeomans, J.C., and Beauchamp, E.G., 1982b, Sulfur in acetylene inhibition of nitrous oxide reduction by soil microorganisms, Soil Sci. Soc. Amer. J. 46:75.

Yoshinari, T., Hynes, R., and Knowles, R., 1977, Acetylene inhibition of nitrous oxide reduction and measurement of denitrification and nitrogen fixation in soil, Soil Biol. Biochem. 9:177.

Yoshinari, T., and Knowles, R., 1976, Acetylene inhibition of nitrous oxide reduction by denitrifying bacteria, Biochem. Biophys. Res. Commun. 69:705.

Zumft, W.G., and Matsubara, T., 1982, A novel kind of multi-copper protein as terminal oxidoreductase of nitrous oxide resperation in *Pseudomonas perfectomarinus*, FEBS Letters: 148:107.

THE USE OF ACETYLENE FOR THE QUANTIFICATION OF N_2 AND N_2O PRODUCTION FROM BIOLOGICAL PROCESSES IN SOIL

Leif Klemedtsson, Gunbritt Hansson
and Arvin Mosier*

Department of Microbiology
Swedish University of Agricultral
Sciences, UPPSALA, Sweden
*USDA - ARS, Fort Collins, CO 80522
U.S.A.

INTRODUCTION

Losses of nitrogen from soil have many adverse environmental effects. Supplies of potable water may be contaminated by NO_3^- leached from agricultural soils, and since the 1970s there has also been concern about the effects of gaseous N-compounds produced from NH_4^+ and NO_3^- by micro- organisms. The rate of degradation of the Earth's protective ozone screen is enhanced by N_2O (Crutzen, 1983), and N_2O also has a significant effect on the Earth's thermal balance via the greenhouse effect (Lacis et al., 1981). N_2 is not harmful to the environment, but the losses of fertilizer nitrogen via denitrification to N_2 are economically undesirable. Consequently, it is essential to develop methods for quantitative studies of the nitrogen cycle in soil in order to find ways of minimizing N losses in agriculture and forest management.

The production of N_2O occurs primarily through denitrification and autotrophic nitrification, while N_2 is produced only during denitrification (Firestone and Davidson, 1989). During autotrophic nitrification, the oxidation of NH_4^+ to NO_3^- by chemoautotrophic bacteria of the family Nitrobacteraceae, N_2O is produced in the first step of the oxidation (Schmidt, 1982) by a denitrification-like process in which NO_2^- serves as an electron acceptor (Poth and Focht, 1985). N_2O can also be produced during NO_3^--respiration by non-denitrifying bacteria (Smith & Zimmerman, 1981), during heterotrophic nitrification (Hynes and Knowles, 1982), and during assimilation of NO_3^- by yeasts and other fungi (Bleakley and Tiedje, 1982). Only during denitrification can significant amounts of N_2O be consumed.

The relative importance of denitrification and autotrophic nitrification to the total production of N_2O in the soil of different ecosystems has not yet been clarified. Denitrification is often assumed to dominate N_2O production in soil but nitrification may be the dominating source of N_2O in certain ecosystems (Martikainen 1985) and in soils fertilized with NH_4^+ or urea (Bremner and

Denitrification in Soil and Sediment, Edited by N. P. Revsbech and
J. Sørensen, Plenum Press, New York, 1990

Blackmer, 1978). In forest ecosystems, fungi may contribute to the N_2O production through heterotrophic nitrification. Robertson and Tiedje (1987) found that only 20 % of the N_2O emission from a forest soil originated either from autotrophic nitrification or anaerobic dissimilatory processes such as denitrification. The mechanisms for N_2O production from microorganisms other than denitrifiers and autotrophic nitrifiers are unclear (Tiedje, 1988) and the importance of these organisms as N_2O producers is unknown.

The Acetylene Inhibition Method, Inhibition of the N_2O Reductase of Denitrifying Bacteria

Fedorova et al. (1973) discovered that acetylene inhibits the reduction of N_2O to N_2 in the denitrification process, and their observation formed the basis of the acetylene inhibition method of measuring denitrification (Yoshinari and Knowles, 1976; Balderston et al., 1976; Klemedtsson et al., 1977; Yoshinari et al., 1977; Smith et al., 1978). The method simply involves treating the atmosphere of soil with 1-10 kPa acetylene (Fig. 1) and analysing the N_2O produced by gas chromatography (Kaspar and Tiedje, 1980; Klemedtsson et al., 1986). Since the reduction of N_2O to N_2 is blocked, the N_2O production will represent the total production of N_2O and N_2 from denitrification. With the acetylene inhibition method low denitrification rates can be measured in ambient atmosphere since the background concentration of N_2O is low, unlike that of N_2. Several variations of the technique have been used to measure the natural denitrification activity in soil, involving both *in situ* systems with enclosures placed on the soil surface and introduction of acetylene into the soil below (Ryden et al., 1979; Colbourn et al., 1984) and incubation of soil cores with acetylene (Aulakh et al., 1982; Rice and Smith, 1982; Parkin et al., 1984; Burton and Beauchamp, 1984; Svensson et al., 1985). The acetylene inhibition method has been reviewed in detail by Tiedje et al. (1989).

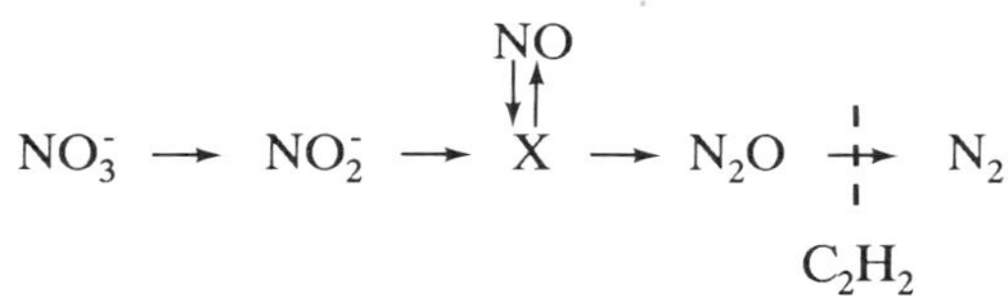

Figure 1. The last step of the denitrification process, the reduction of N_2O to N_2, is inhibited by acetylene.

Inhibition of the Ammonium Monooxygenase of Autotrophic Nitrifying Bacteria by Acetylene

Acetylene inhibits nitrification both in pure cultures of *Nitrosomonas europaea* (Hynes and Knowles, 1978), and in soil (Walter et al., 1979; Mosier, 1980; Berg et al., 1982). In pure cultures of NH_4^+ oxidizers, inhibition is complete within 10-15 minutes, and the inhibition is probably instantaneous, the time-lag representing the time needed for acetylene to diffuse through the medium (Hynes and Knowles, 1978). Berg et al. (1982) found that nitrification was totally inhibited by between 1 and 10 Pa acetylene in soil, and partially inhibited by 0.5 Pa. Both N_2O and NO are produced during NH_4^+ oxidation (Poth and Focht, 1985), and since acetylene inhibits the first step of NH_4^+ oxidation, the production of both compounds will be inhibited by acetylene

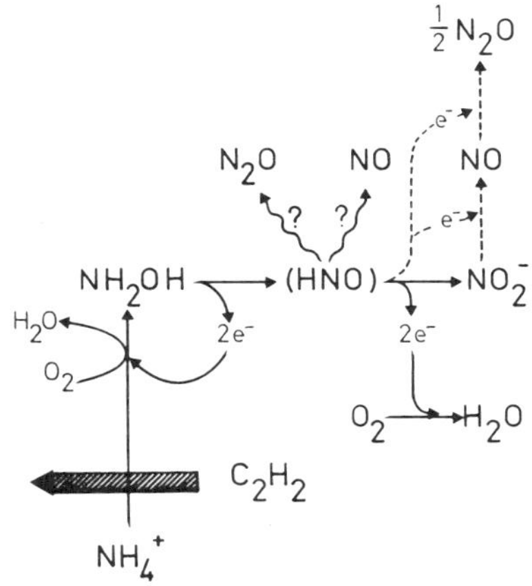

Figure 2. The first step of the autotrophic NH_4^+ oxidation is inhibited by acetylene (diagram of the NH_4^+ oxidation according to Conrad, chapter 7 of this book).

(Fig.2). As the N_2O production from the heterotrophic nitrifier Arthrobacter is not inhibited by acetylene (Hynes and Knowles, 1982), it is possible that heterotrophic nitrification is not affected by acetylene.

The basis of the PPM method (the use of ppm levels of acetylene to inhibit nitrification) is the fact that the N_2O production from nitrification and the reduction of N_2O to N_2 by denitrification are inhibited at different partial

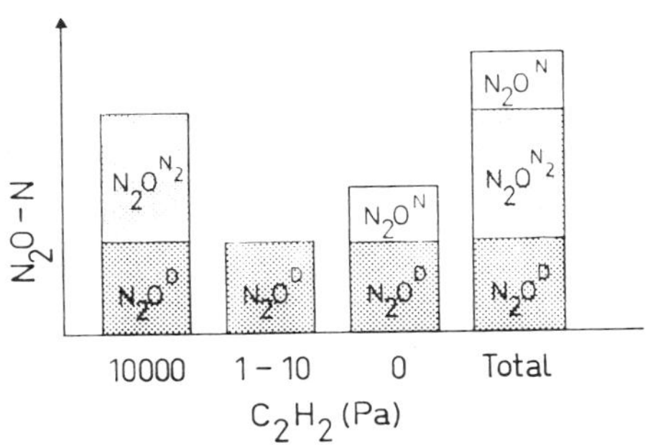

Figure 3. The PPM method: The inhibition of nitrification by low concentrations of acetylene. N_2O^D = N_2O produced by denitrification under natural conditions when also N_2 is formed. $N_2O^{N_2}$ = increase in N_2O production because of acetylene inhibition of nitrous oxide reductase. N_2O^N = N_2O produced by autotrophic nitrification.

pressures of acetylene in soil. N_2O is normally produced in soil from denitrification and from autotrophic nitrification. Partial pressures of acetylene between 1 and 10 Pa inhibit only the N_2O production from autotrophic nitrification. Partial pressures between 1,000 and 10,000 Pa inhibit both autotrophic nitrification and the reduction of N_2O to N_2 by denitrification, and the N_2O produced represents total denitrification which would normally give both N_2 and N_2O. The use of different partial pressures of acetylene thus makes it possible to determine: 1) total denitrification rates, 2) the N_2O production from denitrification, 3) the N_2O production from autotrophic nitrification (Fig. 3). The PPM method has been used to inhibit autotrophic nitrification in several studies: the effect of plants on N_2O production in soil cores from experimental field plots (Svensson et al., 1985) and in a greenhouse experiment (Klemedtsson et al.,1987); the effect of water on various N_2O producing processes in soil in a laboratory experiment (Klemedtsson et al., 1988a); the effect of earthworm casts on N_2O losses from soil (Svensson et al., (1986). The method has also been used to determine the relative importance of autotrophic nitrification and denitrification in a forest soil (Davidson et al., 1986).

THE ACETYLENE INHIBITION METHOD

Advantages

The acetylene method has several advantages: (1) the cost per sampling area is small, (2) the method is relatively simple, (3) the detection limit is low (1 g N ha^{-1} d^{-1}), (4) with the acetylene inhibiton method, denitrification measurements can be made in unfertilized soil without any addition of nitrogen, unlike methods involving the use of ^{15}N fertilizers.

Disadvantages of the Acetylene Inhibiton Method

A number of problems caused by the introduction of acetylene into soil have been discovered. Knowles (1984) described the effects of acetylene on soil microorganisms. Many of the problems he discussed are applicable to the use of the acetylene method in soil systems and are disscussed in this section. The inhibition of autotrophic NH_4^+ oxidizers must be seen as a disadvantage in this context, since a lack of nitrification can lead to a depletion of NO_3^- during long-term incubations of soil, which in turn will lead to decreasing denitrification rates.

Acetylene may be oxidized under both aerobic (Kanner and Barta, 1979) and anaerobic conditions (Culbertson et al., 1981). The existence of acetylene-consuming microorganisms in soil, for example *Nocardia* spp. and *Arthrobacter* spp., makes acetylene unsuitable for long-term experiments and repeated short-term experiments in the same soil since the desired inhibition of denitrification will gradually be lost if a build-up of an acetylene-consuming population of microorganisms occurs. Easily available carbon compounds stimulate the consumption of acetylene in soil (Terry and Duxbury, 1985) and the activity of the acetylene-consuming microorganisms may also increase the O_2 consumption rate, thereby stimulating the denitrification activity. Furthermore, Topp and Germon (1986) found that acetylene-consuming microorganisms oxidize acetylene to ethanol, acetate and acetaldehyde; compounds which may stimulate denitrification by supplying denitrifiers with an energy source. In a study by

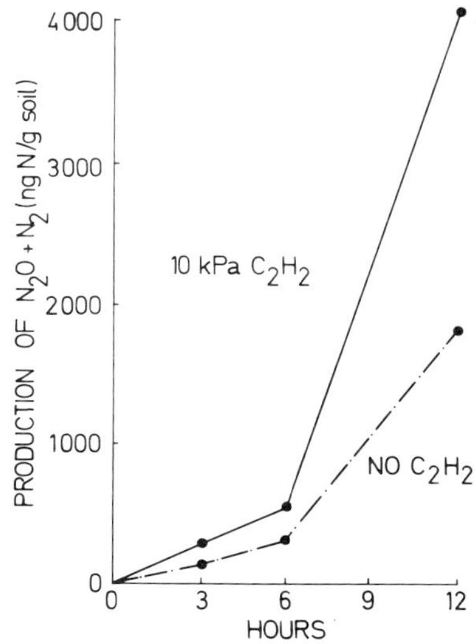

Figure 4. Effect of acetylene on short-term production of N_2O + N_2 from a nitrate-amended and water-saturated soil under ambient atmosphere.

Haider et al. (1983a) acetylene inhibited soil carbon mineralization when NO_3^- was limited but accelerated carbon mineralization when a sufficient amount of NO_3^- was supplied. After three days' exposure to acetylene the gas was removed by purging, but up to 28% of the added acetylene carbon still remained in the soil in a readily available form which was rapidly oxidized aerobically and also anaerobically when NO_3^- was added. The innate capacity of soils for acetylene consumption should always be determined before experiments involving the use of acetylene are performed.

The study by Haider et al. (1983a) shows that acetylene may disappear more rapidly from the soil atmosphere during prolonged incubations when NO_3^- is present, and that acetylene can accelerate NO_3^- reduction. The cause of the accelerated NO_3^- reduction might be that fewer electrons are needed for the reduction of NO_3^- to N_2O than for the reduction of NO_3^- to N_2, and that more NO_3^- thus will have to be consumed when the last step of the denitrification process is inhibited (Smith and Tiedje, 1979; Cho, 1982).

The stimulatory effect of acetylene on denitrification has also been shown in short-term experiments performed by Haider et al (1983b), where determinations of the production of N_2 and N_2O from ^{15}N-nitrate were made with and without acetylene (Fig. 4). These data and those presented by Haider et al. (1983a) show that in this soil NO_3^- reduction was rapidly and significantly increased by the addition of acetylene. The effect persisted for at least a week.

SEPARATION OF N_2O PRODUCED FROM DENITRIFICATION AND NITRIFICATION BY ACETYLENE INHIBITON (PPM METHOD)

Advantages

Acetylene has several advantages as a nitrification inhibitor. For instance, acetylene can be used during short-term incubations of soil since it diffuses rapidly into soil; it moves rapidly through air-filled pores in soil. Nitrapyrin (N-Serve), perhaps the most commonly used nitrification inhibitor, is only moderately volatile, and its water solubility is low (0.04 g/kg H2O at 25°C; Goring, 1962). Compared with acetylene, which is highly soluble in water (0.928 ml/ml H_2O at 760 mm Hg and 25 C), nitrapyrin moves slowly in soil. Since acetylene is a gas, its addition to soil will neither alter the water content nor cause any displacement of water soluble compounds in soil, such as NO_3^- and organic compounds.

The inhibitory effect of nitrapyrin varies depending on the strain of nitrifier present in the soil, and a selection of more nitrapyrin-tolerant nitrifiers may take place during long-term experiments. The inhibitory concentration of nitrapyrin also varies between different types of soil. The organic content and the pH-level influence the inhibition (Goring, 1962).

Nitrapyrin may also stimulate N_2O production from the denitrification of NO_3^- (Klemedtsson et al., 1988b). In this experiment, soil respiration was increased, and the enhanced oxygen consumption may have stimulated denitrification. This stimulation was probably caused by the solvent, xylene, rather than by nitrapyrin. Notton et al. (1979), however, found that nitrapyrin in itself gave rise to an increase in NO_3^- reduction in soil. Unlike nitrapyrin, acetylene does not affect soil respiration to any great extent at low partial pressures (Klemedtsson et al., 1988a) which means that the oxygen status of soil will not be altered by acetylene.

Like acetylene, nitrapyrin may be degraded during prolonged incubation and the degradation rate of nitrapyrin is strongly influenced by soil temperature; the half-life of nitrapyrin can be up to 80 % lower at 20°C than at 10°C (Herlihy and Quirke, 1975).

Disadvantages

The activity of acetylene-consuming microorganisms is of particular importance when the partial pressure of acetylene is low, as when acetylene is used to inhibit nitrification. Furthermore, when acetylene is used to separate the N_2O production from nitrification and denitrification a delicate balance has to be maintained between partial pressures inhibiting nitrification and those inhibiting the N_2 production by denitrifiers. Acetylene inhibits the reduction of N_2O to N_2 by denitrifiers in pure culture by about 10 % at a partial pressure of 1 Pa (Yoshinari and Knowles, 1976), at which nitrification is totally blocked (Berg et al. 1982).

In the soil used in Klemedtsson et al. (1988b) acetylene was not consumed unless glucose (1 mg g^{-1} soil) was added, but even when glucose was added the inhibition of N_2O production from autotrophic nitrification lasted for 2 days at 2.5 Pa acetylene and for 5 days at 5 Pa which indicates a very low acetylene consumption rate.

In contrast with the data presented by Yoshinari and Knowles (1976) we did not find any significant effect of 1 Pa acetylene on the N_2O production in a sandy loam at 100 % and 90 % water-filled pore space during the first 6 days of incubation (Klemedtsson et al., 1988b). Because of the diffusion constraints which exist in soils at high water contents, the partial pressure of acetylene has to be higher than 1 Pa during short-term incubations. Davidson et al., (1986) found that 10 Pa acetylene had no significant effect on the reduction of N_2O in a forest soil during 8 hours of anaerobic incubation when N_2O was added to the soil after it had been almost depleted of NO_3^-. Denitrifiers are less sensitive to inhibition by acetylene at low levels of NO_3^-, and determinations of the inhibition of N_2O production from cultures and soils which have been depleted of NO_3^- might give misleading results (Firestone et al., 1979; Oremland et al., 1984).

An experiment similar to Davidson's et al. (1986) was performed in order to test the method under more natural conditions. The soil was incubated with an air-filled headspace, and at 100 % of WHC to stimulate denitrification (Klemedtsson et al., 1988b). NO_3^- and glucose were added to the soil. The production of N_2O did not increase at a partial pressure of 2.5 Pa (as compared with the control) during 15 hours of incubation. The N_2O-production did not increase significantly at 5 Pa during the first 9 hours of incubation, but then became higher than in the control. This shows that 2.5-5 Pa acetylene can be used during short-term experiments, even when the NO_3^- level is high.

These experiments show that the drawbacks of using acetylene to separate the production of N2O from nitrification and denitrification may be negligible in some soils, but they still have to be considered when experiments are designed since they might be much more apparent in other soils.

APPLICATIONS OF THE ACETYLENE INHIBITION METHOD IN SOIL

Methods where acetylene is used to determine natural denitrification rates involve either in situ systems in which cylinders placed in the soil or enclosures placed over the soil surface are used (Ryden et al., 1979; Burton and Beauchamp, 1984; Colbourn et al., 1984) or incubation in the laboratory or in the field of soil cores (Aulakh et al., 1982; Rice and Smith, 1982; Burton and Beauchamp, 1984; Parkin et al., 1984; Svensson et al., 1985; McConnaughey and Duxbury, 1986). In situ is used only to denote undisturbed soil, and soil cores denote both soil cores incubated in the field and those incubated in the laboratory. Tiedje et al. (1989) have reviewed the acetylene inhibiton method in detail.

In situ measurements with soil covers

Measurements of denitrification with soil covers can be made: (1) with open covers where acetylene flows through the covers and the N_2O produced can be measured either directly in the gas flow at the outlet from the chamber (Denmead, 1979) or be collected on a molecular sieve for later analysis (Ryden et al., 1979), (2) with closed chambers where the N_2O is allowed to accumulate inside the chamber (McConnaughey and Duxbury, 1986). The advantages and disadvantages of the two cover techniques to measure N_2O emission from soil have been discussed in detail by Mosier and Heinemeyer (1985) and in combination with the acetylene inhibition method by Tiedje et al. (1989). Mosier and Heinemeyer (1985) concluded that both methods provide reliable results when used with adequate care.

Acetylene can be introduced into the soil by the method used by Ryden et al. (1979), i.e. before the covers are placed over the soil. The gas flows through perforated probes placed in the soil around the intended position of the chamber. The holes containing the probes are slightly larger than the probes themselves and soil is packed over the tubes at the soil surface. Acetylene will diffuse into the volume of soil over which the cover is later placed. Ryden and Dawson (1982) and McConnaughey and Duxbury (1986) later modified the method devised by Ryden et al. (1979) so that acetylene could diffuse into the soil without any positive pressure and mass flow being exerted.

Advantages. In situ measurements of the denitrification process in soil have the following advantages: (1) minimal physical disturbance of the soil, (2) minimal physical disturbance of plants in cropped ecosystems, which allows the natural daily variation in plant activity to affect the denitrification activity, (3) minimal disturbance of the temperature profile in the soil and the daily variation of the soil temperature during short-term incubations.

Disadvantages. The diffusion of acetylene into and N_2O out of wet soils poses a serious problem when the in situ method is used, particularly in clay soils. Acetylene has to be transported rapidly into the soil over which the cover is placed to ensure a short incubation time and to minimize the effects of acetylene on other soil processes. At the same time, the application of acetylene has to be done without creating a mass flow into the soil atmosphere. If a mass flow occurs, it will alter the oxygen concentration of the soil which in turn will alter the denitrification activity. The system for the introduction of acetylene into the soil must not allow O_2 diffusion into the soil. The N_2O solved in the soil water must be determined since the gas is highly soluble in water. This can be done if the gas is assumed to be at equilibrium with the soil-water phase (Tiedje, 1982). Since this is unlikely, underestimations may be made. Jury et al. (1982) concluded that the possible lag in gas diffusion of acetylene and N_2O may result in N_2O not being a reliable measure of denitrification rates determined with in situ chamber techniques. N_2O solved in the soil water may be lost due to transport downwards in the soil profile or to transport through cracks, macropores or root channels to the soil surface outside. When acetylene is used nitrification will also be inhibited. This can be a problem in grassland ecosystems, where nitrogen is initially present as organic N and ammonium N, and where the denitrification activity is dependent on the nitrification activity (Ryden et al.,1987). Finally, denitrification can be increased and the inhibition of the N_2O reductase lost, as discussed earlier, due to an increased population of acetylene decomposers.

Soil core methods

Intact soil cores can be incubated statically in containers with acetylene (Aulakh et al., 1982; Burton and Beauchamp, 1984; Svensson et al., 1985) or with soil air and acetylene recirculating in the macropores of the core (Parkin et al., 1984).

Advantages. The advantages of the core methods compared with the in situ methods are, according to Ryden et al. (1987): (1) a less complex series of procedures in the field and laboratory, (2) an integration of rates of denitrification during a 24-hour period, (3) denitrification can be studied in systems where farm operations would normally restrict the installation of more permanent equipment, (4) fewer problems associated with the inhibition of nitrification by acetylene, (5) the possibility of using the soil cores for additional analyses of mineral N and water contents. Finally, the problem caused by acetylene-decomposing microorganisms may be minimized during short-term incubations.

Disadvantages. When the core techniques are used, the denitrification activity is determined only in limited parts of the soil profile. It is therefore important to know where the denitrification activity is located in order to estimate the overall denitrification activity correctly. When soil cores are sampled from planted soil, roots will be damaged. This may result in increased availability of carbon for the denitrifiers. The daily variation in root activity, for example root exudation and water uptake, is lost with the core technique. Plants are truncated before the soil cores are taken, and this may affect the denitrification rates. Scaglia et al. (1985) showed that the denitrification activity was lower in soil with truncated maize plants than from soil with intact plants. The use of soil core systems may cause compaction of the soil, something that has been shown to increase denitrification rates (Mosier & Heinemeyer, 1985), most likely because of decreased oxygen diffusion rates. Compaction may also result in lower N_2O fluxes due to lowered gas diffusion rates (Letey et al., 1980). If soil cores are incubated in jars with air and acetylene, the soil in the cores might be exposed to a higher oxygen concentration than that of undisturbed soil. The effect of O_2 will depend on the soil texture and the O_2 concentration in the soil at the time of sampling. The diffusion rates of acetylene and N_2O will also be affected when soil cores are incubated in sealed jars. When soil is near 100 % of water-filled porosity, both the diffusion of acetylene into the soil and the diffusion of N_2O from the soil will be limited in a static core system, i.e. the denitrification rates will be underestimated, particularly in clay soils.

The incubation technique, with soil cores in a recirculating gas phase, eliminates the problem of acetylene and N_2O diffusion into and out of the soil, since pumps are used to recirculate acetylene and soil air through the macropores of the soil (Parkin et al., 1984). The oxygen content of the recirculated air is adjusted to match the oxygen content of the soil macropores. The anaerobic microsites in the soil are assumed to be preserved, and the acetylene distribution and N_2O recovery are rapid enough to allow determinations of short-term denitrification activities within 1-2 hours (Parkin et al., 1984). The forced flow of gas through the cores may, however, disturb the oxygen status of the microsites, since it is not likely that the oxygen concentration in the macropores is uniform in wet soils. Furthermore, errors in the measurements of the oxygen content of soil macropores and its natural variation at the sites where the cores are sampled may give invalid results. The method is not applicable in heavy clay soils at high water contents in which a flow of gas cannot be obtained (Parkin et al., 1984).

COMPARISONS OF TECHNIQUES

Burton and Beauchamp (1984) compared two extracted core techniques (encased cores and loose cores) with two in situ cylinder techniques (single wall or double wall cylinders), which differ less from field chamber techniques. The difference between encased and loose cores is that the encased ones are incubated in the jars within the plastic or steel tubes with which the cores are extracted. The tubes may be perforated to facilitate gas diffusion. The core techniques gave lower denitrification rates than the in situ techniques and the loose core technique gave the lowest values. Burton and Beauchamp (1984) suggest that the exposure of the soil cores to O_2 may explain the lower rates. A double-wall cylinder where acetylene could diffuse into the soil from the outer space of the cylinder gave the highest denitrification values and the least variation. The acetylene introduction and the N_2O accumulation were separated in the double-wall cylinders, which gave a limited change of the N_2O

concentration gradient. Due to the high variability of denitrification rates, significant differences among means for all techniques could only be shown at the 0.75 probability level. Ryden et al. (1987) compared an in situ incubation of soil using enclosures (0.5 x 0.15 m) placed over the soil surface and a field incubation of loose soil cores in jars. The agreement between the two methods was close for well drained soils. In wet soils with a heavy texture, on the other hand, denitrification was underestimated with the enclosure technique because acetylene did not diffuse rapidly enough throughout the soil to inhibit N_2O reduction. This will complicate denitrification measurements in clay soil with a high water content. On the other hand, small soil cores may be compacted when extracted, which would give an increase in denitrification activity. Ryden et al., (1987) concluded that the core technique gave the most realistic results in poorly-drained soils, but no comparisons have been made with rates determined with ^{15}N.

The in situ acetylene inhibition method with covers have been evaluated with ^{15}N by Rolston et al. (1982). The two methods had a correlation factor of 0.6 for samples taken at the same time and the total denitrification losses over a period of 40 days were about equal. Mosier et al. (1986b) directly compared the acetylene method and the ^{15}N method for quantifying N-gas evolution in small field plots. Gas fluxes were measured 2 to 4 times each day after an irrigation until flux rates declined to nearly ambient. The total N-gas flux measured from the acetylene-treated plots averaged 8950 g N ha^{-1} during the week following a 5 cm irrigation but only 6440 g N ha^{-1} in the plots not exposed to acetylene. The two methods of analysis gave virtually identical results when gases evolved from the acetylene plots were analyzed by both methods. These data suggest that the two methods provide the same analytical results when gases from the same plots are analyzed, but that N-gas flux rates were higher from the acetylene plots compared with plots that had not been treated with acetylene.

Denitrification rates measured with the soil core method were compared with rates in microplots determined by mass balance ^{15}N analysis by Parkin et al. (1985). The soil core method gave 1.5-3 times lower denitrification mean values than the ^{15}N method, but the higher water content of the microplots might have stimulated denitrification. The differences were not significant, however, due to the large variation in denitrification rates for both methods. Haider et al. (1983b) evaluated the effect of acetylene on denitrification in soil incubated in the laboratory using ^{15}N analysis (Fig. 4), but to our knowledge no similar studies of soil core techniques have been made.

The PPM method is not suitable for in situ techniques (soil covers) since diffusion aspects have to be considered to a greater extent than with the acetylene inhibition method. If the PPM method is used in soil cores, incubation in a gas flow system is to prefer, since this makes possible the maintainance of low concentrations of acetylene throughout the soil cores.

SUMMARY

Like all methods involving the use of compounds which inhibit soil processes, the acetylene inhibition method and the PPM method have several drawbacks. Diffusion constraints for acetylene and N_2O in soil have caused major problems when the methods have been applied in field experiments. There will always be a time-lag before acetylene has diffused to all active sites of nitrification/denitrification. Since denitrification occurs when the water content of soil is fairly high the diffusion constraints are presumably more severe in soils with high denitrification activity. No particular application can be recommended in general. The choice of method must depend on the location of the research site, the soil management and the soil type. No good application of the acetylene inhibition method has yet been published for poorly drained or heavy clay soils.

The acetylene inhibition methods have not been proved to be perfectly reliable for measuring field losses of N and the large spatial variability of denitrification rates makes it difficult to detect significant differences between ^{15}N and acetylene techniques but, as Tiedje et al. (1989) point out: "Since both methods are sound, it is more important to move forward and investigate underlying principles of the denitrification process than to dwell on denitrification methodology". Provided the limitations of the methods based on acetylene inhibition are given due consideration, these methods are powerful tools for quantifying the production of N_2 and N_2O from various microbial processes.

REFERENCES

Aulakh, M. S., Rennie, D. A. and Paul, E. A., 1982, Gaseous nitrogen losses from cropped and summer fallowed soils. Can. J. Sci., 62:187.

Balderston, W. L., Sherr, B. and Payne, W. J., 1976, Blockage by acetylene of nitrous oxide reduction in Pseudomonas perfectomarinus. Appl. Environ. Microbiol., 31:504.

Berg, P., Klemedtsson, L. and Rosswall, T., 1982, Inhibitory effect of low partial pressures of acetylene on nitrification. Soil Biol. Biochem., 14:301.

Bleakley, B. H. and Tiedje, J. M., 1982, Nitrous oxide production by organisms other than nitrifiers or denitrifiers. Appl. Environ. Microbiol., 44:1342.

Bremner, J. M. and Blackmer, A. M., 1978, Nitrous oxide: Emission from soils during nitrification of fertilizer nitrogen. Science, 199:295.

Burton, D. L. and Beauchamp, E. C., 1984, Field techniques using the acetylene blockage of nitrous oxide reduction to measure denitrification. Can. J. Soil Sci., 64:555.

Cho, C. M., 1982, Oxygen comsumption and denitrification kinetics in soil. Soil Sci. Soc. Am. J., 46:756.

Colbourn, P., Iqbal, M. M. and Harper, I. W., 1984, Estimation of the total gaseous nitrogen losses from clay soils under laboratory and field conditions. J. Soil Sci., 35:11.

Crutzen, P. J., 1983, Atmospheric interactions - homogenous gas reactions of C-, N- and S-containing compounds. in: The major biogeochemical cycles and their interactions, Bolin, B. and Cook, R. B. eds., SCOPE Report 21. Wiley & Sons, Chichester.

Culbertson, C. W., Zehnder, A. J. B. and Ormeland, R.S., 1981, Anaerobic oxidation of acetylene by estuarine sediments and enrichment cultures. Appl. Environ. Microbiol., 41:396.

Davidson, E. A., Swank, W. T. and Perry, T.O., 1986, Distinguishing between nitrification and denitrification as sources of gaseous nitrogen production in soil. Appl. Environ. Microbiol., 62:1280.

Denmead, O. T., 1979, Chamber systems for measuring nitrous oxide emissions from soils in the field. Soil Sci. Soc. Am. J., 43:89.

Fedorova, R. I., Melekhina, E. I. and Ilyuchina, N. I., 1973, Evaluation of the method of "gas metabolism" for detecting extra terrestrial life. Identification of nitrogen-fixing organisms. Izvestia Akademii Nauk. SSSER. Ser. Biol., 6:797

Firestone, M. K. and Davidson, E. A., 1989, Microbial basis of NO and N_2O production and consumption. in: Exchange of Trace Gases between Terrestrial Ecosystems and the Atmosphere, Andreae, M. O. and Schimel, D. S. eds., Dahlem Conference, Chichester, John Wiley & Sons.

Firestone, M. K., Smith, M. S., Firestone, R. B. and Tiedje, J. M., 1979, The influence of nitrate, nitrite and oxygen on the composition of gaseous products of denitrification in soil. Soil Sci. Soc. Am. J., 43:1140.

Goring, C. A. I., 1962, Control of nitrification by 2-chloro-6-(trichloro-metyl)pyridine. Soil Sci., 93:211.

Haider, K., Mosier, A. R. and Heinemeyer, O., 1983a, Side effects of acetylene on the conversion of nitrate in soil. Z. Pflanzenerhaehr Bodenk., 146:623.

Haider, K., Mosier, A. R. and Heinemeyer, O., 1983b, Side effecs of acetylene on the conversion of nitrate in soil. Agronomy Abstracts., 75:159.

Herlihy, M. and Quirke, W., 1975, The persistence of 2-chloro-6(trichloromethyl)pyridine in soil. Comm. Soil Sci. Pl. Anal., 6:513.

Hynes, R. K. and Knowles, R., 1978, Inhibition by acetylene of ammonia oxidation in Nitrosomonas europaea. FEMS Microbiol. Lett., 4:319.

Hynes, R. K. and Knowles, R., 1982, Effect of acetylene on autotrophic and heterotrophic nitrification. Can J. Microbiol., 28:334.

Jury, W. A., Letey, J. and Colleus, T., 1982, Analysis of chamber methods used for nitrous oxide production in the field. Soil Sci. Soc. Am. J., 46:250.

Kanner, D. and Bartha, R., 1979, Growth of Nocardia rhodochromum on acetylene gas. J. Bacteriol., 139:225.

Kaspar, H. and Tiedje, J. M., 1980, Responses of electron-capture detectors to hydrogen, oxygen, nitrogen, carbon dioxide, nitric oxide and nitrous oxide. J. Chrom., 193:142.

Klemedtsson, L., Svensson, B. H., Lindberrg, T. and Rosswall, T., 1977, The use of acetylene inhibition of nitrous oxide reductase in quantifying denitrification in soils. Swedish J. Agric. Res., 7:179.

Klemedtsson, L., Svensson, B. H. and Rosswall, T., 1987, Dinitrogen and nitrous oxide produced by denitrification and nitrification in soil with and without barley plants. Plant Soil., 99:303.

Klemedtsson, L., Simkins, S. and Svensson, B. H., 1986, Tandem thermo-conductivity and electron-capture detectors and non-linear calibration curves in quantitative nitrous oxide analysis. J. Chrom., 361:107.

Klemedtsson, L., Svensson, B. H. and Rosswall, T., 1988a, Relationships between soil moisture content and nitrous oxide production during nitrification and denitrification. Biol. Fertil. Soils, 6:106.

Klemedtsson, L., Svensson, B. H. and Rosswall, T., 1988b, A method of selective inhibition to distinguish between nitrification and denitrification as sources of nitrous oxide in soil. Biol. Fert. Soils, 6:112.

Knowles, R., 1984, Some effects on inhibitors on nitrogen transformations. in: Proc. Int. Symp. Nitrogen in the Environment, Malik, K. A., Naqvi, S. H. M. and Aleem, M. I. H., NIAB, Faisalabad.

Lacis, A., Hanson, G., Lee, P., Mitchell, T. and Lebedoff, S., 1981, Greenhouse effects of trace gases, 1970-1980. Geophys. Res. Lett., 8:1035.

Letey, J., Jury, W. A., Hadas, A. and Valois, N., 1980, Gas diffusion as a factor in laboratory incubation studies on denitrification. J. Environ. Qual., 9:223.

Martikainen, P.J., 1985, Nitrous oxide emission associated with autotrophic ammonium oxidation in acid coniferous forest soil. Appl. Environ. Microbiol., 50:1519.

McConnaughey, P. K. and Duxbury, T.R., 1986, Introduction of acetylene into soil for measurement of denitrification. Soil Sci. Soc. Am. J., 50:210.

Mosier, A. R. and Mack, L., 1980, Gas chromatography system for precise, rapid analysis of N_2O. Soil Sci. Soc. Am. J., 44:1121.

Mosier, A. R., 1980, Acetylene inhibition of ammonium oxidation in soil. Soil Biol. Biochem., 12:443.
Mosier, A. R., Guenzi, W. D. and Schweizer, E. E., 1986b, Field denitrification estimation by nitrogen-15 and acetylene inhibition techniques. Soil Sci. Soc. Am. J., 50:831.
Mosier, A. R. and Heinemeyer, O., 1985, Current methods used to estimate N_2O and N_2 emissions from field soils. in: Denitrification in the nitrogen cycle. Golterman, E. L. ed., NATO Conference Series I: Ecology, Vol 9, Plenum Press, New York.
Notton, B. A., Walson, E. F. and Hewitt, E. J., 1979, Effects of N-serve (2-chloro-6-(trichloromethyl)pyridine) formulations on nitrification and loss of nitrate in a sand culture experiment. Plant Soil, 51:1.
Oremland, R. S., Umberger, C., Culbertson, C. W., Smith, R. L., 1984. Denitrification in San Fransisco Bay intertidal sediments. Appl. Environ. Microbiol., 47: 1106-1112.
Parkin, T. B., Kaspar, H. F., Sextone, A. J. and Tiedje, J. M., 1984, A gas-flow soil core method to measure field denitrification rates. Soil Biol. Biochem., 16:323.
Parkin, T. B., Sexton, A. J. and Tiedje, J. M., 1985, Comparison of field denitrification raes determined by acetylene-based soil core and nitrogen-15 methods. Soil. Sci. Soc. Am. J., 49:94.
Poth, M. and Focht, D. D., 1985, ^{15}N kinetic analysis of N_2O production by Nitrosomonas europaea: an examination of nitrifier denitrification. Appl. Environ. Microbiol., 49:1134.
Rice, C. W. and Smith, M. S., 1982, Denitrification in no-till and ploughed soils. Soil Sci. Soc. Am. J., 46:1168.
Robertson, G. P. and Tiedje, J. M., 1987, Nitrous oxide sources in aerobic soil: nitrification, denitrification and other biological processes. Soil Biol. Biochem., 19:187.
Rolston, D. E., Sharpley, A. N., Toy, D. W. and Broadbent, F. E., 1982, Field measurement of denitrification: II. Rates during irrigation cyles. Soil Sci. Soc. Am. J.,46:289.
Ryden, J. C. and Dawson, K.P., 1982, Evaluation of the acetylene inhibition technique for field measurements of denitrification in grassland soils. J. Sci Food Agric., 1197.
Ryden, J. C., Lund, L. J., Letey, J. and Focht, D. D., 1979, Direct measurement of denitrification loss from soils: II. Development and application of field methods. Soil Sci. Soc. Am. J., 43:110.
Ryden, J. C., Skinner, J. H. and Nixon, D. J., 1987, Soil core incubation systems for the field measurement of denitrification using acetylene inhibition. Soil Biol. Biochem., 19:753.
Scaglia, J., Lensi, R. ans Chalamet, A., 1985, Relationships between photosyntesis and denitrification in planted soil. Plant Soil, 84:37.
Schmidt, E. L., 1982, Nitrification in soil. in: Nitrogen in Agricultural Soils, Stevenson, F. J. ed., Am. Soc. Agro. Inc, Soil Sci. Soc. Am. Inc, Madison.
Smith, M. S., Firestone, M. K. and Tieje, J. M., 1978 The acetylene inhibition method for short-term measrurement of soil denitrification and its evaluation using nitrogen-13. Soil Sci. Soc. Am. J., 42:611.
Smith, M. S. and Tiedje, J. M., 1979, Phases of denitrification followed by oxygen depletion in soil. Soil Biol. Biochem., 11:261.
Smith, M. S. and Zimmerman, K., 1981, Nitrous oxide production by non-denitrifying soil nitrate reducers. Soil Sci Soc. Am. J., 43:865.
Svensson, B. H., Klemedtsson, L. and Rosswall, T., 1985, Preliminary field denitrification studies of nitrate-fertilized and nitrogen-fixing crops. in:

Denitrification in the Nitrogen Cycle. Golterman, E. L. ed., NATO Conference Series I: Ecology, Vol 9, Plenum Press, New York.

Svensson, B. H., Boström, U. and Klemedtsson, L., 1986, Potential for higher rates of denitrification in earthworm casts than in the surrounding soil. Biol. Fertil. Soil, 2:147.

Terry, R. E. and Duxbury, J. M., 1985, Acetylene decomposition in soils. Soil Sci. Soc. Am. J., 49:90.

Tiedje, J. M., 1982, Denitrification. in: Methods of Soil Analysis, part 2: Chemical and Microbial Properties (2nd ed) Page, A. L., Miller, R. H. and Keeney, D. R. eds., Am. Soc. Agro. Inc, Soil Sci. Soc. Am. Inc, Madison.

Tiedje, J. M., 1988, Ecology of denitrification and dissimilatory nitrate reduction to ammonium. in: Biology of Anaerobic Microorganisms, Zehnder, J. B. ed., Wiley, New York.

Tiedje, J. M., Simkins, S., Groffman, P. M., 1989. Perspectives on measurement of denitrification in the field including recommended protocols for acetylene based methods. in: Ecology of Arable Land, Perspectives and Challenges. Clarholm, M., Bergström, L. eds., Kluwer Academic Publishers, Dordrecht, Holland, pp. 217-240.

Topp, E and Germon, J. C., 1986, Acetylene metabolism and stimulation of denitrification in an agricultural soil. Appl. Environ. Microbiol., 52:802.

Walter, H. M., Keeney, D. R. and Fillery, I. R., 1979, Inhibition of nitrification by acetylene. Soil Sci. Soc. Am. J., 43:195.

Yoshinari, T. and Knowles, R., 1976, Acetylene inhibiton of nitrous oxide reduction by denitrifying bacteria. Biochem. Biophys. Res. Comm., 69:705.

Yoshinari, T., Hynes, R. and Knowles, R., 1977, Acetylene inhibiton of nitrous oxide reducation and measurement of denitrification and nitrogen fixation in soil. Soil Biol. Biochem., 9:177.

MEASURING DENITRIFICATION IN SOILS USING ^{15}N TECHNIQUES

David D. Myrold

Department of Soil Science
Oregon State University
Strand Agriculture Hall 202
Corvallis, OR 97731-2213
U.S.A.

INTRODUCTION

Soil scientists who were involved in some of the pioneering studies of denitrification in soils used ^{15}N methodology (Nômmik, 1956; Wijler and Delwiche, 1954). At that time, thirty years ago, ^{15}N or the use of artificial atmospheres were the only tools available to scientists interested in understanding the workings of denitrification in soil systems. Because of the expense of ^{15}N, limited access to mass spectrometer facilities, and the analytical difficulties of working with ^{15}N, few scientists were able to study denitrification. Despite these limitations, however, many fundamental insights were gained into the denitrification process and useful techniques were developed that were adapted several years later (Hauck and Bouldin, 1961; Hauck et al., 1958). The pace of denitrification research quickened greatly following the discovery in the mid-1970's that C_2H_2 blocks the reduction of N_2O to N_2 (Balderston et al., 1976; Yoshinari and Knowles, 1976). The C_2H_2 inhibition method made the study of denitrification accessible to many more scientists, as shown by the explosive increase in the number of articles about denitrification published since the introduction of the C_2H_2 blockage technique. This new technique also stimulated comparisons of the efficacy of ^{15}N versus C_2H_2 inhibition methods (Focht, 1985; Hauck, 1986; Keeney, 1986; Rolston, 1986).

The purpose of this paper is to describe and evaluate the various ways that ^{15}N can be used in studies of denitrification in soil. Denitrification studies using ^{15}N can be grouped into five categories: (1) variations in ^{15}N natural abundance, (2) mass balances using ^{15}N-labelled fertilizers, (3) use of $^{15}NO_3^-$ in isotope dilution and modeling of N cycle processes, (4) additions of $^{15}NO_3^-$ and measurement of ^{15}N-gas emissions, and (5) $^{15}N_2$ isotope dilution studies. In the course of this discussion, comparisons of ^{15}N methodology with the C_2H_2 inhibition technique will be made.

VARIATIONS IN ^{15}N NATURAL ABUNDANCE

There has been a recent resurgence of interest in the use of variations in

Denitrification in Soil and Sediment, Edited by N. P. Revsbech and J. Sørensen, Plenum Press, New York, 1990

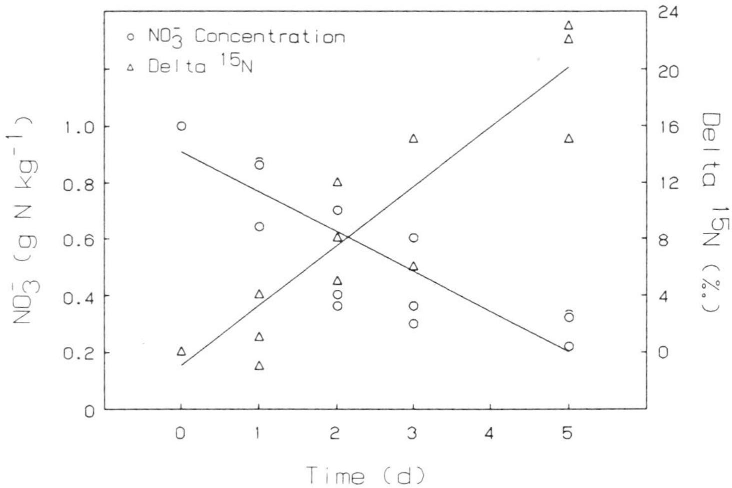

Fig. 1. Isotope effect of denitrification during an anaerobic incubation of soil slurries amended initially with 1 g N kg^{-1} as KNO_3 at natural abundance. Data from Blackmer and Bremner (1977).

the natural abundance of stable isotopes for studying ecosystem processes (Peterson and Fry, 1987; Rundel et al., 1989). The isotopic composition of soil N pools varies because N cycle processes discriminate slightly between ^{14}N and ^{15}N. Generally speaking, the lighter isotope (^{14}N) reacts faster than the heavier isotope (^{15}N), leaving more of the heavier isotope behind. In denitrification, the NO_3^- pool becomes somewhat enriched with $^{15}NO_3^-$. This is known as the isotope effect (Hauck and Bremner, 1976). The amount of discrimination is small and is usually of no consequence in ^{15}N tracer studies, however, the small variations in natural abundance of N pools in soil may provide clues about the magnitudes of soil N transformations.

Variations in ^{15}N natural abundance are usually expressed in terms of $\delta^{15}N$, which is defined as:

$$\delta^{15}N = 1000 * \frac{(\text{atom \% } ^{15}N_{\text{sample}} - \text{atom \% } ^{15}N_{\text{standard}})}{\text{atom \% } ^{15}N_{\text{standard}}}$$

Usually 0.3663 atom % ^{15}N, the natural abundance of ^{15}N in the atmosphere, is used as the standard.

Laboratory studies using soils incubated under anaerobic conditions have found $\delta^{15}N$ values for NO_2^- of 11 to 33‰ (Mariotti et al., 1982) and for NO_3^- of 6.5 to 23‰ (Blackmer and Bremner, 1977; Chien et al., 1977). An example of these data are shown in Fig. 1, which shows the increase in $\delta^{15}N$ of the remaining NO_3^- as NO_3^- is denitrified. Isotope fractionation values increase with decreasing incubation temperature, i.e., lower denitrification rates (Mariotti et al., 1982), and decrease with the addition of C, i.e., higher denitrification rates (Chien et al., 1977; Mariotti et al., 1982). Such fluctuations in the isotope effect for denitrification may be useful in controlled laboratory studies, but are likely to confuse the interpretation of field results. A further complicating factor is that

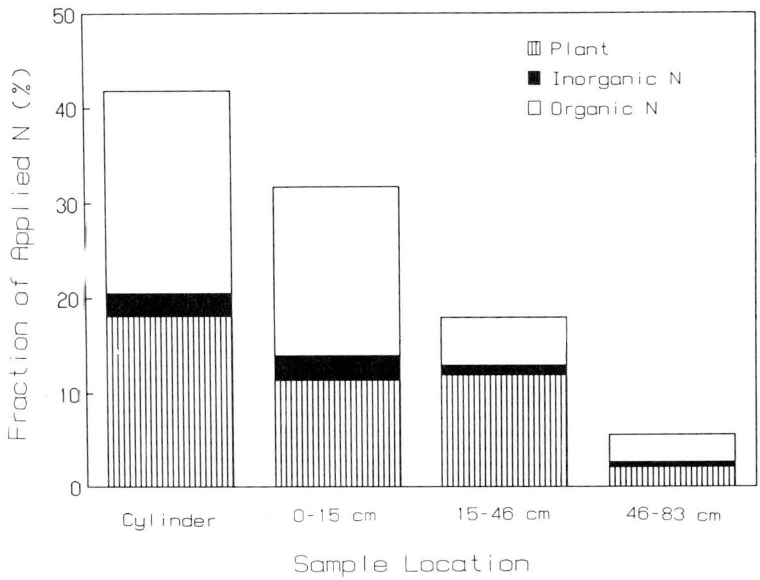

Fig. 2. Lateral movement of ^{15}N from 15-cm deep by 15.2-cm diameter microplots of an irrigated clay loam soil cropped to barley in Colorado. Microplots were fertilized with the equivalent of 200 kg N ha^{-1} of 99 atom % ^{15}N as $(NH_4)_2SO_4$. Soils were sampled to a depth of 90 cm. Data from Mosier et al. (1986a;b).

the NO_3^- pool participates in other N cycle reactions, and isotope discrimination occurs in these reactions, too. For example, nitrification produces NO_3^- slightly depleted in ^{15}N, which directly counters the ^{15}N enrichment of the NO_3^- pool produced by denitrification. Despite the greater utilization of stable isotope techniques for examining variations in natural abundance, the complexities of the soil N cycle and the many factors that influence denitrification rates are likely to prohibit the application of ^{15}N natural abundance techniques for soil denitrification studies.

^{15}N-LABELED FERTILIZERS IN MASS BALANCE STUDIES

Losses of N by denitrification were first estimated by using a budgeting approach and mass balance studies remain the standard against which other denitrification methods are measured. Although N budgets can be constructed without the use of ^{15}N (Legg and Meisinger, 1982), greater sensitivity is obtained by applying ^{15}N. When ^{15}N is used, it is possible to differentiate between the fate and cycling of added N compared to that of native soil N (Barraclough et al., 1984; Meyer et al., 1989). For example, in studies with plants, plant N uptake and the proportion of fertilizer N taken up by the plant can be calculated.

The basic approach of mass balance studies is to add ^{15}N to the soil and subsequently sample the several soil and plant N pools after the desired time interval has elapsed. In the case of denitrification, inorganic ^{15}N fertilizer is often added, all other inputs and outputs of N and changes in soil and plant N

Table 1. Example of the use of ^{15}N mass balance data for measuring denitrification losses from a cropped system. Data are the mean ± standard deviation for three replicate lysimeters, each of which was fertilized with 120 kg N ha^{-1} as $CaNO_3$. Data from Dowdell and Webster (1984) and Dowdell et al. (1984).

Nitrogen pool	Recovery as % of applied ^{15}N
Grain and straw	52.0 ± 4.5
Roots and soil	24.6 ± 1.3
Drainage water	6.5 ± 0.2
Unaccounted for (Denitrification)	16.7 ± 3.4

are measured, and denitrification is determined by difference. The difficulty comes with measuring all the appropriate N pools with the necessary degree of sensitivity. For example, Hauck (1986) points out that an analytical error of 2% in total Kjeldahl N analysis for a soil containing 0.2% total N is equivalent to an error of 90 kg N ha^{-1}.

In the laboratory, some of the problems related to sensitivity and sampling errors are largely overcome by mixing the isotope homogeneously throughout the soil and incubating as a closed system where inputs and outputs are regulated or measured (Aulakh and Rennie, 1987; Heinemeyer et al., 1988; Meyer et al., 1989). Field ^{15}N balance studies have attempted to reduce variability by similar means: defining the boundaries of the system and measuring outputs from the system. Of course, as the design of the microplots becomes more restrictive, the microplots become poorer models of the real field situation.

Unrestricted microplots are the most natural experimental system, but the extent of lateral movement of ^{15}N is undefined. Sampling soil and plant material outside the microplot borders can provide some information about the importance of lateral movement of the tracer but including this fate further increases variability (Jokela and Randall, 1987; Recous et al., 1988a). For example, Mosier et al. (1986a;b) found that more than half of the $(^{15}NH_4)_2SO_4$ added to irrigated microplots was recovered outside the microplots (Fig. 2). Although unconfined (or very shallow confined microplots) are undoubtedly less accurate than confined microplots, unconfined microplots would be useful in studies with trees, because rooting patterns would not be disturbed.

Confined microplots limit lateral movement of ^{15}N making sampling easier. Restriction of lateral flow may, however, alter soil moisture relations and thereby confound interpretation of denitrification rates estimated in this way (Parkin et al., 1985). Neither confined nor unconfined microplots are equipped to directly measure leaching or gaseous losses of N. By sampling the distribution of ^{15}N with depth, it is possible to indirectly determine if ^{15}N has been lost by leaching. For example, Parkin et al. (1985) found less than 0.1% as much $^{15}NO_3^-$ at the

Table 2. Accumulation of error in estimating denitrification losses by difference using the ^{15}N mass balance method for a loam soil in California. Data are mean ± 95% confidence intervals from duplicate plots of each treatment. Plots were maintained at a soil water potential of -1 kPa. Data from Rolston et al. (1979).

Nitrogen pool	Manure	Cropped	Uncropped
	------------- kg N ha^{-1} -------------		
Fertilizer N added	301 ± 1	296 ± 1	305 ± 1
Plant uptake	---	34 ± 1	---
Inorganic soil N	1 ± 0	0	1 ± 0
Organic soil N	25 ± 8	52 ± 17	27 ± 9
Leaching loss	37 ± 48	194 ± 203	267 ± 203
Unaccounted for	238 ± 49	16 ± 206	10 ± 203

30-40 cm depth compared to the 0-10 cm depth after one growing season, suggesting that NO_3^- leaching was not an important loss of N in their study. When considerable $^{15}NO_3^-$ is found at the lowest sampling depth, it is impossible to accurately estimate the relative importance of leaching and denitrification losses (Fredrickson et al., 1982).

More elaborate field designs, such as lysimeters (Dowdell and Webster, 1984; Dowdell et al., 1984), which allow for direct measurement of leaching losses, have also been used. Dowdell and coworkers (Dowdell and Webster, 1984; Dowdell et al., 1984) used replicated lysimeters of a silt loam soil cropped to barley in England to measure the fate of ^{15}N-labeled NO_3^- over a four-year period. Representative data for the first year of the study are summarized in Table 1. The results are fairly typical in that the crop took up about half of the applied N, about one quarter of the ^{15}N remained in the soil, and the remaining quarter was lost by leaching and denitrification. The denitrification loss represents 20 kg N ha^{-1} yr^{-1}. Although data on the variability of the measured N pools was not reported on a per lysimeter basis, having triplicate lysimeters provided an indication of the overall variability of denitrification in this experiment.

Calculation of denitrification by difference using an ^{15}N budgeting approach leads to the accumulation of additive errors. The potential impact of additive errors on the quality of the denitrification measurement is shown in Table 2, which uses data of Rolston et al. (1979). The large confidence intervals for these data are primarily the result of the large uncertainty in leaching losses. Nevertheless, the denitrification estimates obtained from the mass balance approach were in reasonable agreement with direct measurements of evolved N_2O and $^{15}N_2$, which were 218±91, 40±12, and 10±4 for the manure, cropped, and uncropped treatments, respectively (Rolston et al., 1979). According to Legg and Meisinger (1982), the standard deviation of the denitrification estimates in Table 2 may be overestimated because covariance between measured N pools was not considered in calculating the variance of the unaccounted for N.

Although the mass balance method is best suited for determining the amount of added N that is denitrified, estimates of total denitrification can be made if the average atom % ^{15}N of the NO_3^- pool that has undergone denitrification over the sampling interval is known (Meyer et al., 1989). Total denitrification is equal to the amount of ^{15}N lost, as calculated from the mass balance approach, divided by the average atom % ^{15}N of the NO_3^- pool. This calculation assumes that the rate of denitrification is constant over the sampling interval, that there is one homogeneously labeled NO_3^- pool, and that the atom % ^{15}N of the NO_3^- pool is fairly constant. In reality, none of these assumptions is likely to be true, but errors can be minimized by thoroughly mixing the isotope into the soil and sampling at more frequent time intervals.

One of the major advantages of the ^{15}N mass balance approach for measuring denitrification rates is that it integrates biological activity over time and space. No other methods of quantifying denitrification in the field can easily be adapted to large plot sizes or low sampling frequency. Because the ^{15}N budgeting approach requires destructive sampling of soils and plants, it is poorly adapted to resolving the temporal dynamics and spatial variability of denitrification. It is also an expensive technology because of the larger plot size required and the large number of N pools that must be measured.

$^{15}NO_3^-$ ISOTOPE DILUTION AND N CYCLE PROCESS MODELS

Studies by Koike and coworkers (Koike and Hattori, 1978; Nishio et al., 1983) in coastal sediment systems and flooded paddy soils (Hasebe et al., 1987) demonstrated that nitrification and denitrification rates could be determined by using $^{15}NO_3^-$ as a tracer and measuring the temporal changes in the atom % ^{15}N and concentration of the NO_3^- pool. This technique is known as isotope dilution or pool dilution, because the nitrification rate is determined from the dilution of the ^{15}N tracer enriched product pool with NO_3^- at natural abundance. To date, applications of the principle of isotope dilution to soils have concentrated on N mineralization-immobilization dynamics and nitrification (Barraclough and Smith, 1987; Bjarnason, 1988; Nishio et al., 1985), although Myrold and Tiedje (1986) have shown that denitrification can be measured in soils using this method. The principle of this technique is shown by the compartmental model in Fig. 3 and the following equations:

$$\frac{dC_n}{dt} = R_n - R_d \tag{1}$$

$$\frac{dA_n}{dt} = \frac{R_n(A_a - A_n)}{C_n} \tag{2}$$

where C_n is the concentration of the NO_3^- pool, A_a is the atom % ^{15}N of the NH_4^+ pool, A_n is the atom % ^{15}N of the NO_3^- pool, R_n is the nitrification rate, and R_d is the denitrification rate. This model assumes that there is only one homogeneously labeled NO_3^- pool, that there are no other sources (heterotrophic nitrification of organic-N) or sinks of NO_3^- (plant uptake, leaching, dissimilatory reduction to NH_4^+, or immobilization), that R_n and R_d are constant over the sampling interval, and that A_a remains constant (probably at natural abundance). Given these assumptions and approximating A_n and C_n by their mean values over the time of the measurement interval, the model can be solved algebraically

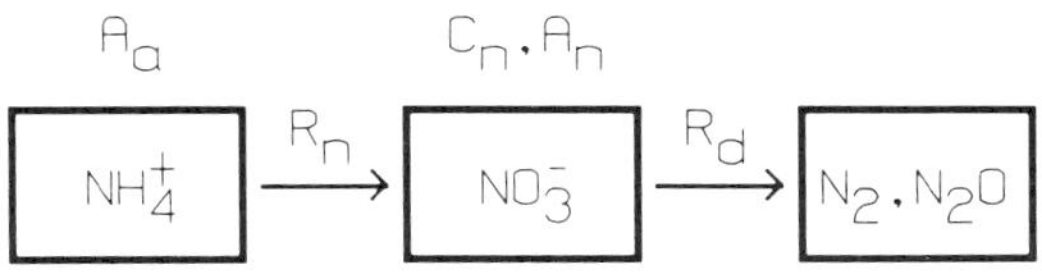

Fig. 3. Simplified compartmental model of soil denitrification, which is applicable for ^{15}N isotope dilution experiments. Variables are explained in the text.

because there are two equations and only two unknowns. An example of such a solution is given in Table 3.

When A_a, R_n, and R_d are constant, Equations [1] and [2] can be integrated analytically (Kirkham and Bartholomew, 1955). This exact solution gives slightly higher values of 7.66 and 3.66 for the nitrification and denitrification rates of the example given in Table 3, because A_n decreases curvilinearly with time, which makes the arithmetic mean of A_n too large. The assumption that process rates are constant implies zero-order kinetics. If this is not true, or if A_a is not constant, then more complete and complicated models are necessary and nonlinear parameter estimation methods must be used to determine N cycle rates (Myrold and Tiedje, 1986). For example, Myrold and Tiedje (1986) fit a first-order kinetic model with five parameters to ^{15}N dilution data obtained from laboratory incubation of a clay loam soil and estimated a rate constant for denitrification of 0.0132 day^{-1}. This rate constant translated to denitrification fluxes of 0.54 to 0.96 mg N kg^{-1} day^{-1}, which were about four times greater than field rates for the same soil measured in soil cores by the C_2H_2 inhibition method (Parkin et al., 1985).

The $^{15}NO_3^-$ dilution method has the advantages of being integrative in time and space and in estimating rates of other N cycle processes. It can be shown that the sensitivity for determining denitrification rates with this technique is

Table 3. Example calculation for determining simultaneous nitrification and denitrification rates by the isotope dilution method.

Given that: $A_a = 0.37$
At time 0, $C_{n0} = 8$ and $A_{n0} = 1.74$
At time 1, $C_{n1} = 12$ and $A_{n1} = 1.00$

Therefore: the mean C_n $(C_{nm}) = 10$ and
the mean A_n $(A_{nm}) = 1.37$

Rearranging Eqn. [2] gives:

$$R_n = C_{nm}(A_{n1} - A_{n0})/(A_a - A_{nm}) = 10(1.00 - 1.74)/(0.37 - 1.37) = 7.4$$

Rearranging Eqn. [1] gives:

$$R_d = R_n - (C_{n1} - C_{n0}) = 7.4 - (12 - 8) = 3.4$$

mainly a function of the sensitivity of measuring changes in soil NO_3^- concentrations (R_d only appears in Equation [1]). For example, if differences in soil NO_3^- concentrations of 0.1 mg N kg^{-1} are detectable, this translates to a change of about 250 g ha^{-1} assuming a bulk density of 1.25 Mg m^{-3}. Another serious disadvantage is that no other fates for NO_3^- are allowed by this procedure, which is quite unrealistic under field situations. To overcome this last drawback it would be necessary to include more compartments and fluxes in the model and require monitoring those compartments for changes in ^{15}N as done by Myrold and Tiedje (1986).

DIRECT MEASUREMENTS OF $^{15}N_2$ AND $^{15}N_2O$ EVOLUTION

Denitrification rates can be determined most directly by measuring the products of the reaction, N_2 and N_2O, in a closed system. It is quite easy to measure N_2O evolution rates, because the ambient concentration of N_2O in air is low (about 0.3 cm^3 m^{-3}) and N_2O can be measured with great sensitivity using simple gas chromatography. Quantitatively assaying increases in N_2, however, is difficult because the atmosphere is 78% N_2. The inherent insensitivity in measuring N_2 evolution can be overcome by measuring the evolution of $^{15}N_2$ following application of $^{15}NO_3^-$ to soil using any of several mass spectrometry techniques (Cho and Sakdinan, 1978; Focht et al., 1980; Siegel et al., 1982).

Four basic approaches have been used to measure direct ^{15}N-gas losses: (1) determination of N_2O by gas chromatography and $^{15}N_2$ by mass spectrometry after first removing interfering gases (Colbourn et al., 1984; Mosier et al., 1986a; b; Rolston et al., 1978, 1979, 1982); (2) determination of $^{15}N_2O$ and $^{15}N_2$ by gas chromatographic separation followed by mass spectrometry (Focht et al., 1980; Iimura and Ito, 1984; Ryden et al., 1979), (3) removal of contaminating gases, such as O_2 and CO_2, and cryogenic removal of $^{15}N_2O$, measurement of $^{15}N_2$, followed by subsequent reduction of $^{15}N_2O$ to $^{15}N_2$ for measurement (Mulvaney and Kurtz, 1982, 1984); and (4) conversion of $^{15}N_2O$ and $^{15}N_2$ to randomly labelled ^{15}N oxides, chemical conversion to $^{15}NH_4^+$ salts, which are oxidized to $^{15}N_2$ by the standard LiOBr method (Buresh and Austin, 1988; Craswell et al., 1985). The reader is urged to consult the cited references for complete details of these methodologies, but two points are important to make here. First, isotope ratio mass spectrometers give about 100-fold greater precision than gas chromatograph mass spectrometer systems and therefore greater sensitivity for measuring denitrification. Second, the procedure of Craswell et al. (1985) randomizes the isotopic composition of the N in the gas sample, which invalidates calculations based on the assumption that distribution of ^{14}N and ^{15}N in the gases evolved via denitrification are different than the isotopic distribution of atmospheric N_2 at natural abundance (Hauck and Bouldin, 1961; Hauck et al., 1958; Mulvaney, 1984; Mulvaney and Boast, 1986; Siegel et al., 1982).

Changes in the atom % ^{15}N of the N-gases in the enclosed atmosphere can be used to calculate a denitrification rate by isotope dilution, analogous to the determination of a nitrification rate (Table 3). The appropriate expression is given by Equation 3, where C_d and A_d are the concentration and atom % ^{15}N of the N-gas pool (N_2 plus N_2O). This calculation procedure can be used with data collected using any of the analytical procedures listed previously, but suffers from the disadvantage of requiring information about the atom % ^{15}N of the NO_3^- pool. If the isotopic labeling of the NO_3^- pool is measured or inferred, e.g., from the atom % ^{15}N of plants growing at the site (Focht and Stolzy, 1978), then the calculation of the denitrification rate is straightforward.

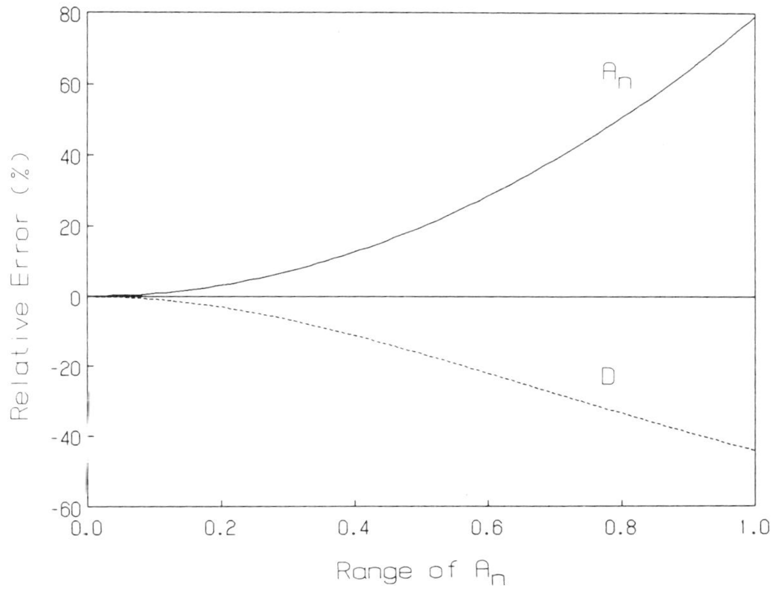

Fig. 4. Relative errors in A_n (isotopic composition of NO_3^-)and D (fraction of N-gas from denitrification) predicted for two pools of NO_3^- using different ranges of ^{15}N enrichment about a constant mean (A_n=0.5). For example, a range of 0.2 refers to two pools of NO_3^- at 0.4 and 0.6 mole fraction ^{15}N. Nitrogen from the NO_3^- pool was 0.01% of N initially present in the gas phase.

$$\frac{dA_d}{dt} = \frac{R_d(A_n - A_d)}{C_d} \tag{3}$$

Hauck et al. (1958) derived an alternative method of calculating denitrification rates by taking advantage of the fact that the isotopic distribution of the evolved denitrification gases is different than that present in the atmosphere. The usefulness of this procedure is that it allows for the calculation of both the isotopic composition of NO_3^- pool, A_n, assuming homogeneous mixing of labeled and native NO_3^-, and the fraction of the N-gas coming from denitrification, D. (Hauck and others (Hauck and Bouldin, 1961; Hauck et al., 1958; Mulvaney, 1984; 1988; Siegel et al., 1982) use the symbols $^{15}X_N$ and d, respectively.) This is made possible by measuring two different ion current ratios with a mass spectrometer and relating each of these ratios to A_n and D using the underlying assumption that $^{28}N_2^+$, $^{29}N_2^+$, and $^{30}N_2^+$ conform to a binomial distribution. If the ion current ratios of $^{29}N_2/^{28}N_2$, r, and $^{30}N_2/(^{28}N_2 + ^{29}N_2)$, r', are measured, the following equations result:

$$A_n = 2.015(r'/r)/(1 + (2.015(r'/r)) \tag{4}$$

$$D = r'/(A_n)^2 \tag{5}$$

The complete derivation of these equations, or more recent refinements that take advantage of newer mass spectroscopy methodology, have been published elsewhere (Hauck and Bouldin, 1961; Hauck et al., 1958; Mulvaney, 1984; Mulvaney and Boast, 1986; Siegel et al., 1982). The denitrification rate is obtained by multiplying D by the total amount of N gas in the sampling chamber.

Initial work using ^{15}N-gas evolution methods reported sensitivities of 0.1 to 1 kg N ha^{-1} day^{-1} (Focht and Stolzy, 1978; Rolston et al., 1978, 1982) but this has been improved by using highly enriched ^{15}N materials and newer mass spectrometers with greater precision. Siegel et al. (1982) reported a detection limit of 5 g N ha^{-1} day^{-1}. A second advantage is that these methods are nondestructive, although the addition of $^{15}NO_3^-$ could potentially affect the kinetics of denitrification by altering the concentration of the soil NO_3^- pool. Relatively good spatial and temporal resolution can be achieved, because of practical limitations on how large a soil cover can be and the relatively high sensitivity, which allows for short sampling periods.

The complex gas sampling and gas inlet systems associated with this methodology (Mulvaney and Kurtz, 1982; Siegel et al., 1982) are a practical limitation to the use of this technique. More important considerations, however, are the need to address the issues of the homogeneity of the labeling of the NO_3^- pool and the physical limitation of N-gas flux from the soil. Focht (1985) and Duxbury (1986) suggested that rates of denitrification would be underestimated using Equations [4] and [5]. A recent series of papers by Mulvaney and coworkers (Boast et al., 1988; Mulvaney and Vanden Heuvel, 1988) have proven mathematically and with computer simulations that Equations [4] and [5] will underestimate denitrification if more than one isotopically distinct NO_3^- pool exists (Fig. 4). Laboratory and field studies, however, suggest that soil and sampling variability and analytical errors are probably larger than the error introduced by heterogeneous label distribution (Mulvaney, 1988; Vanden Heuvel et al., 1988). It should be noted also that the problem of nonuniform distribution of added ^{15}N will introduce bias into all methods that use ^{15}N to calculate denitrification.

Accumulation of ^{15}N-labeled gases in the headspace of a soil cover is a measure of N-gas flux from the soil surface and may not represent the actual rate of denitrification. The flux is a function not only of the production rate but also transport processes of mass flow and diffusion. This was recognized in some of the initial studies using ^{15}N-gas evolution that attempted to correct for gas diffusion (Colbourn et al., 1984; Letey et al., 1980; Rolston et al., 1978, 1979, 1982), but such corrections are rarely applied. Although problems of gas entrapment and diffusional constraints occur in well aerated soils, the problem is most clearly illustrated in flooded systems. For example, under laboratory conditions Lindau et al. (1988) found that over 50% of the $^{15}N_2$ evolved in flooded soil columns was trapped within the column.

Comparison of the ^{15}N balance and direct ^{15}N-gas methods for measuring denitrification have been made in three studies (Buresh and Austin, 1988; Heinemeyer et al., 1988; Rolston et al., 1979). Rolston et al. (1979) measured denitrification in the field under twelve different treatments that varied in organic matter input, season of the year, and soil moisture. Although there was no statistical difference between the two methods, the ^{15}N-gas evolution usually gave lower estimates of denitrification than the ^{15}N mass balance (median of 42%).

Because Rolston et al. (1979) corrected for limitations in gas diffusion, the underestimation was probably because of nonuniform mixing of added $^{15}NO_3^-$ with soil NO_3^-. Under laboratory conditions, the ^{15}N-gas evolution method gave significantly lower estimates of denitrification than the ^{15}N mass balance for well aerated (Heinemeyer et al., 1988) and flooded (Buresh and Austin, 1988) soils. Both nonhomogeneity and diffusional constraints may have been important factors contributing to the underestimation of denitrification rates observed in these laboratory experiments. Losses of N as NO were not accounted for by any of the direct ^{15}N-gas methods used in these three studies.

$^{15}N_2$ ISOTOPE DILUTION

An alternative approach to measuring the increase in atom % ^{15}N of N-gases by adding $^{15}NO_3^-$, is to create an ^{15}N-enriched atmosphere and measure its dilution from N-gases arising from NO_3^- at natural abundance.

Rates of denitrification can be calculated by integrating simultaneously the following equation, which assumes that the rate of denitrification is constant and that there is no N_2-fixation:

$$\frac{dC_d}{dt} = R_d \tag{6}$$

with Equation (3) to yield:

$$R_d = C_{d0}(A_d - A_{d0})/(A_n - A_d) \tag{7}$$

where C_{d0} and A_{d0} are the initial concentration and atom % ^{15}N of the N-gas pool, respectively.

Although this type of approach has the same disadvantages in terms of homogeneity of labeling and N-gas entrapment as the ^{15}N-gas evolution method, it is less disruptive to the soil and can be easily moved from location to location in the field. The biggest drawback with using $^{15}N_2$ isotope dilution is its insensitivity. This was recognized by Limmer et al. (1982) in the only published work using this method. In order to achieve a sensitivity of 17 g N ha^{-1} day^{-1}, they needed to use a gas lysimeter that could be maintained at subambient concentrations of 1% N_2. A comparison of the relative sensitivities of the two methods is shown in Fig. 5. The reason that the ^{15}N-gas evolution method is about 100-fold more sensitive than the ^{15}N-gas isotope dilution approach is that with the same relative precision, the atom % ^{15}N difference in the numerator of Equation [7] is a smaller absolute number when N-gases at natural abundance are enriched with ^{15}N-gases than when an N-gas pool of high atom % ^{15}N is diluted. It might be possible to reverse the sensitivity of the method if ^{14}N was viewed as the tracer in a highly ^{15}N-enriched atmosphere. Difficulties would likely be encountered, however, in the design of mass spectrometer collector amplification circuitry, which is normally designed to be most sensitive to $^{30}N_2$ and least sensitive to $^{28}N_2$. Sample contamination with atmospheric N_2 probably would be a bigger problem.

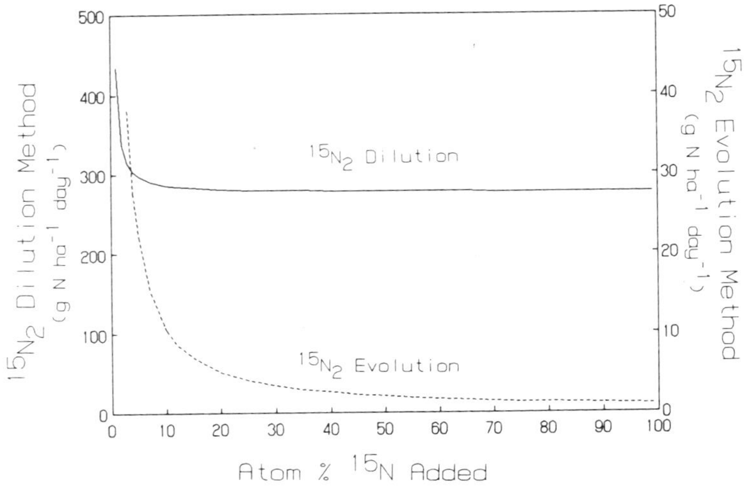

Fig. 5. Relative detection limits of the ^{15}N-gas evolution and ^{15}N isotope dilution methods. The curves were generated assuming a soil cover with a surface area of 66 cm^2 and volume of 400 cm^2 at atmospheric concentrations of N_2, and a relative sensitivity of 0.05% in the measurement of atom % ^{15}N.

COMPARISON OF ^{15}N TECHNIQUES WITH THE C_2H_2 INHIBITION METHOD

Application of the C_2H_2 inhibition method to measuring denitrification rates under field conditions has been done in two ways: using soil covers (Mosier et al., 1986a; b) or using intact soil covers (Parkin et al., 1985). The use of soil covers with C_2H_2 has many of the same problems, such as diffusion, and advantages of using soil cores for ^{15}N-gas evolution measurements. Soil cores are completely closed systems and are more amenable to complete mixing of gases to avoid diffusional limitations, however, they are more disruptive of soil structure. Biological affects of C_2H_2 may also affect the measurement of denitrification rates by blocking nitrification, which could result in underestimation if NO_3^- is limiting, or serving as a substrate for denitrifying bacteria, which would result in overestimation if C is limiting to denitrification (Duxbury, 1986; Haider et al., 1983; Keeney, 1986; Rolston, 1986). A more detailed review of the C_2H_2 technique is provided by R. Knowles in Chapter 9, this book.

Generally, the C_2H_2 inhibition method has given estimates of denitrification that agree with rates determined by the ^{15}N-gas evolution technique (Table 4). Rates measured by the C_2H_2 inhibition method are often slightly higher than those obtained by the ^{15}N-gas evolution method, which may be due to the extra pair of electrons that would have been used to reduce N_2O to N_2 were used instead to reduce more NO_3^- (Tiedje, 1978), although numerous other explanations are possible. The most important point, however, is that the similar rates obtained from the two methods that measure gaseous products are likely because both methods are subject to similar physical transport processes.

As with the ^{15}N-gas evolution method, the C_2H_2 inhibition technique often underestimated the amount of denitrification when compared to the ^{15}N balance approach (Table 5); although good agreement has been obtained with Canadian

Table 4. Comparison of ^{15}N-gas evolution and C_2H_2 inhibition methods for measuring denitrification rates in soils.

Study	^{15}N-gas evolution	C_2H_2 inhibition	^{15}N-gas evolution / C_2H_2 inhibition
Ryden et al. (1979)[a]			
loam	37.5	35	1.07
sandy loam	69.2	72.8	0.95
loam	68.5	53.1	1.29
Rolston et al. (1982)[b]			
3x per week, 24 days	4.1	4.3	0.95
1x per week, 42 days	3.2	3.4	0.94
1x per 2 weeks, 42 days	1.9	2.7	0.70
Mosier et al. (1986b)[c]			
5 cm, June, 3 days	1.86	2.15	0.87
3 cm, July, 5 days	0.75	1.00	0.75
5 cm, July, 8 days	6.44	8.95	0.72

[a]Laboratory study; rates are mg N kg^{-1} over 200 hr.
[b]Field study; loam soil; three irrigation frequencies; rates are kg N ha^{-1} for the time period.
[c]Field study; clay loam; three irrigation events; rates are kg N ha^{-1} for the time period.

prairie soils (Aulakh et al., 1983, 1984). The extreme underestimation in the data of Recous et al. (1988a,b) is probably because of incomplete accounting of the lateral movement of ^{15}N from the unconfined microplots and the infrequent measurement of N_2O production. In the studies of Germon (1985) and Parkin et al. (1985), denitrification measured with the C_2H_2 inhibition technique was 30-70% of that by the ^{15}N mass balance. This may have been caused by restricted diffusion of C_2H_2 into (or N_2O out of) anaerobic microsites, which is supported by the closer agreement between the two methods in the sandy loam soil. It also should be noted that the C_2H_2 inhibition method does not measure any N losses as NO.

SUMMARY

Over the past decade, significant progress has been made in the use of ^{15}N to study denitrification. Despite problems with sensitivity, the ^{15}N balance method remains the standard for comparison. Efforts to reduce the measurement errors associated with parts of the mass balance would prove to be useful, both for getting better estimates of annual losses of N and for improving its utility as a calibration standard for other methods.

Recent studies on entrapment of N-gases in soils has again emphasized the need to incorporate the expertise of soil physicists in denitrification studies. The

Table 5. Comparison of ^{15}N mass balance and C_2H_2 inhibition methods for measuring denitrification rates in soils.

Study	^{15}N mass balance	C_2H_2 inhibition	C_2H_2 inhibition / ^{15}N mass balance
Germon (1985)[a]	77.3	27.1	0.35
Aulakh et al. (1983)[b]			
clay loam	2.3	1.8	0.78
clay loam	1.2	1.2	1.00
Aulakh et al. (1984)[c]			
clay loam, CT, no straw	4.62	4.28	.93
clay loam, CT + straw	6.65	7.25	1.09
clay loam, NT no straw	11.2	11.9	1.06
clay loam, NT + straw	20.9	22.5	1.08
Parkin et al. (1985)[d]			
clay loam	6.75	2.18	0.32
sandy loam	1.77	1.14	0.66
Recous et al. (1988ab)[e]			
silt loam, urea	33	0.5	0.02
silt loam, NH_4NO_3	36	1.4	0.04

[a]Soil cores on field lysimeters; mg N per lysimeter for a season.
[b]Soil cores and confined miniplots; 50 kg N ha^{-1} as $(NH_4)_2SO_4$; rates as kg N ha^{-1} over 98 days.
[c]Soil cores and confined miniplots; 100 kg N ha^{-1} as $(NH_4)_2SO_4$; CT--conventional tillage, NT--no tillage; straw added at 3 Mg ha^{-1}; rates as kg N ha^{-1} over 92 days.
[d]Soil cores and confined miniplots; rates as mg N kg^{-1} over 30 days.
[e]Soil cores and unconfined miniplots; 160 kg N ha^{-1} added; rates as kg N ha^{-1} over 154 days.

influence of gas movement out of soils on the denitrification estimates obtained by the direct ^{15}N-gas evolution method needs to be evaluated. Much progress has been made in understanding the effects of nonhomogeneous mixing of added $^{15}NO_3^-$ and further experimentation should demonstrate whether the theoretical inaccuracies are meaningful when applied to field situations. Finally, evidence is mounting to show that NO flux from soils may represent a significant and usually unmeasured loss of N (R. Conrad, Chapter 7, this book).

As mass spectrometers become more sensitive and precise, it will be important to reevaluate the potential of the $^{15}N_2$ isotope dilution method. Although it suffers from some of the same shortcomings as the ^{15}N-gas procedure, it has the advantage of not disrupting the native NO_3^- pool. No comparisons of the $^{15}N_2$ isotope dilution study with other methods have been made.

Finally, there is the need to compare all of the ^{15}N methods and the C_2H_2 inhibition technique under similar conditions. The data to date suggest that the direct ^{15}N-gas evolution and C_2H_2 inhibition method are comparable and that both underestimate denitrification rates compared to the ^{15}N mass balance, but all three have not been examined in one study.

ACKNOWLEDGEMENTS

I thank S. Christensen and J.P. Myrold for their reviews of this paper. This work was supported in part by a Presidential Young Investigator Award, BSR-8857269, from the National Science Foundation and Forest Research Laboratory of the College of Forestry.

REFERENCES

Aulakh, M.S., Rennie, D.A., and Paul, E.A., 1983, Field studies on gaseous nitrogen losses from soils under continuous wheat versus wheat-fallow rotation, Plant Soil, 75:15.

Aulakh, M.S., Rennie, D.A., and Paul, E.A., 1984, The influence of plant residues on denitrification rates in conventional and zero tilled soils, Soil Sci. Soc. Am. J., 48:790.

Aulakh, M.S., and Rennie, D.A., 1987, Effect of wheat straw incorporation on denitrification of N under anaerobic and aerobic conditions, Can. J. Soil Sci., 67:825.

Balderston, W.L., Sherr, B., and Payne, W.J., 1976, Blockage by acetylene of nitrous oxide reduction in Pseudomonas perfectomarinus, Appl. Environ. Microbiol., 31:504.

Barraclough, D., and Smith, M.J., 1987, The estimation of mineralization, immobilization, and nitrification in nitrogen-15 field experiments using computer simulation, J. Soil Sci., 38:519.

Barraclough, D., Geens, E.L., and Maggs, J.M., 1984, Fate of fertilizer nitrogen applied to grassland. II. Nitrogen-15 leaching results, J. Soil Sci., 35:191.

Bjarnason, S., 1988, Calculation of gross nitrogen immobilization and mineralization in soil, J. Soil Sci., 39:393.

Blackmer, A.M., and Bremner, J.M., 1977, Nitrogen isotope discrimination in denitrification of nitrate in soils, Soil Biol. Biochem., 9:73.

Boast, C.W., Mulvaney, R.L., and Baveye, P., 1988, Evaluation of nitrogen-15 tracer techniques for direct measurement of denitrification in soil: I. Theory, Soil Sci. Soc. Am. J., 52:1317.

Buresh, R.J., and Austin, E.R., 1988, Direct measurement of dinitrogen and nitrous oxide flux in flooded rice fields, Soil Sci. Soc. Am. J., 52:681.

Chien, S.H., Shearer, G., and Kohl, D.H., 1977, The nitrogen isotope effect associated with nitrate and nitrite loss from waterlogged soils, Soil Sci. Soc. Am. J., 41:63.

Cho, C.M., and Sakdinan, L., 1978, Mass spectrometric investigation on denitrification, Can. J. Soil Sci., 58:443.

Colbourn, P., Harper, I.W., and Iqbal, M.M., 1984, Denitrification losses from ^{15}N-labelled calcium nitrate fertilizer in a clay soil in the field, J. Soil Sci., 35:539.

Craswell, E.T., Byrnes, B.H., Holt, L.S., Austin, E.R., Fillery, I.R.P., and Strong, W.M., 1985, Nitrogen-15 determination of nonrandomly distributed dinitrogen in air, Soil Sci. Soc. Am. J., 49:664.

Dowdell, R.J., and Webster, C.P., 1984, A lysimeter study of the fate of fertilizer nitrogen in spring barley crops grown on a shallow soil overlying Chalk: denitrification losses and the nitrogen balance, J. Soil Sci., 35:183.

Dowdell, R.J., Webster, C.P., Hill, D., and Mercer, E.R., 1984, A lysimeter study of the fate of fertilizer nitrogen in spring barley crops grown on a shallow soil overlying Chalk: crop uptake and leaching losses, J. Soil Sci., 35:169.

Duxbury, J.M., 1986, Advantages of the acetylene method of measuring denitrification, in: "Field measurement of dinitrogen fixation and denitrification," R.D. Hauck and R.W. Weaver, eds., Soil Science Society of America Special Publication 18, Soil Science Society of America, Madison, WI.

Focht, D.D., 1985, Differences in nitrogen-15 enrichments of evolved nitrous oxide and dinitrogen and the question of a uniform nitrate-15 pool, Soil Sci. Soc. Am. J., 49:786.

Focht, D.D., and Stolzy, L.H., 1978, Long-term denitrification studies in soils fertilized with $(^{15}NH_4)_2SO_4$, Soil Sci. Soc. Am. J., 42:894.

Focht, D.D., Valoras, N., and Letey, J., 1980, Use of interfaced gas chromatography-mass spectrometry for detection of concurrent mineralization and denitrification in soil, J. Environ. Qual., 9:218.

Fredrickson, J.K., Koehler, F.E., and Cheng, H.H., 1982, Availability of ^{15}N-labeled nitrogen in fertilizer and in wheat straw to wheat in tilled and no-till soil, Soil Sci. Soc. Am. J., 46:1218.

Germon, J.C., 1985, Denitrification in cropped soils, Fert. Agric., 89:3.

Haider, K., Mosier, A.R., and Heinemeyer, O., 1983, Side effects of acetylene on the conversion of nitrate in soil, Z. Pflanzenernaehr. Bodenk., 146:623.

Hasebe, A., Koike, I., Ohmori, M., and Hattori, A., 1987, Variations in the process of nitrification and nitrate reduction in submerged paddy soils as measured by ^{15}N isotope dilution technique, Soil Sci. Plant Nutr., 33:201.

Hauck, R.D., 1986, Field measurement of denitrification--an overview, in: "Field measurement of dinitrogen fixation and denitrification," R.D. Hauck and R.W. Weaver, eds., Soil Science Society of America Special Publication 18, Soil Sci. Soc. Am., Madison, WI.

Hauck, R.D., and Bouldin, D.R., 1961, Distribution of isotopic nitrogen in nitrogen gas during denitrification, Nature, 191:871.

Hauck, R.D., and Bremner, J.M., 1976, Use of tracers for soil and fertilizer nitrogen research, Adv. Agron., 28:219.

Hauck, R.D., Melsted, S.W., and Yankwich, P.E., 1958, Use of N-isotope distribution in nitrogen gas in the study of denitrification, Soil Sci., 86:287.

Heinemeyer, O., Haider, K., and Mosier, A., 1988, Phytotron studies to compare nitrogen losses from corn-planted soil by the ^{15}N balance or direct dinitrogen and nitrous oxide measurements, Biol. Fertil. Soils, 6:73.

Iimura, K., and Ito, J., 1984, Direct measurement of N_2 denitrified from soil and fertilizer in paddy fields using ^{15}N as a tracer, Jap. Agric. Res. Q., 18:27.

Jokela, W.E., and Randall, G.W., 1987, A nitrogen-15 microplot design for measuring plant and soil recovery of fertilizer nitrogen applied to corn, Agron. J., 79:322.

Keeney, D.R., 1986, Critique of the acetylene blockage technique for field measurement of denitrification, in: "Field measurement of dinitrogen fixation and denitrification," R.D. Hauck and R.W. Weaver, eds., Soil Science Society of America Special Publication 18, Soil Science Society of America, Madison, WI.

Kirkham, D., and Bartholomew, W.V., 1955, Equations for following nutrient transformations in soil, utilizing tracer data: II., Soil Sci. Soc. Am. Proc., 19:189.

Koike, I., and Hattori, A., 1978, Simultaneous determination of nitrification and nitrate reduction in coastal sediments by a ^{15}N dilution technique, Appl. Environ. Microbiol., 35:853.

Legg, J.O., and Meisinger, J.J., 1982, Soil nitrogen budgets, in: "Nitrogen in agricultural soils," F.J. Stevenson, ed., Agronomy Monograph 22, American Society of Agronomy, Madison, WI.

Letey, J., Jury, W.A., Hadas, A., and Valoras, N., 1980, Gas diffusion as a factor in laboratory incubation studies on denitrification, J. Environ. Qual., 9:223.

Limmer, A.W., Steele, K.W., and Wilson, A.T., 1982, Direct field measurement of N_2 and N_2O evolution from soil, J. Soil Sci., 33:499.

Lindau, C.W., Patrick, W.H., Jr., Delaune, R.D., Reddy, K.R., and Bollich, P.K., 1988, Entrapment of nitrogen-15 dinitrogen during soil denitrification, Soil Sci. Soc. Am. J., 52:538.

Mariotti, A., Germon, J.C., and Leclerc, A., 1982, Nitrogen isotope fractionation associated with the $NO_2^- \rightarrow N_2O$ step of denitrification in soils, Can. J. Soil Sci., 62:227.

Meyer, M.L., Bloom, P.R., and Grava, J., 1989, Transformation and losses of applied nitrogen-15 labeled ammonium in a flooded organic soil, Soil Sci. Soc. Am. J., 53:79.

Mosier, A.R., Guenzi, W.D., and Schweizer, E.E., 1986a, Soil losses of dinitrogen and nitrous oxide from irrigated crops in northeastern Colorado, Soil Sci. Soc. Am. J., 50:344.

Mosier, A.R., Guenzi, W.D., and Schweizer, E.E., 1986b, Field denitrification estimation by nitrogen-15 and acetylene inhibition techniques, Soil Sci. Soc. Am. J., 50:831.

Mulvaney, R.L., 1984, Determination of ^{15}N-labeled dinitrogen and nitrous oxide with triple-collector mass spectrometers, Soil Sci. Soc. Am. J., 48:690.

Mulvaney, R.L., 1988, Evaluation of nitrogen-15 tracer techniques for direct measurement of denitrification in soil: III. Laboratory studies, Soil Sci. Soc. Am. J., 52:1327.

Mulvaney, R.L., and Boast, C.W., 1986, Equations for determination of nitrogen-15 labeled dinitrogen and nitrous oxide by mass spectrometry, Soil Sci. Soc. Am. J., 50:360.

Mulvaney, R.L., and Kurtz, L.T., 1982, A new method for determination of ^{15}N-labeled nitrous oxide, Soil Sci. Soc. Am. J., 46:1178.

Mulvaney, R.L., and Kurtz, L.T., 1984, Evolution of dinitrogen and nitrous oxide from nitrogen-15 fertilized soil cores subjected to wetting and drying cycles, Soil Sci. Soc. Am. J., 48:596.

Mulvaney, R.L., and Vanden Heuvel, R.M., 1988, Evaluation of nitrogen-15 tracer techniques for direct measurement of denitrification in soil: IV. Field studies, Soil Sci. Soc. Am. J., 52:1332.

Myrold, D.D., and Tiedje, J.M., 1986, Simultaneous estimation of several nitrogen cycle rates using ^{15}N: theory and application, Soil Biol. Biochem., 18:559.

Nishio, T., Kanamori, T., and Fujimoto, T., 1985, Nitrogen transformations in an aerobic soil as determined by a $^{15}NH_4^+$ dilution technique, Soil Biol. Biochem., 17:149.

Nishio, T., Koike, I., and Hattori, A., 1983, Estimates of denitrification and nitrification in coastal and estuarine sediments, Appl. Environ. Microbiol., 45:444.

Nômmik, H., 1956, Investigations on denitrification in soil, Acta Agric. Scand., 6:195.

Parkin, T.B., Sexstone, A.J., and Tiedje, J.M., 1985, Comparison of field denitrification rates determined by acetylene-based soil core and nitrogen-15 methods, Soil Sci. Soc. Am. J., 49:94.

Peterson, B.J., and Fry, B., 1987, Stable isotopes in ecosystem studies, Ann. Rev. Ecol. Syst., 18:293.

Recous, S., Fresneau, C., Faurie, G., and Mary, B., 1988, The fate of labelled ^{15}N urea and ammonium nitrate applied to a winter wheat crop. I. Nitrogen transformations in soil, Plant Soil, 112:205.

Recous, S., Machet, J.M., and Mary, B., 1988, The fate of labelled ^{15}N urea and ammonium nitrate applied to a winter wheat crop. II. Plant uptake and N efficiency, Plant Soil, 112:215.

Rolston, D.E., 1986, Limitations of the acetylene blockage technique for field measurement of denitrification. in: "Field measurement of dinitrogen fixation and denitrification," R.D. Hauck and R.W. Weaver, eds., Soil Science Society of America Special Publication 18. Soil Science Society of America, Madison, WI.

Rolston, D.E., Broadbent, F.E., and Goldhamer, D.A., 1979, Field measurement of denitrification: II. Mass balance and sampling uncertainty, Soil Sci. Soc. Am. J., 43:703.

Rolston, D.E., Hoffman, D.L., and Toy, D.W., 1978, Field measurement of denitrification: I. Flux of N_2 and N_2O, Soil Sci. Soc. Am. J., 42:863.

Rolston, D.E., Sharpley, A.N., Toy, D.W., and Broadbent, F.E., 1982, Field measurement of denitrification: III. Rates during irrigation cycles, Soil Sci. Soc. Am. J., 46:289.

Rundel, P.W., Ehleringer, J.R., and Nagy, K.A., 1989, Stable isotopes in ecological research, Ecological Studies, Vol. 68, Springer-Verlag, New York.

Ryden, J.C., Lund, L.J., and Focht, D.D., 1979, Direct measurement of denitrification loss from soils: I. Laboratory evaluation of acetylene inhibition of nitrous oxide reduction, Soil Sci. Soc. Am J., 43:104.

Ryden, J.C., Skinner, J.H., and Nixon, D.J., 1987, Soil core incubation system for the field measurement of denitrification using acetylene-inhibition, Soil Biol. Biochem., 19:753.

Siegel, R.S., Hauck, R.D., and Kurtz, L.T., 1982, Determination of $^{30}N_2$ and application to measurement of N_2 evolution during denitrification, Soil Sci. Soc. Am. J., 46:68.

Tiedje, J.M., 1978, Denitrification in soil, in: "Microbiology 1978," W.D. Schlessinger, ed., American Society of Microbiology, Washington, D.C.

Vanden Heuvel, R.M., Mulvaney, R.L., and Hoeft, R.G., 1988, Evaluation of nitrogen-15 tracer techniques for direct measurement of denitrification in soil: II. Simulation studies, Soil Sci. Soc. Am. J., 52:1322.

Wijler, J., and Delwiche, C.C., 1954, Investigations on the denitrifying process in soil, Plant Soil, 5:155.

Yoshinari, T., and Knowles, R., 1976, Acetylene inhibition of nitrous oxide reduction by denitrifying bacteria, Biochem. Biophys. Res. Commun., 69:705.

RHIZOSPHERE DENITRIFICATION; A MINOR PROCESS BUT INDICATOR OF DECOMPOSITION ACTIVITY

Søren Christensen[1], Peter Groffman[2], Arvin Mosier[3] and Donald R. Zak[4]

1: Dep. Population Biology, Copenhagen Univ.
Universitetsparken 15, DK-2100 Copenhagen Ø, Denmark
2: Coll. Resource Dev., Univ. Rhode Island
Kingston, RI 02881, USA
3: US Department of Agriculture, ARS
P.O.Box E, Fort Collins, CO 80522, USA
4: School of Natural Resources
Univ. of Michigan, Ann Arbor, MI 48109, USA

INTRODUCTION

The primary inputs of organic matter driving heterotrophic processes in soil are of plant origin. Microbial activity increases more than an order of magnitude when moving from the bulk soil through the rhizosphere towards the root surface (Brown, 1975; Helal and Sauerbeck, 1986). The composition of the soil microbial population (Rouatt et. al., 1960) and its denitrifying capability (Smith and Tiedje, 1979) show dramatic changes over this small distance in soil. Decomposition of recalcitrant soil organic matter can be stimulated in the rhizosphere (Helal and Sauerbeck, 1986) and grazing on soil microbes by soil animals is greatly increased in this environment (Clarholm, 1985). As members of the rhizosphere microbial community, the activity of denitrifying bacteria is expected to be dramatically affected by the presence of plants. Since the comprehensive work by Woldendorp (1963), several investigations in arable soil have shown a substantial influence of plants on denitrification (Bakken, 1988, Christensen 1985, Klemendtsson et. al., 1987, Von Rheinbaben and Troldenier, 1984 are more recent examples). Plants in waterlogged soil (Broadbent and Tusnem, 1971; Garcia, 1975; Reddy and Patrick, 1986) and algae in marine sediments (Andersen et. al., 1984) can also have a marked influence on denitrification. Some studies found no or weak influence of plants on denitrification (Haider et al., 1985; 1987).

Plant effects on denitrification can be analyzed by asking how the three main determinants of denitrification (NO_3^-, organic C and O_2 partial pressure) are influenced by the presence of plants. Plants can directly influence nitrate availability through uptake and assimilation making it unavailable to denitrification (Fig. 1). An indirect effect on NO_3^- levels arises from the supply of organic matter of root origin; mineralization and nitrification of this material can potentially provide more NO_3^- for denitrification and conversely immobilization

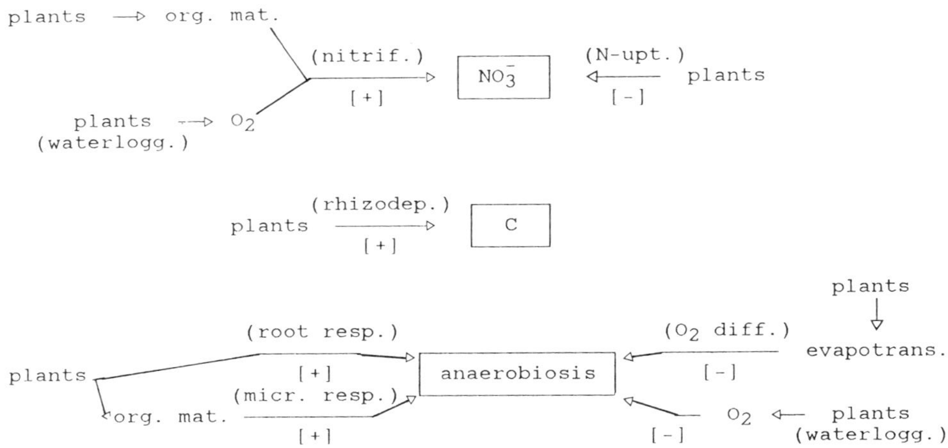

Figure 1. Theoretical examination of mechanisms whereby plants influence the primary determinants of soil denitrification. Plant effects increasing denitrification to the left, negative effects to the right.

can reduce NO_3^- levels. Another indirect effect is the ability of water-logged root systems to supply O_2 to the rhizosphere thereby promoting nitrification.

Moving to carbon, plants are the main source of C substrate to denitrifyers through rhizodeposition. Presumably, however, it is mainly the soluble fraction comprising 1-9 % of the total rhizodeposition (Barber and Martin, 1976; Martin, 1977) that can be directly utilized by denitrifying bacteria. Denitrifiers are probably unable to decompose the structural carbohydrates of root cells.

Plant effects on anaerobiosis are complex. The utilization of O_2 for plant respiration is presumably of minor importance since root respiration is normally considered of smaller magnitude compared to microbial respiration (Barber and Martin, 1976; Johnen and Sauerbeck, 1977). Microbial degradation of root derived organic matter in the rhizosphere is an important process during plant growth (Helal and Sauerbeck, 1987; Mercks et. al., 1985) and of course the only O_2 consuming process when dead plant remains are to be decomposed. The opposite effect can be caused by plants increasing evapotranspiration and soil gas exchange, thereby increasing soil oxygen. The inhibitory effect of roots supplying O_2 to the rhizosphere of waterlogged soils is probably of minor importance to denitrification compared to the positive effect of this extra O_2, stimulating nitrification and aerobic decomposition of complex organic molecules in this environment (Broadbent and Tusnem, 1971).

To summarize, the most important negative effects of plants on denitrification in arable soils are decreased soil water content and/or removal of NO_3^- from the soil solution. The major positive effect is rhizodeposition giving organic C to the rhizosphere microbes thereby decreasing the O_2 partial pressure during decomposition.

The purpose of this paper is twofold: (i) to present data showing the relative magnitude of N-transformation rates as influenced by the presence of plants and (ii) to demonstrate how the large spatial variability in soil denitrification can be driven by the supply of plant derived organic matter.

CONVERSION RATES OF N IN THE RHIZOSPHERE

We used ^{15}N to directly measure rates of the different N cycle processes in soil and to determine the significance of denitrification in the soil N budget. $(^{15}NH_4)_2SO_4$ or $NH_4{}^{15}NO_3$ was added to undisturbed soil cores in the field. The cores were sampled after two days to identify the fate of the added isotope depending on nitrogen form and presence of plants.

Field Experiment

The experiment was conducted on a poorly drained soil in an old field on the Michigan State University campus. The soil had a pH of 7.82±0.04 and contained 4.74±2.74 and 0.23±0.13 % total C and N respectively. Sixteen PVC cores 25 cm long and 10 cm diameter were hammered 19 cm into the ground within a 15 m^2 area. Half of the rings were placed around goldenrod plants (*Solidago* sp.) the remainder were located on bare soil. During the next three weeks all new seedlings were removed. In the three weeks preceeding this experiment, the soil contained 0.8-0.9 ppm NO_3^--N and 3-6 ppm NH_4^+-N and had a water content of 34-36 %. On 28 April 1986 about 10 mg ^{15}N as 99.5 % enriched $(^{15}NH_4)_2SO_4$ or $NH_4{}^{15}NO_3$ was applied to the rings giving four treatments in four replicates. Ten 5 ml portions of isotope solution containing 10 mg of ^{15}N in total were distributed evenly by injection with a syringe in the 0-4 cm soil layer.

During the next two days beginning about six hours after N addition, the 0.5 l headspace above the soil was enclosed four times with a lid for 1 h. Five ml of headspace gas was sampled after 20, 40 and 60 min and stored in 3.3 ml Venoject vials. Total N-gas flux from the rings was determined from the amount and isotopic composition of the gases (Mosier et. al., 1986).

Laboratory Analysis

Two days (49 h) after the ^{15}N application the cores were taken out of the soil. The 0 - 6 cm top segments (520 g D.W. avg.) were homogenized after removal of root material (1.8 g D.W. avg.) from cores with plants. Samples were extracted with 2 M KCl and analyzed for NH_4^+ and $NO_2^-+NO_3^-$ on a Technicon Autoanalyzer II. Total N in soil and plant material was determined by Kjeldahl digestion followed by NH_4^+ measurement on the autoanalyzer. Total N was determined on soil used for inorganic N extractions, but was additionally leached with excess 2 M KCl and deionized water before Kjeldahl digestion. Plant content of N was determined on shoots and roots separately. Soil biomass N was determined by the fumigation-incubation procedure (Clark and Paul, 1988). Aliquots of fumigated and control soil were placed in sealed jars at room temperature. Nitrogen and carbon mineralized during a 10 day incubation was determined by measuring accumulation of inorganic N and CO_2 in the flasks. Carbon mineralization was determined as the increase in headspace CO_2 of the flasks measured on gas chromatograph. The k_N factor (fraction of biomass N mineralized during incubation) was determined according to the formula $k_N = 0.8 * (C_f/N_f)^{-0.43}$ (Clark and Paul, 1989). C_f and N_f are the amount of CO_2-C and mineral N, respectively, liberated in fumigated minus non-fumigated soil. For mineral N and total N in soil and plant material two replicate analyses were performed for each core and the coefficient of variation among the four core averages determined. Single determinations of N mineralized from fumigated and

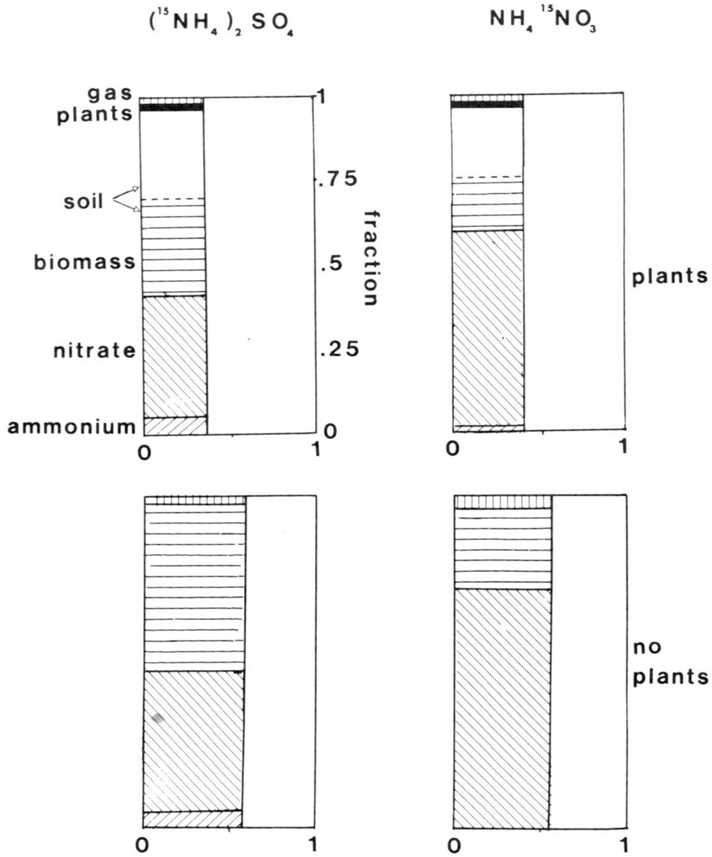

Figure 2. Budget for the addition of ^{15}N to an organic soil. Width of bars on the X-axis indicates the recovery of label in the 0-6 cm layer after two days. The Y-axis is divided into the fractions in which the recovered label was found. The unshaded part is the fraction of soil N (^{15}N not extractable with 2 M KCl) that does not belong to the biomass pool.

unfumigated soil were performed on the four replicate cores and pooled before they were subtracted from each other and divided by k_N to obtain a figure for microbial biomass.

The $^{15}N/^{14}N$ ratio was determined on all the fractions for which N-content was measured. Samples were prepared by NH_3 diffusion in sealed containers (MacKown et. al., 1987). Aliquots of soil extract or soil or plant digest were placed in 120 ml plastic containers and MgO was added to convert NH_4^+ into NH_3 which volatilized and was trapped in a test tube containing dilute H_2SO_4. During the two-week diffusion time the containers were constantly shaken. After collection of the NH_3 from NH_4^+ in soil extracts, Devardas alloy was added to convert NO_3^- into NH_3. A Micromass 622 mass spectrometer was used for determination of $^{15}N/^{14}N$ ratios in soil and plant material. For all samples where less than 500 μg N was present, 0.5 mg N as unenriched $(NH_4)_2SO_4$ was added as a spike. The ^{15}N content in the spike was subtracted from sample measurements to give actual ^{15}N in the sample.

Results and discussion

Recovery of the label was low in all the treatments, ranging from 37 to 58 % of the added N-15 (Fig. 2). The most likely explanation for the low recovery is leaching of N below the 6 cm sampling depth at labelling since addition of the label solution probably favored saturated flow. The lower recovery in planted than in unplanted soil supports this since vertical movement of water will be facilitated by root channels. Because of the low recovery, the amount of ^{15}N and ^{14}N in NH_4^+ and NO_3^- at the start of the experiment was not known. For this reason it was not possible to estimate the conversion of the different soil N pools, but only to follow conversion of the added N-15.

Large amounts of label were found in the NO_3^- pool in NH_4^+-labelled treatments indicating high nitrification activity. A major process occurring with both N sources was the formation of labelled compounds not extractable with 2 M KCl (soil fraction in Fig. 2) suggesting a considerable immobilization. In the unplanted treatments biomass ^{15}N determined by the fumigation incubation procedure ($k_N=0.33$) was similar to the amount of label not extractable with 2 M KCl within 5 %. In the planted treatments some ^{15}N not extractable with 2 M KCl did not belong to the biomass pool either: Biomass ^{15}N only amounted to half of the non-extractable ^{15}N (Fig. 2). This ^{15}N not extracted with 2 M KCl and not belonging to the biomass in planted soil could be label in small roots not removed from the soil at the termination of the experiment or microbial metabolites/dead cells. The ^{15}N concentration in root tissue was 1.8 and 2.7 times higher than the shoot concentration for addition of NH_4^+ and NO_3^- label, respectively. Considering those concentration gradients of ^{15}N within the plants and the low amount of ^{15}N found in plants (Fig. 2) it is improbable that the nonextractable ^{15}N not belonging to the microbial biomass was contained in tiny root segments left over in the soil. Therefore the non-extractable, non-biomass ^{15}N is considered of microbial origin. Moreover, the biomass ^{15}N pools might be overertimated which implies that the non-extractable ^{15}N pools outside the microbial biomass are even larger: The ^{15}N fraction of the biomass is probably more readily decomposed than the the average biomass N resulting in a higher k_N for ^{15}N than the gross k_N of 0.33. ^{15}N in plants and gaseous form did not account for more than 2-4 % of the labelled pools after two days.

Denitrification rates were high at the first sampling and decreased steadily during the two days of incubation (Fig. 3). Denitrification in unamended soil sampled in small soil cores (2.3 cm diameter) at the site of the labelled field cores was determined in the laboratory with the acetylene inhibition method. The average rate from 3 samplings during the days of the labelling experiment was 94 ng N g^{-1} soil day^{-1} (60-70 ng N m^{-2} sec^{-1}). These core rates are well in accordance with the ^{15}N determination of total denitrification at the later samplings (Fig. 3). The abovementioned core rates were increased 50 times by addition of water; the decreasing rates during incubation are probably caused by decreased soil water content in the days following addition of isotope solution. The denitrification rates and pattern of response to water addition is in accordance with results from irrigation experiments (Mosier et. al., 1986, Rolston et al. 1982, Terry et al. 1986).

The denitrification rate in cores amended only with NH_4^+ was almost as high as in NO_3^- supplied cores, again suggesting a high nitrification rate. The presence of plants seemed to decrease denitrification rates slightly. The net effect of plants on denitrification was therefore either increased soil aeration caused by presence of roots or decreased soil NO_3^- due to plant uptake. NO_3^- competition

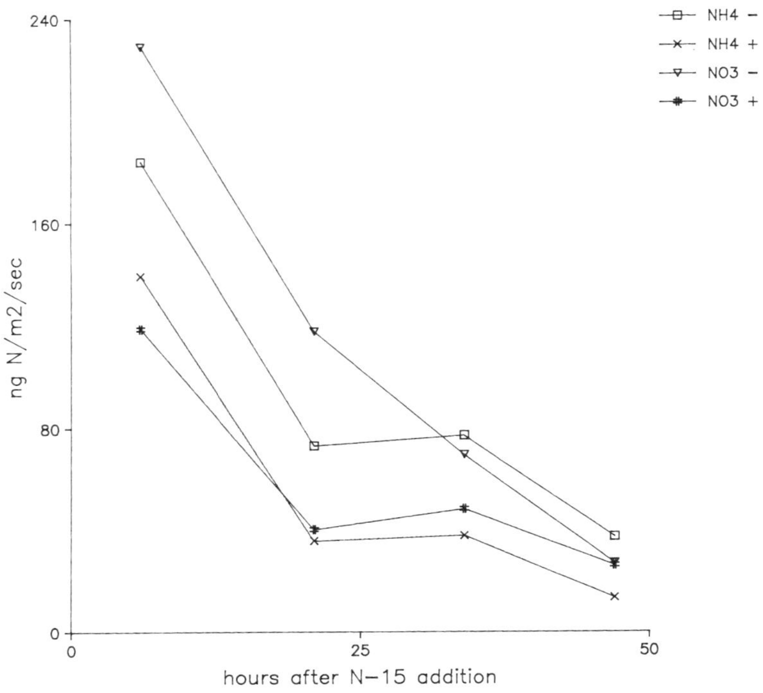

Figure 3. Denitrification during the two days of N-15 labelling as measured by amount of labelled and unlabelled N accumulating under a lid during 60 min. Bars are the means from four cores per treatment per day. Treatments are ^{15}N label added in NH_4^+ and NO_3^- with (+) and without (-) plants.

does not seem likely since the very low plant N uptake probably did not alter the soil pool much and the NO_3^- concentration in the soil solution was well above 1 mM. This concentration is more than 100 times above the half saturation constant (K_m) for denitrifying bacteria (Christensen and Tiedje, 1988) and therefore unlikely to be limiting (Myrold and Tiedje, 1985). The same result was obtained at variable or unlimited NO_3^- supply (Haider et. al., 1985; 1987). For those reasons increased aeration due to root channels is the primary negative plant effect outweighing any positive effects due to rhizodeposition.

Making the assumption that all loss of ^{15}N occurred at the start of the experiment, i.e., only label in the form added and not reaction products disappeared, reaction rates for the added ^{15}N can be calculated (Table 1). A common feature for all treatments is that rates of processes removing N from the soil (plant uptake and denitrification) are low relative to internal soil N transformations.

Since we added large amounts of ^{15}N to follow isotope incorporation in the large pool of soil organic matter we were not able to determine turnover rates at natural N concentrations, (indigenous pools of NH_4^+ and NO_3^- are increased by more than 2 and 10 times respectively). On the other hand, potential rates with respect to N may be determined at least for NO_3^- label. Only 40 % of the added NO_3^- label was reduced to organic form, gases or NH_4^+ during the two days of incubation whereas almost the entire pool of added NH_4^+ label (94-96%) disappeared (Table 1 last column). It is likely that some $^{15}NH_4^+$ previously

Table 1. Fates of ^{15}N applied to cores in the field as NH_4^+ in $(^{15}NH_4)_2SO_4$ or as NO_3^- in $NH_4{}^{15}NO_3$. Field incubation lasted 49 h.

	Immobilization			Nitrification rate	Plant uptake rate	Denitrification rate[&]	Inorganic N	
	rate	biomass					added	turnover[*]
		amount	turnover[@]					
	ng N/g/d	ng N/g	d^{-1}	---	ng N/g/d	---	ng N/g	d^{-1}
NH_4^+ no pl.	3420[#]	7330	0.47	2830	-	108	13400	0.48
NO_3^- no pl.	1350	2750	0.49	-	-	107	11400	0.14
NH_4^+ plants	2370	2520	0.94	1570	42	62	8600	0.47
NO_3^- plants	1620	1410	1.15	-	28	60	8600	0.20

#: Coefficient of variation: Immobilization 32-37 %, nitrification 15-26 %, plant uptake 21-90 %, denitrification 48-90 %.
*: Assuming that all ^{15}N not accounted for was lost at time 0 h.
@: Biomass turnover; immobilization rate divided by amount of biomass.
&: Assuming that all activity occurred in the 0-6 cm soil layer.

immobilized would be remineralized to inorganic N during the two days of incubation. Measured rates of NH_4^+ conversion can therefore cover both saturated and unsaturated situations with respect to N and will probably underestimate the potential rates. The average immobilization rate for all treatments of 2.2 μg N g^{-1} soil day^{-1} is 15 times below the respiration rate of 33.5 μg C g^{-1} soil day^{-1} determined in soil cores. Since the C/N ratio of the soil organic matter is 20.6 this suggests that N mineralization rate is of about the same magnitude as the immobilization rate.

The estimated turnover rate for ^{15}N incorporated in microbial biomass was similar for label source NH_4^+ and NO_3^- but was twice as high in planted compared to unplanted soil (Table 1). This is consistent with models that show that grazers may be stimulated in the rhizosphere, increasing turnover of microbial populations (Clarholm 1985; Robinson et. al., 1989). Considering the low yield efficiency of the microfauna when feeding on microbes (Coleman et. al., 1978) one would have expected to find the residual ^{15}N not belonging to the microbial biomass in the inorganic ^{15}N pools. The most likely explanation is the formation of microbial metabolites and/or partly decomposed microbes in the rhizosphere eventually mediated by microfaunal grazing.

The average ^{15}N content in the microbial biomass was 1.6 atom-%. The overall mean N-15 content during the incubation in pools labelled (NH_4^+ or NO_3^-) was 42 % still assuming that all label unaccounted for was lost at time zero. This means that 1.6/0.42 = 3.8 % of the biomass N was taken up during the incubation. Assuming that the rest of the biomass N was inactive, turnover rates of all biomass N is estimated as 6.6 yr^{-1} and 14.5 yr^{-1} for unplanted and planted soil, respectively. These turnover rates are probably overestimated due to the considerations above. The figures are in the upper end of the range of

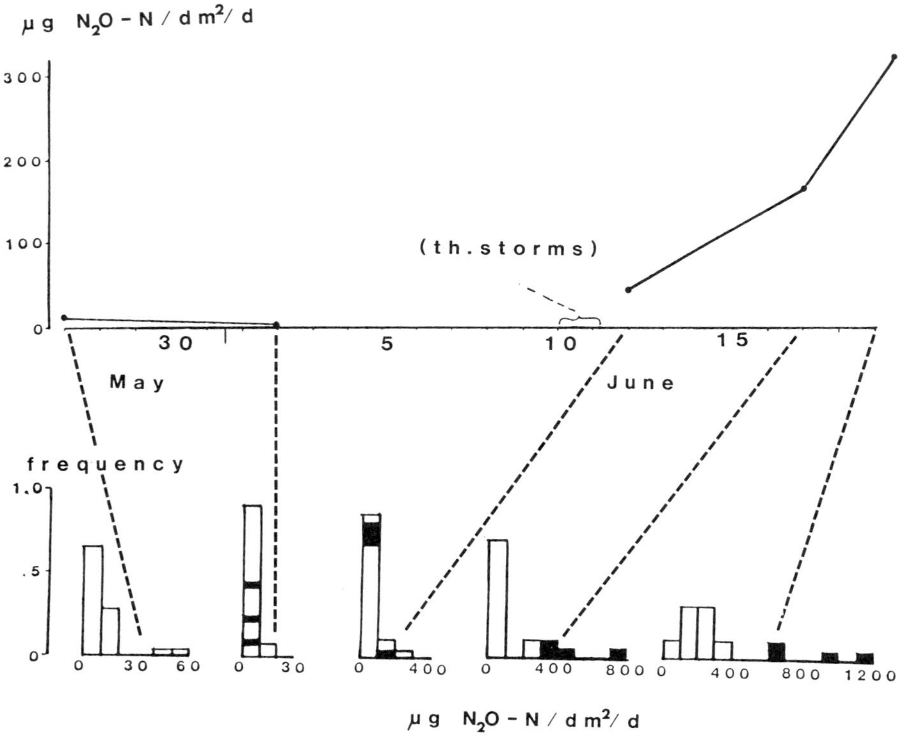

Figure 4. Denitrification as effected by heavy thunderstorms. (a), Mean rates of N_2O production (μg N/dm^2/d = g N/ha/d); (b), frequency plots of the mean rates in (a). Cores with highest rates at the last measurement are labelled black. These are shown at the relevant position in the frequency columns at the preceeding days. (Christensen et. al., 1990a).

values calculated by models of 0.4, 0.7 and 5.5 yr^{-1} obtained by Jenkinson and Ladd 1981, McGill et al. (1981) and Parton et al. (1988), respectively.

MAPPING OF PLANT-INDUCED DENITRIFICATION HOT SPOTS

Denitrification activity in soil often show a dramatic spatial variability (Folorunso and Rolston, 1984; Parkin 1987). This variability is for a large part due to the occurrence of denitrification hot spots that can be mediated by decomposing plant material (Parkin 1987).

To investigate the change in denitrification activity in these spots with time, we need to measure the process several times at the very same sites, i.e. nondestructive sampling is required. Choosing an acid soil that only produced N_2O by denitrification allowed us to determine the denitrification rate as the accumulation rate of N_2O under soil covers. Thirty permanent rings of 10 cm diameter arranged in a 5 by 6 grid with 1 m spacing were sampled as described above for ^{15}N gas at 0, 10 and 20 min (Christensen et. al., 1990a). Measurements were made under conditions of no NO_3^- limitation and with soil water contents at field capacity or above.

The results presented below are part of a study dealing with spatial and temporal variation in field denitrification and its causes. The presentation here focus on the effect of decaying plants on denitrification in the field (Christensen et. al., 1990a) and on laboratory experiments with artificial hot spots of denitrification created by addition of a carbon substrate to simulate the plant effect in the field (Christensen et. al., 1990b).

Field Study

The study was performed on a sandy loam soil of pH 3.8 at Michigan State University Farms. The plant cover on the area was dominated by Hieracum pilosella and mosses but also contained grasses and other herbs.

Denitrification activity in this area was generally very low, less than 20 ug N dm^{-2} day^{-1} (20 g N ha^{-1} day^{-1}). At very few occations e.g. following 1-2 days of heavy thunderstorms (130 mm rain) denitrification increased more than 10 times (Figure 4a). Plotting the rates in a frequency diagram shows that the data follow a skewed distribution (Figure 4b). The frequency plot also show that four points in the field were responsible for much of the increase in the mean rates and that these hot spots had been active for several days.

Whereas no correlation was observed between these spots of high denitrification activity and soil pH or water content, changes in plant appearance were in accordance with the denitrification hot spots. Following the heavy rain, plants started to decompose within three of the field cores. These three locations were all among the four denitrification hot spot cores. The four cores of high denitrification activity were brought to the laboratory and sliced into successively smaller sections. In two of the four hot spot cores (dominated by Hieracum) the whole denitrification activity could be traced to occur within 4-8 % of the core volume (Table 2) whereas the activity could not be recovered upon sectioning of

Table 2. Denitrification in whole soil cores and sections thereof. At the first slicing cores were divided in two top and two bottom sections amounting to 14 % and 36 % of the original soil mass, (marked a and b respectively). At the second sectioning, the top sections (a), were divided in three giving subsections of 4 % of the original soil mass. Values are μg N_2O-N/core section/day (Christensen et. al., 1990a).

	Whole core		First sectioning	Second sectioning		
	Field	Laboratory				
core 1	152	126	104^a	0	0	126
			34^a	0	0	66
			0^b			
			0^b			
core 2	129	93	98^a	0	83	27
			0^a	0	0	0
			0^b			
			0^b			

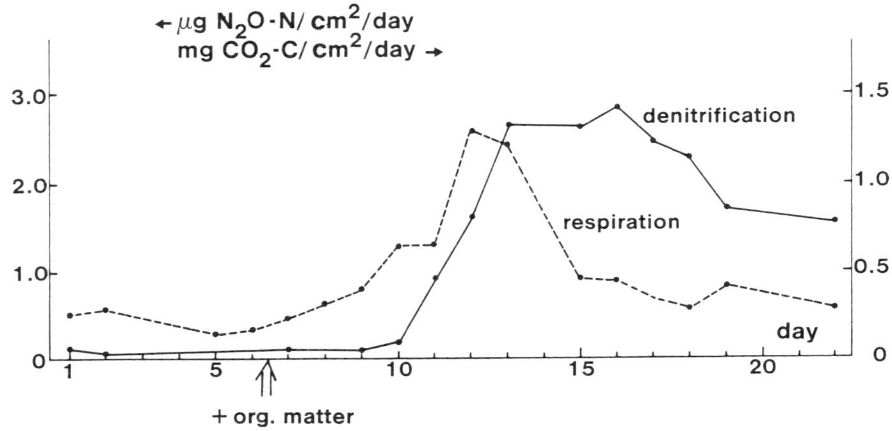

Figure 5. Respiration and denitrification in a soil core in the laboratory following injection of organic matter (Christensen et. al., 1990b).

the last two cores (dominated by grasses and almost bare soil, respectively, Christensen et al. 1990a). This observation is in accordance with an earlier study tracing the whole denitrification activity in a soil core to less than 1 % of the soil mass (Parkin 1987). The rate of denitrification in these hot spots was close to the potential rate (Christensen et. al., 1990a) as suggested in a recent model suggesting denitrification to occur at the potential rate in a few spots distributed in a soil matrix of no activity (Parkin and Robinson, 1989).

Artificial hot spots in the laboratory

To better understand the events occurring during the formation of denitrification hot spots, experiments with artificial hot spots could be advantageous. Figure 5 shows the changes in respiration and denitrification occurring in one soil core with the same soil as above following injection of 9 ml of particulate organic matter (dead E. coli cells, 18.8 % DM) to a depth of 4 cm (Christensen et. al., 1990b). The respiration starts to increase right after the addition of POM. Denitrification, however, did not respond to the addition until respiration had increased for several days. Applying the same carbon source to anaerobic soil resulted in an immediate increase in denitrification at POM addition compared to unamended soil. (Christensen et. al., 1990b). The indigenous denitrifying enzymes in this soil were thus able to respond to the addition given a favourable environment. The lag observed in the cores with atmospheric air (Fig 5) must therefore be due to the time required to create an anaerobic zone at the point of injection of organic matter. At conditions of no NO_3^- limitation denitrification activity can therefore serve as a marker of anaerobic soil volumes, i.e. when and where high microbial activity occur.

SUMMARY

The organic matter driving soil decomposition processes and thus denitrification is of plant origin. A short term field experiment with ^{15}N on a poorly drained soil showed that even though denitrification rate was relatively high it only amounted to 1-2 % of the immobilization plus nitrification rates.

Moreover the experiment indicated that turnover rate for biomass ^{15}N was 2 and 2.3 times higher in planted than unplanted soil for NH_4^+ and NO_3^- label respectively. This impact of active roots on microbial turnover in the soil might ultimately be very important to the functioning of the soil system. Since denitrification is a process of minor quantitative importance, plant effects on this process will not have much influence on the state of the soil environment.

Denitrification variability indicates spatial and temporal variability in the dominating processes in the plant soil system, i.e. mineralization of organic matter. In an acid sandy loam soil, the occurrence of denitrification hot spots was often related directly to variations in supply of organic matter of plant origin. These observations suggest that measurement of denitrification can be used as a bioassay for the spatial and temporal variability in the main processes occurring in the plant-soil system even though denitrification in itself is of minor importance.

REFERENCES

Andersen, T.K., Jensen, M.B., and Sørensen, J., 1984, Diurnal variation of nitrogen cycling in coastal, marine sediments I. Denitrification. Mar. Biol. 83, 171- 176.

Bakken L.R. 1988. Denitrification under different cultivated plants: effects of soil moisture tension, nitrate concentration, and photosynthetic activity. Biol. Fert. Soils 6, 271-278.

Barber, D.A., and Martin, J.K., 1976, The release of organic substances by cereal roots into soil. New Phytol. 76, 69-80.

Broadbent, F.E., and Tusnem, M.E., 1971, Losses of nitrogen from some flooded soils in tracer experiments. Soil Sci. Soc. Am. J. 35, 922-926.

Brown, M.E., 1975, Rhizosphere microorganisms opprotunists, bandits or benefactors, in "Soil Microbiology," N. Walker, ed., pp 21-38, Butterworth Scientific, London.

Christensen, S., 1985, Denitrification in an acid soil: effects of slurry and potassium nitrate on the evolution of nitrous oxide and on nitrate reducing bacteria. Soil Biol. Biochem. 17, 757-764.

Christensen, S., and Tiedje, J.M., 1988, Oxygen control prevents denitrifiers and barley plant roots from directly competing for nitrate. FEMS Microbiol. Ecol. 53, 217-221.

Christensen, S., Simkins, S., and Tiedje, J.M., 1990a, Spatial variation in soil denitrification: Occurrence of activity centers (hot spots) as influenced by the soil environment. subm. Soil Sci. Soc. Am. J.

Christensen, S., Simkins, S., and Tiedje, J.M., 1990b, Activity centers (hot spots) of denitrification in the field. Mechanisms causing their formation. subm. Soil Sci. Soc. Am. J.

Clark, F.E., and Paul, E.A., 1989, Soil microbiology and biochemistry. Academic Press, San Diego, USA.

Clarholm, M., 1985, Possible roles of roots, bacteria, protozoa and fungi in supplying nitrogen to plants, in "Ecological interactions in soil," A.H.Fitter, ed., Blackwell Scientific Publications, Oxford.

Coleman, D.C., Anderson, R.V., Cole, C.V., Elliott, E.T., Woods, L., and Campion, M.K., 1978, Trophic interactions in soils as they affect energy and nutrient dynamics. IV. Flows of metabolic and biomass carbon. Microb. Ecol. 4, 373- 380.

Folorunso, O.A., and Rolston, D.E., 1984, Spatial variability of field-measured denitrification gas fluxes. Soil Sci. Soc. Am. J. 48, 1214- 1219.

Garcia, J.L., 1975, Effet rhizosphere du riz sur la denitrification. Soil Biol. Biochem. 7, 139-141.

Haider, K., Mosier, A., and Heinemeyer, O., 1985, Phytotron experiments to evaluate the effect of growing plants on denitrification. Soil Sci. Soc. Am. J. 49, 636-641.

Haider,K., Mosier, A., and Heinemeyer, O., 1987, The effect of growing plants on denitrification at high soil nitrate concentrations. Soil Sci. Soc. Am. J. 51, 97-102.

Helal, H.M., and Sauerbeck, D., 1986, Effect of plant roots on carbon metabolism of soil microbial biomass. Zeitsschr. Pflanzenernahr. Bodenk. 149, 181-188.

Helal, H.M., and Sauerbeck, D., 1987, Direct and indirect influences of plant roots on organic matter and phosphorus turnover in soil. Intecol Bull. 15, 49-58.

Jenkinson, D.S., and Ladd, J.N., 1981, Microbial biomass in soil: Measurement and turnover, in "Soil Biochemistry vol. 5," E.A.Paul and J.N.Ladd, eds., Marcel Dekker Inc., New York and Basel.

Johnen, B.G., and Sauerbeck, D.R., 1977, A tracer technique for measuring growth, mass and microbial breakdown of plant roots during vegetation, in "Soil organisms as Components of Ecosystems," Ecol. Bull. (Stockholm) 25, 366-373.

Klemendtsson, L., Svensson, B.H., and Rosswall, T., 1987, Dinitrogen and nitrous oxide produced by denitrification and nitrification in soil with and without barley plants. Plant Soil 99, 303-319.

Mackown, C.T., Brooks, P.D., and Smith, M.S., 1987, Diffusion of nitrogen-15 Kjeldahl digests for isotope analysis. Soil Sci. Soc. Am. J. 51, 87-90.

Martin, J.K., 1977, Factors influencing the loss of organic carbon from wheat roots. Soil Biol. Biochem. 9, 1-7.

Mercks, R., Hartog, A.D., and vanVeen, J.A., 1985, Turnover of root derived material and related microbial biomass formation in soils of different texture. Soil Biol. Biochem. 17, 565-569.

McGill, W.B., Hunt, H.W., Woodmansee, R.G., and Reuss, J.O., 1981, Phoenix, a model of the dynamics of carbon and nitrogen in grassland soils, in, "Terrestrial nitrogen cycles," F.E.Clark and T.Rosswall, eds., Ecol. Bull. (Stockholm) 33.

Mosier, A.R., Guenzi, W.D., and Schweitzer, E.D., 1986, Soil losses of dinitrogen and nitrous oxide from irrigated crops in northeastern Colorado. Soil Sci. Soc. Am. J. 50, 344-348.

Myrold, D.D., and Tiedje, J.M., 1985, Diffusional constraints on denitrification in soil. Soil Sci. Soc. Am. J. 49, 651-657.

Parkin, T.B., 1987, Soil microsites as a source of denitrification variability. Soil Sci. Soc. Am. J. 51, 1194-1199.

Parkin, T.B., and Robinson, J.A. 1989, Stochastic model of soil denitrification. Appl. Env. Microbiol. 55, 72-77.

Parton, W.J., Schimel,D.S., and Cole,C.V., 1988, Dynamics of C, N, P and S in grassland soils: a model. Biogeochem. 5, 109- 131.

Reddy, K.R., and Patrick, W.H., 1986, Fate of fertilizer nitrogen in the rize root zone. Soil Sci. Soc. Am. J. 50, 649-651.

Robinson, D., Griffiths, B.S., Ritz, K., and Wheatley, R.E., 1989, Root driven N mineralization: A theortetical approach. Plant Soil.

Rolston, D.E., Sharpley, A.N., Toy, D.W. and Broadbent, F.E., 1982, Field measurement of denitrification: III. Rates during irrigation cycles. Soil Sci. Soc. Am. J. 46, 289-296.

Rouatt, J.W., Katznelson, H., and Payne, T.M.B., 1960, Statistical evaluation of the rhizosphere effect. Soil Sci. Soc. Am. Proc. 24, 271- 273.

Smith, M.S., and Tiedje, J.M., 1979, The effect of roots on soil denitrification. Soil Sci. Soc. Am. J. 43, 951-955.

Terry, R.L., Jellen, E.N., and Breakwell, D.P., 1986, Effect of irrigation method and acetylene exposure on filed denitrification measurements. Soil Sci. Soc. Am. J. 50, 115-120.

von Rheinbaben, W., and Troldenier, G., 1984, Influence of plant growth on denitrification in relation to soil moisture and potassium nutrition. Zeitsschr. Pflanzenernähr. Bodenk. 147, 730-738.

Woldendorp, J.W., 1963, The influence of living plants on denitrification. Medelingen Landbouwhogeschole Wageningen 63, 1-100.

CHARACTERIZING THE VARIABILITY OF SOIL DENITRIFICATION

Timothy B. Parkin

U.S. Department of Agriculture-ARS
Environmental Chemistry Laboratory
Beltsville Agricultural Research Center-West
Beltsville, Maryland 20705, USA

INTRODUCTION

It is generally recognized that quantification of the spatial variability associated with soil processes is important, and over the past 5-10 years a wide variety of fairly novel statistical techniques have been applied to the study of soil variables in an attempt to characterize variability including i) geostatistics (Burgess and Webster, 1980; Vieira et al., 1983; Vauclin et al., 1983), ii) spectral and autocorrelation analysis (Kachanoski et al., 1985a; 1985b), iii) the state space approach (Morkoc et al., 1985), and iv) bootstrapping (Dane et al., 1986). The spatial variability exhibited by soil denitrification is among the highest reported for soil processes, with coefficients of variation in the range of 100-500% (Folorunso and Rolston, 1984; 1985; Parkin et al., 1987; Myrold, 1988). While there has been much emphasis on applying novel techniques to characterize variability, less emphasis has been placed on trying to answer the question, "How does characterizing variability really help us understand the process under study?"

The factors which are important in controlling denitrification are well known from laboratory studies with both soil and pure cultures. Figure 1 outlines these factors and illustrates some of the ways these factors interact in soil to control denitrification. The three factors which exert a direct effect on soil denitrification are organic carbon, oxygen, and nitrate (depicted by the heavy arrows). The effects of these three primary factors are influenced by a number of secondary interactions. Soil water content exerts diffusional constraints on soil oxygen (A). Soil water also influences the availability of organic carbon released through wetting and drying cycles (B). Organic carbon has an effect on nitrate availability which is manifested through mineralization/immobilization reactions (C). Nitrate availability is also subjected to diffusional constraints imposed by soil water content (D). Soil water, by affecting general microbial activity in soil, controls the amount of available organic carbon (E). Finally, microbial respiration of the available organic carbon can have a dramatic effect on oxygen levels at the microsite level (F).

Although not explicitly indicated, soil structure imposes an additional layer of control upon the interactions depicted in Fig. 1. Thus, while these interactions are, conceptually, not complex, the heterogeneous nature of the soil matrix makes

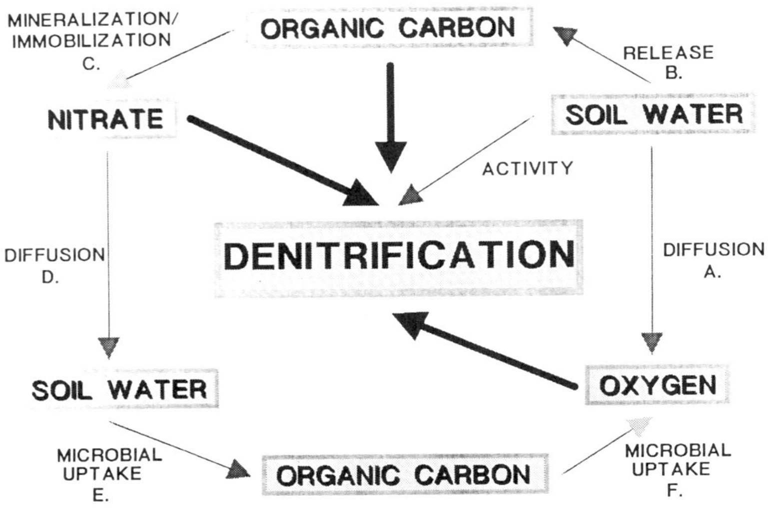

Figure 1. Interactions of soil factors which influence denitrification.

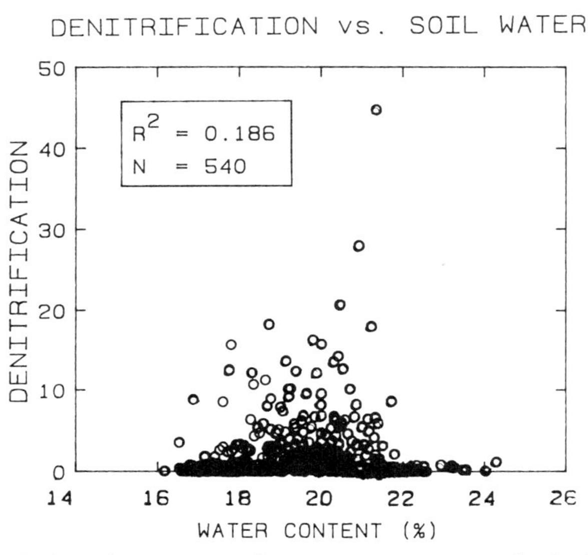

Figure 2. Correlation between soil water content and denitrification rate (ng-N/g/d).

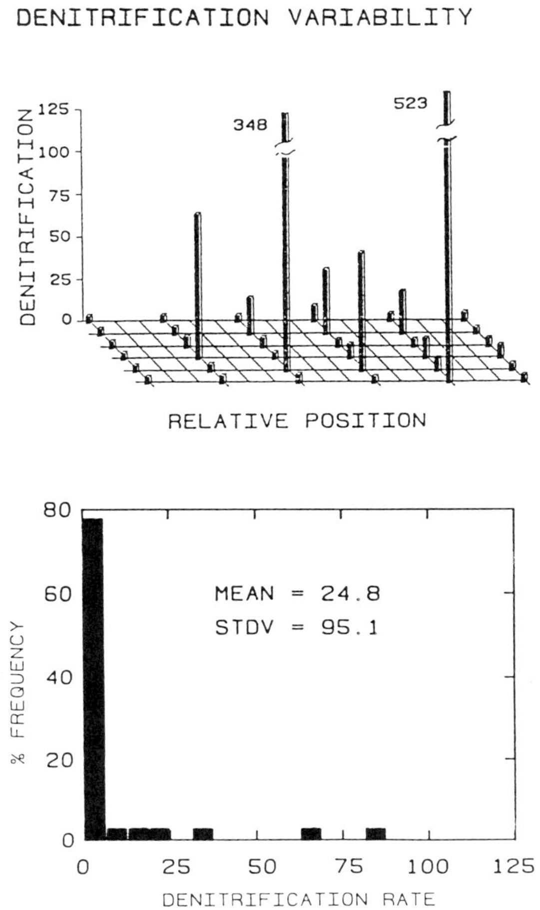

Figure 3. Spatial plot of denitrification rates (ng N/g/d) and sample histogram of 36 intact soil cores.

the relationships between soil factors and denitrification difficult to quantify. For example, we know qualitatively that wetter soils support higher denitrification rates than drier soils, but due to high variability it is difficult to quantify this relationship. Figure 2 shows a scatter plot of water content vs. denitrification rate measured in 540 intact soil cores using the acetylene block technique. Regression analysis of these data resulted in a correlation coefficient of 0.186. The correlation between soil water content and denitrification is significant due to the high degree of replication, but because of the high variability this relationship has poor predictive power.

In order to deal more effectively with high variability, the factors or sources responsible for variability must be identified. The strength of this approach lies in the fact that identification of the factors responsible for the high variability leads to identification of the factors important in controlling denitrification in soil. In addition, the study of variability provides a mathematical or statistical framework which is useful in elucidating, in a quantitative manner, both the interactions involved in controlling denitrification as well as estimating the magnitude of denitrification-N loss from soils.

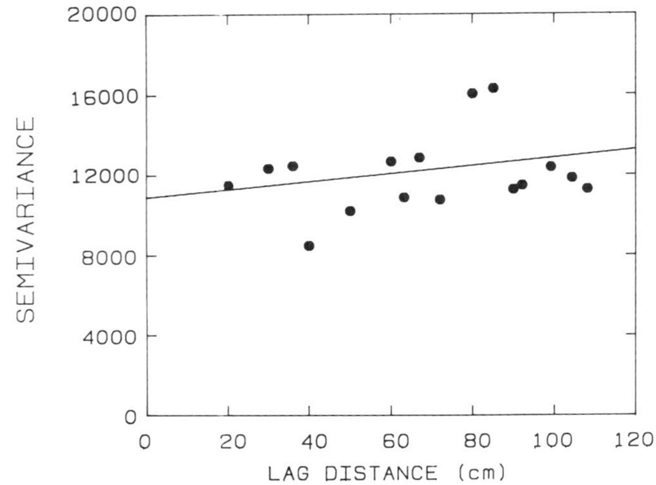

Figure 4. Semivariogram of denitrification data presented in Fig. 3.

The objective of this paper is to discuss some of the factors which are perceived to be important in controlling denitrification variability and to illustrate how characterizing variability can be a useful approach both in quantifying denitrification and in identifying the factors which control denitrification in nature. In one sense denitrification can be viewed as a model system for the study of soil microbial processes, as some of the approaches and techniques discussed here can be applied to investigations of other natural processes.

CHARACTERIZING VARIABILITY

There are several techniques which can be used in characterizing variability, including: i) computing a summary statistic (variance or coefficient of variation), ii) constructing a spatial plot of the data, iii) characterizing the shape of the underlying distribution, and iv) using geostatistics. The use of a single summary statistic such as the variance or coefficient of variation is the most commonly applied technique for estimating the magnitude of the variability. Constructing a spatial plot of the sample data is a simple way of obtaining a visual picture of the magnitude of the spread in a given data set and to observe if any spatial pattern exists. The shape of the underlying distribution is very important piece of information which influences data interpretation as well as decisions concerning the appropriate statistical techniques to use. Finally, geostatistics, a statistical method developed in the mining industry, has recently found application in the soil sciences. The main tool of geostatistics is the semivariogram which quantifies the spatial structure of the variability.

The techniques listed above have been applied to denitrification rate data determined using the intact soil core-acetylene block technique (Fig. 3). The X-Y plane (Fig 3, top) indicates the relative positions of 36 individual soil cores collected in a small 1.2 x 2 meter area of a field, and the Z axis indicates the denitrification rates of the individual samples (Parkin et al., 1987). No apparent spatial pattern is observed, and adjacent samples can exhibit rates which vary by

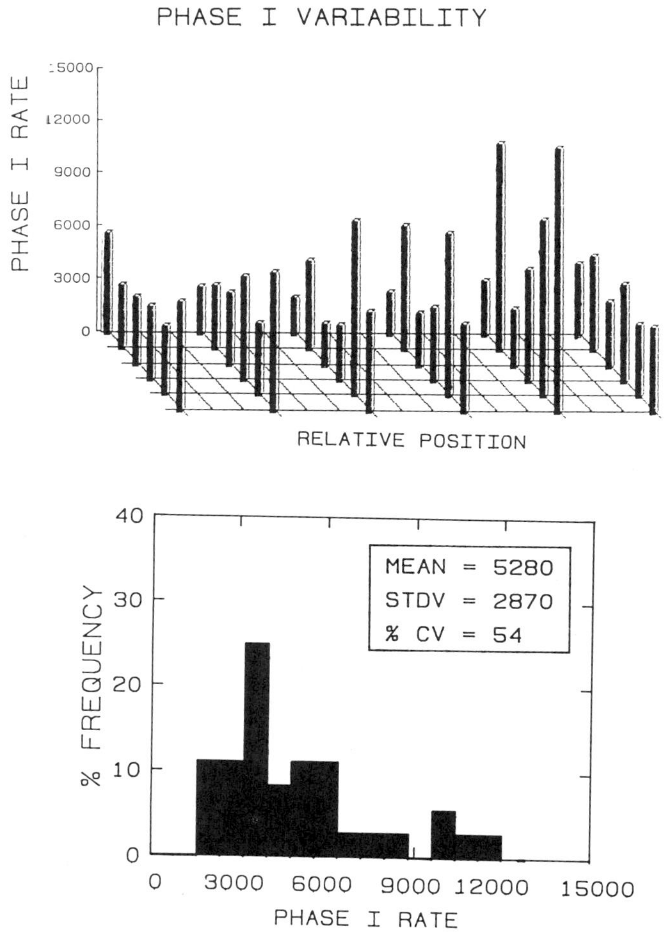

Figure 5. Spatial plot and sample histogram of denitrification enzyme activity (ng N/g/d) associated with core data of Fig. 3.

2-3 orders of magnitude. This data set had a mean denitrification rate of 24.8 ng N/g/d and a coefficient of variation of nearly 400%. The sample histogram (Fig. 3, bottom) indicates that the data are positively skewed, with most samples having very low rates, but a few samples exhibiting very high rates. This sample distribution is well approximated by a lognormal distribution.

Semivariogram analysis of this data set is presented in Fig 4. The relatively flat slope of the regression line indicates that there is little spatial dependence of the variability. Rather, the high value of the Y-intercept suggests that most of the observed variability occurs at the microscale level. This microscale variability (distances <20 cm) accounts for 84% of the total variability observed for this data set. Similar results have been observed by Folorunso and Rolston (1984) who observed no spatial structure of denitrification gas fluxes measured at 1 m interval intervals in field soils.

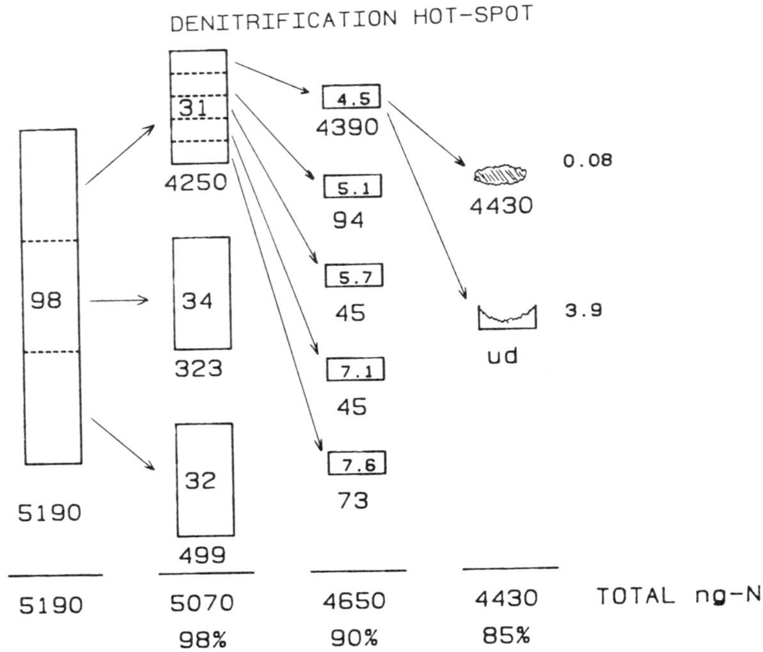

Figure 6. Results from a core segmentation experiment to identify active denitrifying sites in soil (from Parkin, 1987). All incubations were conducted in an air atmosphere with 10% acetylene. Values inside each soil segment refer to the mass (g) of that segment, and the denitrification rate (ng-N/segment/day) appears below each segment. Values at bottom of figure refer to % of original activity observed in the initial incubation of the 15 cm, 98 g soil core.

The high variability observed for this data set is the result of a few samples which exhibit very high rates, prompting the question, "Why are these high rate samples different?". It is unlikely that the answer to this question is due to a heterogeneous spatial dispersion of denitrifying organisms. Figure 5 shows the spatial plot of the phase I denitrifying enzyme activity of the individual samples as determined by the anaerobic slurry technique of Smith and Tiedje (1979). Dispersion of enzyme activity is fairly homogeneous and the rates are, on the average, 20-fold greater than those observed for the individual cores. The fact that high denitrification potential exists in every sample suggests that it is not the heterogeneous dispersion of organisms which causes the high variability, rather the variability is due to factors regulating the expression of this activity.

SOURCES OF DENITRIFICATION VARIABILITY

In an attempt to identify the source of denitrification variability a series of incubations were performed on segmented intact core samples (Fig. 6). The initial incubation of a 98 g intact core revealed a denitrification rate of 5190 ng N / day, (specific rate of 53 ng N/g soil/day). Through repeated incubations of segments of this core, a small piece of organic matter was identified which supported a rate of 4430 ng N/day (specific rate of 55000 ng N/g/day). Thus, 85% of the activity observed in the initial incubation of the intact core was due to the denitrification activity associated with an 0.08 g piece of organic matter.

Table 1. Influence of sample volume on estimates of field denitrification rates.

Core diameter	Core mass	Denitrification		
		Median	Mean	Std.dev.
- cm -	- g -	---	ng-N $g^{-1}d^{-1}$	---
1.73	61.8	0.82[a]	3.14[a]	8.41
2.15	96.8	1.57[ab]	24.8[b]	95.1
3.45	265.0	1.57[ab]	12.8[b]	45.9
4.15	371.0	3.10[bc]	16.6[b]	52.7
5.40	612.0	4.05[c]	26.4[b]	89.2

Means followed by the same letter are not significantly different as determined by overlap of 95% confidence limits.
Medians followed by the same letter are not significantly different as determined by Tukey's LSD on ln-transformed sample data (P>0.05).

These experiments indicate that a possible source of denitrification variability is the patchy dispersion of particulate organic carbon supporting high specific denitrification rates. High denitrification rates associated with particulate plant carbon have also been observed to contribute to the high temporal and spatial variability observed in a sandy loam soil (Christensen et al., 1989).

Microsites of high denitrification may be predominantly associated with particulate organic matter in structureless soils (Parkin, 1987; Christensen et al., 1989) or associated with the anaerobic zones existing in the aggregates of a structured soil (Sexstone et al., 1985; chapters 14 and 15, this book). A computer simulation presented by Parkin (1987) provides supporting evidence that the patchy dispersion of denitrification hotspots may be a significant source of the observed sample variability.

The two predominant characteristics of denitrification variability can be summarized: i) sample data are well described by the lognormal distribution, and ii) a preponderance of the observed variability is expressed at the microscale level.

VARIABILITY CONSIDERATIONS IN QUANTIFYING DEINTRIFICATION

The most important information which will used in quantifying soil denitrification is the shape of the sample frequency distribution. Adoption of the lognormal model has several implications regarding sampling, data analysis, and data interpretation including: i) selection of the appropriate location parameter as the summary variable, ii) selection of appropriate statistics to estimate location and scale, iii) determining sample number requirements, and iv) determining sample size (volume) necessary to obtain a representative sample.

The application of these considerations are illustrated from a series of experiments performed to investigate the influence of sample volume on estimates of field denitrification rates. Table 1 shows denitrification rate estimates obtained using 5 different size soil cores. Thirty-six of each of 5 different sized soil cores were collected from a small area of a field plot (1.2 x 2 m) and denitrification rates were measured using the acetylene inhibition technique (Parkin et al., 1987).

The core samplers used ranged from 1.7 cm to 5.4 cm in diameter, and the average mass of a soil core collected ranged from 62 to 612 g. The median and mean denitrification rates, along with the associated standard deviations for each size are indicated in Table 1.

A number of questions can be raised concerning the data in Table 1. i) Why, for a given sample size, are the mean and median different?, ii) Why does the median tend to increase with increasing sample size?, iii) Why are the mean and variance estimates using the smallest core size different from the other sizes, and iv) Which is the appropriate summary location parameter for estimating field denitrification; the mean or median? We can use our knowledge of the variability of the process to answer these questions.

The answer to question 1 is obvious when we consider that the data can be described by a lognormal distribution. Since the lognormal distribution is asymmetric, the mean and the median have different values. The mean is defined by $\exp(\mu + \sigma/2)$ and the median defined by $\exp(\mu)$, where μ and σ^2 are the mean and variance of the log-transformed variable. These two parameters actually yield different information about the distribution. The mean is the center of mass of the distribution while the median is the center of probability of the distribution. Thus, considering the asymmetric nature of distributions exhibited by denitrification rates, it is expected that the mean and median will have different values.

Question 2, "Why does the median increase with increasing sample volume?", may be addressed by considering the central limit theorem, which states that regardless of the shape of the underlying distribution, the distribution of means approaches normality as sample size (n) increases. This effect can be illustrated using Monte Carlo simulation (Fig. 7). When 1000 samples are drawn from a lognormal distribution having a mean of 400 and a median of 179, the resulting sample mean and median are 391 and 179, respectively (Fig 7, top). When samples are randomly selected from this batch of 1000 samples and averaged in groups of 2, the resulting sample histogram is less skewed and has a mean and median of 391 and 246, respectively (Fig. 7, center). If this bulking process is repeated by averaging groups of 10 samples drawn from the original 1000 samples, the resulting histogram of means is more symmetric and has a mean and median of 391 and 346, respectively (Fig. 7, bottom). Thus, through the process of bulking the distribution approaches symmetry and the value of the median approaches the value of the mean. With regard to the data of Table 1, since a majority of the variability occurs at the microscale level, increasing core size is equivalent to bulking of small samples. Thus, collecting 1 of the large-size cores (612 g) is equivalent to bulking 10 of the smallest size cores (61.8 g), which, as predicted by Fig. 7, will result in a greater value for the median.

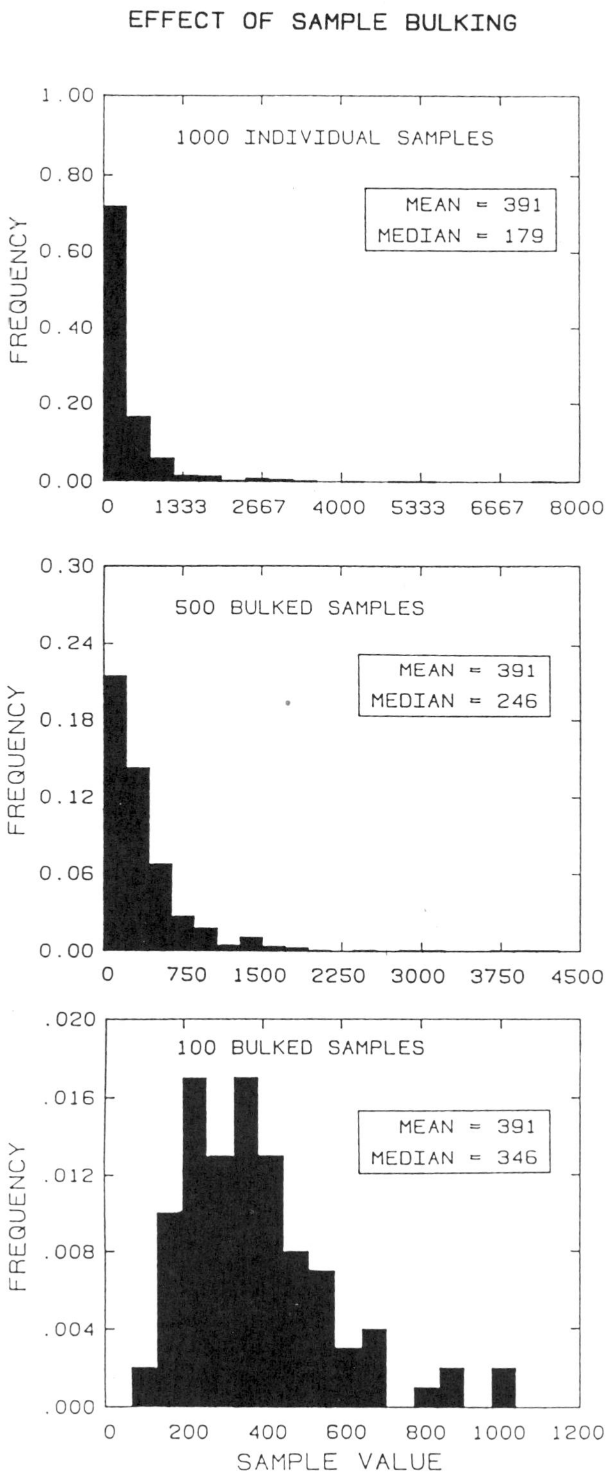

Figure 7. Influence of sample bulking on the mean and median.

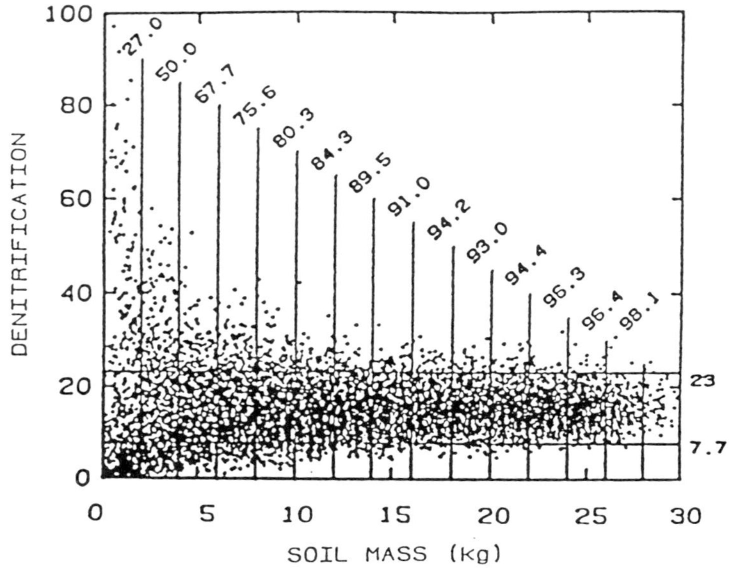

Figure 8. Representative soil mass estimates for denitrification rates (ng-N/g/d), from Parkin et al. (1987).

The previous discussion indicates that the mean should remain constant, however the data in Table 2 indicate that the mean for the smallest core size is significantly less than the means of the other sizes. One possible reason for this result is that 36 of the small-sized core samples is not enough soil to provide a representative sample of the field. This possibility was explored by using computerized random resampling (Parkin et al., 1987). For this evaluation all 180 rates (36 of each of the 5 different sized samples) were pooled and subsamples were randomly selected. The weighted average denitrification rate was calculated for each of these subsamples and plotted against the total soil mass represented by the subsample. Figure 8 shows the results of 10,000 of these subsamplings. The two horizontal lines indicate $\pm$ 50% of the weighted average of the 180 samples (taken to be the best estimate of the population mean), and the vertical lines divide the graph into 2 kg soil segments. The numbers at the top of the graph indicate the percentage of occurrences the mean rate of a given subsample falls within $\pm$ 50% of our estimate of the population mean. It is seen that when only 0-2 kg of soil is collected a representative estimate of the population mean is obtained only 27% of the time. This result indicates that the soil mass represented by 36 of the small size cores (36 of the smallest core size = ca. 2 kg) is not representative of the field and denitrification is underestimated. Figure 8 also indicates that, at this site, collection of 12-16 kg soil will result in an estimate which is within $\pm$ 50% of the population mean approximately 90% of the time. An alternate explaination for the observation of significantly lower denitrification rates observed in the smallest soil cores is the disruption of soil structure (chapters 14 and 15, this book). However, it is unlikely that this is the case for these soils. Parkin et al. (1987) determined denitrification rates on soil samples having different surface area:volume ratios and concluded that spatial heterogeneity of active denitrifying sites were responsible for observed variability, and that altering the surface area:volume ratio of samples did not alter the observed denitrification rates of samples.

Another consequence of bulking or averaging which occurs when larger samples are collected is that the variance should decrease. This is not observed in Table 1, and no significant difference was observed between variances of the sample sizes. It must be recognized, however, that variance estimates from sample data have a high degree of associated uncertainty. The variances of the variance estimates presented in Table 1 (calculated using eq. 24 of Parkin et al., 1988) ranged from 7.9×10^5 to 1.6×10^{11}. Given the high uncertainty associated with the variance estimates in Table 1, it is not surprising that significant differences were not observed.

The question, "Should the mean or median be used as the summary parameter for denitrification?", arises because of the asymmetric distribution exhibited by denitrification. The answer to this question is not straightforward, and really depends upon the objectives of the study. For the objective of quantifying the mass of N-loss by denitrification from soil, the mean is the appropriate summary parameter, as the sample mean is the best estimator of the center of gravity (mass) of the distribution.

Choice of the mean as the summary location parameter raises questions concerning the appropriate statistical methods for estimating the population mean from lognormal sample data. Confusing recommendations exist in the literature concerning parameter estimation as well as the calculation of confidence limits for the mean. For example Sokal and Rolf (1969) suggest that it is appropriate, for lognormally distributed data, to first perform a normalizing transformation by taking the logarithms of the sample data. Then the mean and associated confidence limits of the log-transformed values are computed, followed by a back transformation to obtain an estimate of the mean and confidence limits on the untransformed scale. However, this procedure actually yields an estimate of the population median (recall that for the lognormal distribution the population median = e^{μ}, where μ is the mean of the log-transformed variable). Several statistical procedures for estimating the mean, variance, and coefficient of variation from sample data were recently evaluated where the optimum parameter estimation methods were identified and described (Parkin et al., 1988). In a subsequent study (Parkin et al., in press) five methods for calculating confidence limits for the mean of a lognormally distributed variable were evaluated, and an exact method was identified.

MODELING DENITRIFICATION VARIABILITY

The second objective of this paper is to examine how an understanding of the variability will aid in developing quantitative relationships between denitrification and soil/environmental factors. The ultimate goal underlying this objective is to develop a mathematical model of denitrification. Many deterministic models of denitrification have been developed (Christensen, 1985; Mosier,1985; Rolston et al., 1984). However, it may be unreasonable to expect that a highly variable process such as soil denitrification can be adequately described by a deterministic model.

In a recent study of Parkin and Robinson (1989), probabilistic (stochastic) models were developed to understand how positively skewed distributions for field denitrification rate estimates result from the combined influences of variables known to affect denitrification. The assumption underlying this modelling effort is that the distribution observed for measured denitrification

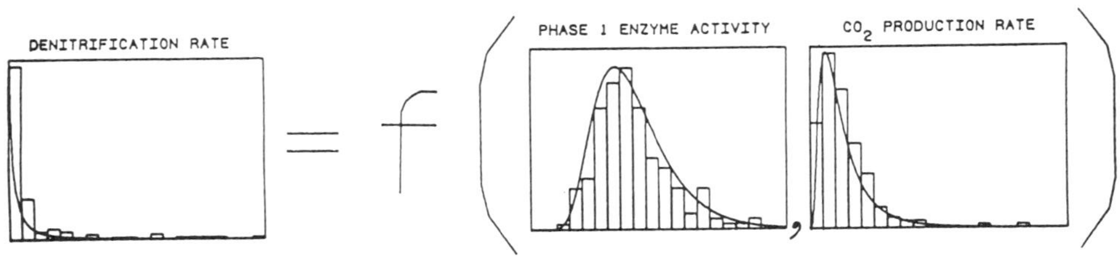

Figure 9. General form of stochastic models (reprinted with permission from Parkin and Robinson, 1989).

rates is functionally related to the distributions of phase I enzyme activity and CO_2 production rates (Fig. 9). The choice of these two driving variables for the model arises from the hotspot concept described earlier.

Experimental results indicate that high denitrification potential is present in every sample, yet only a fraction of this potential is expressed in situ. Therefore, the variability associated with natural denitrification rate estimates is not simply controlled by the presence of denitrifying enzymes, but by factors controlling enzyme expression. Also, earlier work has shown that available carbon as estimated by CO_2 production rates of both bulk soil and individual pieces of particulate organic carbon show positive correlations with denitrification (Rice et al., 1988; Parkin, 1987). A recent study indicates that denitrification rates of soil aggregates are limited by availability of organic carbon (Seech and Beauchamp, 1988). Therefore, CO_2 production is used as a driving variable controlling the expression of denitrification enzyme activity in these models since CO_2 production activity indicates both organic C availability as well as soil aeration status (ie. O_2 consumption activity).

Several function relationships were developed for the general model shown in Fig. 9, including a second order model (Eq. 1) and a second order model with threshold (Eq. 2).

Model 1:

$$\text{Denitrification} = K_1 * \text{phase I} * CO_2 \quad (1)$$

where: K_1 = pseudo second-order rate coefficient

Model 2:

$$\text{Denitrification} = K_1 * \text{phase I} * (CO_2 - K_2) \quad [CO_2 > K_2]$$
$$\text{Denitrification} = 0.05 \text{ ng-N/g/d} \quad [CO_2 \leq K_2] \quad (2)$$

where: K_1 = pseudo second-order rate coefficient
K_2 = threshold rate of CO_2 production
0.05 = the detection limit of the denitrification rate assay

Parameter estimates for these models were obtained through nonlinear regression and the models were validated using 3 large independent data sets.

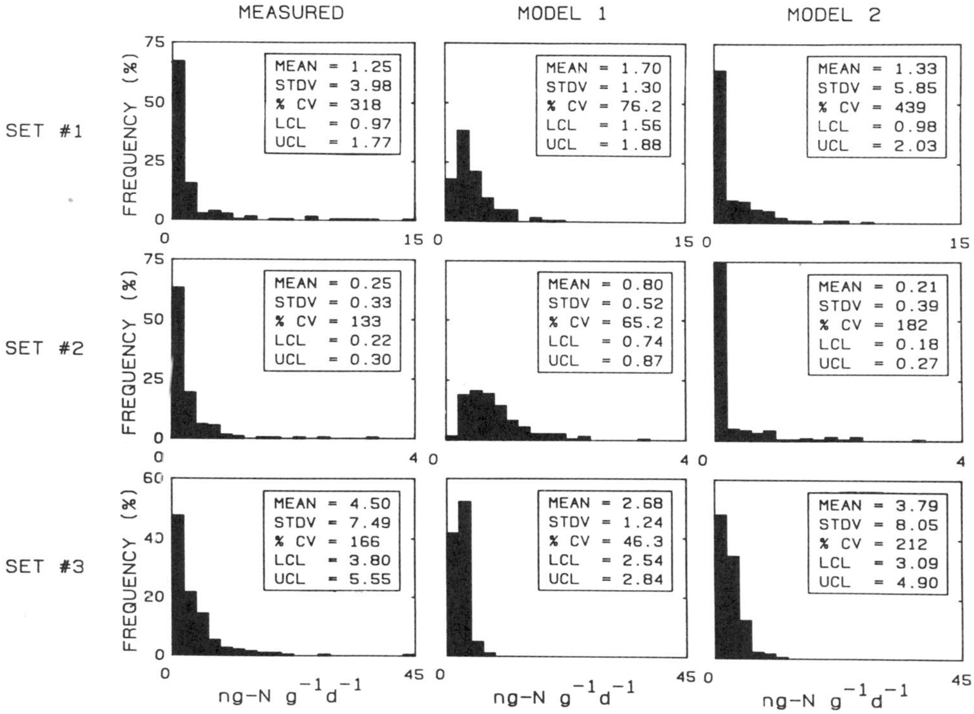

Figure 10. Results of stochastic models. Measured soil core denitrification rates for three validation data sets along with denitrification rate distributions predicted from models 1 and 2 (from Parkin and Robinson, 1989).

In the implementation of the stochastic models, Monte Carlo simulation was used to randomly select variates from the lognormal probability density functions describing the frequency distributions of phase I activity and CO_2 production. These values were then combined according to the functional relationships expressed by Eqs. 1 and 2 to obtain a single predicted denitrification rate. This process was repeated 180 times (n=180) to obtain a distribution of predicted rates for each of the two models. The models were then evaluated by comparing the predicted denitrification rate distributions with the measured denitrification rate distributions obtained for each validation data set (Figure 10).

It is evident from the shape of the histograms that model 2 gave populations of denitrification rates similar to the measured rates. Mean rate estimates obtained from model 2 were not significantly different ($P>0.1$) from the measured rates for all three data sets, as indicated by the overlap of the 95% confidence limits.

One advantage of taking a stochastic approach is that in addition to predictions of the mean denitrification rate, variance estimates are also obtained. Thus, variability can be used as a diagnostic in model discrimination. Model 2 accurately described the variability exhibited by natural denitrification rate estimates. Coefficients of variation of predicted rates were very similar to the actual rates, and variances of the predicted rates from model 2 were not significantly different ($P>0.1$) from the measured denitrification rates. Model 1 only yielded an accurate prediction of mean denitrification for validation data set

1, and variances predicted from model 1 underestimated the measured variances for all three validation data sets.

Due to the high degree of microscale heterogeneity, Fluhler et al. (1976) recommend a statistical approach to describe soil aeration status. The basis for this recommendation is that a single summary parameter such as the mean may be insufficient to describe highly variable soil process. Rather, the entire frequency distribution of the variable under study should be considered. The stochastic approaches of modeling denitrification rates measured in intact soil cores (Parkin and Robinson, 1989), and of modeling denitrification at the microsite level (chapter 15, this book) incorporate the high variability typically exhibited by soil denitrification by considering the frequency distributions of the underlying variables thought to control denitrification in soil.

SUMMARY

A feature common to measurements of many soil processes is a high degree of associated variability. Rather than ignoring this variability, it should be recognized that variability contains information concerning the underlying factors controlling natural processes. The nature of the variability, and in particular, the underlying frequency distribution, provides valuable insights into the process under study. The rationale for this view was early recognized:

> Knowledge of the connection between the form of the curves and the causes to which the form is due may lead us, in some cases, to precious indications about the nature of the true causes (Kapteyn, 1903).

In this paper, an effort has been made to illustrate how an exploration of the factors contributing to denitrification variability can be used to extract important information concerning the mechanisms controlling the denitrification process in soil. Realistically, it may be impossible to completely identify and characterize the factors contributing to the variability of a given process. However, even a partial knowledge of the factors contributing to the underlying variability can aid in: i) sampling and experimental design, ii) data interpretation, iii) making decisions concerning the application of appropriate statistical analyses, and iv) model discrimination. An increased understanding of the factors contributing to the high variability of soil denitrification (such as the form and shape of the frequency distributions associated with individual denitrifying microsites) should have a direct impact on our understanding of the factors which are important in controlling this process as well as on our ability to predict soil denitrification.

REFERENCES

Burges, T.M. and Webster, R., 1980, Optimal interpolation and isarithmic mapping of soil properties. I. The semi-variogram and punctual kriging, J. Soil Sci., 31:315.

Christensen, S. 1985, Denitrification in a sandy loam soil as influenced by climatic and soil conditions, Tidsskr. Planteavl, 89:351.

Dane, J.H., Reed, R.B., and Hopmans, J.W., 1986, Estimating soil parameters and sample size by bootstrapping, Soil Sci. Soc. Am. J., 50:283.

Folorunso, O.A., and Rolston, D.E., 1984, Spatial variability of field-measured denitrification gas flux, Soil Sci. Soc. Am. J., 48:1214.

Folorunso, O.A., and Rolston, D.E., 1985, Spatial and spectral relationships between field-measured denitrification gas fluxes and soil properties. Soil Sci. Soc. Am. J., 49:1087.

Fluhler, H.L., Stolzy, H. and Ardkani, M.S., 1976, A statistical approach to define soil aeration in respect to denitrification, Soil Sci. 122:115.

Kapteyn, J.C. 1903. Skew frequency curves in Biology and Statistics. Astronomical Lab., Groningen: Noordhoff.

Kachanoski, R.G., Rolston, D.E. and De Jong, E., 1985a, Spatial and spectral relationships of soil properties and microtopography: I. Density and thickness of A Horizon, Soil Sci. Soc. Am. J., 49:804.

Kachanoski, R.G., Rolston, D.E. and De Jong, E., 1985b, Spatial and spectral relationships of soil properties and microtopography: II. Density and thickness of B Horizon, Soil Sci. Soc. Am. J., 49:812

Morkoc, F., Biggar, J.W., Nielsen, D.R., and Rolston, D.E., 1985, Analysis of soil water content and temperature using state-space approach. Soil Sci. Soc. Am. J., 49:798.

Mosier, A.R. and Parton, W.J., 1985, Denitrification in a shortgrass prairie: a modeling approach, In D.E. Caldwell et al. (eds) Planetary ecology., Van Nostrand Reinhold Co Inc., New York.

Myrold, D.D. and Tiedje, J.M., 1985, Diffusional constraints on denitrification in soil, Soil Sci. Soc. Am. J., 49:651.

Parkin, T.B. 1985. Automated analysis of nitrous oxide, Soil Sci. Soc. Am. J., 49:273.

Parkin, T.B. 1987. Soil microsites as a source of denitrification variability, Soil Sci. Soc. Am. J., 51: 1194.

Parkin, T.B., Starr, J.L., and Meisinger, J.J., 1987, Influence of sample size on measurement of soil denitrification, Soil Sci. Soc. Am. J., 51:1492.

Parkin, T.B., Meisinger, J.J., Chester, S.T., Starr, J.L., and Robinson, J.A., 1988, Evaluation of statistical estimation methods for lognormally distributed variables. Soil Sci. Soc. Am. J., 52:323.

Parkin, T.B. and Robinson, J.A., 1989, Stochastic models of soil denitrification. Appl. Environ. Microbiol., 55:72.

Parkin, T.B., Chester, S.T. and Robinson, J.A., 1990, Evaluation of statistical methods for calculating confidence intervals for the mean of a lognormally distributed variable. Soil Sci. Soc. Am. J., 54:in press.

Rice, C.W., Sierzega, P.E., Tiedje, J.M., and Jacobs, L.W., 1988, Simulated denitrification in the microenvironment of a biodegradable organic waste injected into soil, Soil Sci. Soc. Am. J., 52:102.

Rolston, D.E., Rao, P.S.C., Davidson, J.M., and Jessup, R.E., 1984, Simulation of denitrification losses of nitrate fertilizer applied to uncropped, cropped, and manure-amended field plots, Soil Sci. 137:270.

Seech, A.G. and Beauchamp, E.G., 1988, Denitrification in soil aggregates of different sizes, Soil Sci. Soc. Am. J., 52:1616.

Sexstone, A.J., Revsbech, N.P., Parkin, T.B., and Tiedje, J.M., 1985, Direct measurement of oxygen profiles and denitrification rates in soil aggregates, Soil Sci. Soc. Am. J., 49:645.

Smith, K.A., 1980, A model of the extent of anaerobic zones in aggregated soils and its potential application to estimates of denitrification, J. Soil Sci., 31:263.

Smith, M.S. and Tiedje, J.M., 1979, Phases of denitrification following oxygen depletion in soil, Soil Biol. Biochem., 11:261.

Sokal, R.R. and Rohlf, F.J., 1969, Biometry, W.H. Freeman & Co., New York.

Vauclin, M., Vieira, S.R., Vachud, G. and Nielsen, D.R., 1983, The use of cokriging with limited field soil observations, Soil Sci. Soc. Am. J., 47:175.

Vieira, S.R., Hatfield, J.L., Nielsen, D.R., and Bigger, J.W., 1983, Geostatistical theory and application to variability of some agronomic properties, Hilgardia, 51:1.

ANAEROBIC ZONES AND DENITRIFICATION IN SOIL: MODELLING AND MEASUREMENT

Keith A. Smith

The Edinburgh School of Agriculture
West Mains Road,
Edinburgh EH9 3JG, UK

INTRODUCTION

The general requirements for denitrification are (i) the presence of microorganisms possessing the metabolic capacity, (ii) suitable electron donors, (iii) anaerobic conditions or at least restricted oxygen availability, and (iv) nitrate, nitrite, nitric oxide (NO), and/or nitrous oxide (N_2O) as terminal electron acceptor (Firestone, 1982). In terms of the spatial distribution and extent of anaerobic conditions in soils and sediments, it is convenient to consider soil environments in three broad categories.

The first includes water-covered sediments and flooded soils which are devoid of plants. All of the biologically active volume is anaerobic, except for an aerobic layer at the surface, which can vary in thickness from the order of 1 millimetre to several centimetres, depending on the oxygen respiration rate (Patrick, 1982).

The second situation differs from the first only in the presence of plants which have adapted to the anaerobic environment by developing a capacity to transport oxygen to root meristems through conducting vessels (aerenchyma) running the length of the roots. An obvious example is the rice paddy. Some of the oxygen transported down the roots of rice and other aerenchymous plants is also released to the surrounding soil, creating narrow aerobic zones around the roots (Armstrong, 1964) which are believed to be important for nitrification and denitrification, as a result of diffusion of ammonium and nitrate ions across the aerobic-anaerobic interface (Patrick, 1982). The presence of more or less cylindrical aerated zones around roots within an otherwise anaerobic layer produces a more complex spatial arrangement than that prevailing in the absence of plants, but it may be expected to change only slowly with time.

The third category comprises soils which are normally predominantly aerobic, but within which anaerobic regions -- large or small -- may develop where denitrification can occur. The occurrence of these regions is characterised by great spatial and temporal variability (Chapters 13 and 15 of this book, by Arah and Parkin, respectively). This paper is addressed primarily to a consideration of the factors which contribute to the development of these anaerobic regions, which are the sources of denitrification activity in soils. As is explained below, among these factors are soil structure and spatial distribution of air-filled pores, and methods of modelling the effects of these variables on the extent of anoxia are outlined. The consequences of soil structural effects for the reliability of popular methods of measurement of denitrification are also discussed.

OXYGEN CONSUMPTION IN THE SOIL

In soils of the temperate zone, the net flux of oxygen from the atmosphere into the soil profile necessary to maintain the aerobic respiration of roots and microorganisms varies from over 20 g m^{-2} d^{-1} in cropped soils in summer to less than one-tenth of this amount in mid-winter (Table 1), with somewhat higher values being found in tropical environments.

The principal mechanism by which oxygen enters the soil is by diffusion (Smith, 1977). The diffusion coefficient for oxygen is four orders of magnitude greater in air-filled pores than in water-filled ones, so the entry of oxygen into the soil is heavily dependent on the maintenance of a continuous system of air-filled pores from the surface down into the soil profile. When the soil is saturated, diffusion from the atmosphere virtually ceases, and the supply of oxygen to respiring roots and microorganisms becomes limited to that which is dissolved in the soil water. As the solubility of oxygen is only about 0.006 ml ml^{-1} in water in equilibrium with air, the mere replacement of air in a pore with air-saturated water reduces the oxygen reserves by a factor of over 30. The small amount that is present in solution is usually consumed within a period ranging from a few hours to a few days (Jackson and Campbell, 1976; Drew and Lynch, 1980), and the soil becomes anaerobic. Such conditions are readily identified by the visible effects on growing plants and by characteristic soil colour changes resulting from the reduction of ferric iron (gleying) (Marshall and Holmes, 1988; Rowell, 1988), as well as by chemical analysis.

In soils which are not completely saturated, anaerobic regions may occur within a surrounding matrix of aerated soil. These regions may occupy, at one extreme, the greater part of the soil volume, e.g. in clayey soil with a massive

Table 1. Oxygen Consumption in Soil (after Payne and Gregory, 1988)

Location/experimental conditions		O_2 consumption (g m^{-2} d^{-1})
Rothamsted, England		
soil under kale	July	24
soil under kale	January	2.0
bare soil	July	12
bare soil	January	0.7
Sweden, forest soils		13-20
Puerto Rico, tropical forest soils		8-40*

*Derived from measurements of CO_2 release, assuming RQ of unity.

Fig. 1. Massive gleyed peds excavated from clay soil (Denchworth Series, near Stratford-on-Avon, England) in summer, after development of shrinkage cracks. The stem of the gas sampling probe also shown in the photograph is 20 mm in diameter.

structure, such as the Denchworth Series soil from the Midlands of England depicted in Fig. 1. In such soils, drainage, aeration and root penetration may be confined to inter-ped fissures and the surface layers of the peds. At the other extreme, anaerobic regions may develop because of intense microbial activity in a small volume, which may be of millimetre or sub-millimetre dimensions. In such circumstances the oxygen concentration gradient through the water film between adjacent air-filled pores and the centre of microbial activity is extremely steep.

Such a region of intense activity may be a microsite within a small structural unit (Dowdell and Smith, 1974; Fluhler et al., 1976; Tiedje et al., 1984), or it may be associated with the decomposition of a discrete piece of organic material. An example of the latter was provided by Parkin (1987), who found that virtually all the denitrification activity in a soil core was associated with a fragment of plant leaf within the uppermost 1 cm of the core. In this soil there was no visible aggregation of soil particles into structural units.

It is important to recognise that, as far as the creation of anaerobic regions is concerned, there is no fundamental distinction between "microsites" within soil aggregates and "hotspots" of microbial activity associated with organic fragments. Both result from a localised demand for oxygen which exceeds the supply. The anaerobic region within an aggregate may have quite a modest oxygen demand per unit volume of soil, the inadequate oxygen supply being caused by a relatively thick diffusive barrier of mainly water-filled pores, whereas the organic hotspot is likely to have a respiration rate orders of magnitude higher, sufficient to make diffusion through even thin water films inadequate to maintain aerobic conditions. Thus the principal factors responsible for the two situations may differ, and the geometries of the sites may vary from near-spherical to laminar, but the governing processes are the same.

The capacity of readily degradable organic material to stimulate first aerobic respiration and then anaerobic denitrification activity was demonstrated in an interesting experiment by Christensen and Tiedje (1988). They introduced dead E. coli cells to a small region within soil cores at field capacity in the laboratory. Respiration was stimulated immediately, whereas a four-day lag period preceded an increase in denitrification. By contrast, when a similar addition was made to an already anaerobic soil slurry, denitrification increased immediately. It was concluded that formation of local anaerobic conditions was the factor limiting denitrification when a new organic source became available in the soil. As these authors pointed out, in nature large amounts of organic matter will often be available in one spot, e.g. when a plant root or soil animal dies.

SIGNIFICANCE OF SOIL STRUCTURE

Early mathematical treatments of gas diffusion in soil treated it as a homogeneous porous medium (Marshall and Holmes, 1988). For light sands and sandy loams, and loessial soils with low organic matter, for example, such a concept is useful even if not rigorously applicable. However, many soils show heterogeneity of pore size distribution, e.g. those with a highly developed natural ped structure, and cultivated soils with crumbs and clods (Currie, 1965). These have distinct zones of crumb, or intra-aggregate, pores, separated by a more continuous system of inter-aggregate pores. In such soils, diffusion to depth will occur preferentially in inter-aggregate pores, but the paths to the respiring sites will be completed within crumb pores that are partly or even wholly filled with water. As Currie (1965) put it, "any attempt to describe the aeration pattern in soils possessing such a bimodal pore size distribution needs some knowledge of gas diffusion within the crumbs themselves".

That gas diffusion may be very slow indeed, particularly in soils with a high clay content and consequent high proportion of very fine pores. It has been shown (Avery, 1964; Thomasson and Robson, 1967) that in such soils aggregates may remain virtually waterlogged for extended periods after the pores between them have drained.

Currie (1961) and Greenwood and Berry (1962) presented expressions to describe the radial diffusion of oxygen towards the centres of physically and biologically homogeneous soil aggregates (idealised by assuming spherical shape), which made it possible to calculate what fraction of such an aggregate was anaerobic, if its radius, and the external oxygen concentration, diffusion coefficient, solubility of oxygen, and respiration rate were all known. For any given values of these parameters, there is a critical radius r_c below which no anaerobic zone exists at the centre of an aggregate:

$$r_c = (6DC/Q)^{1/2} \quad (1)$$

where D is the diffusion coefficient, C is the oxygen concentration, and Q is the respiration rate per unit volume of aggregate.

It is well established that in soils in the field aggregate sizes are log-normally distributed. A simple test of this property is to plot the percentage

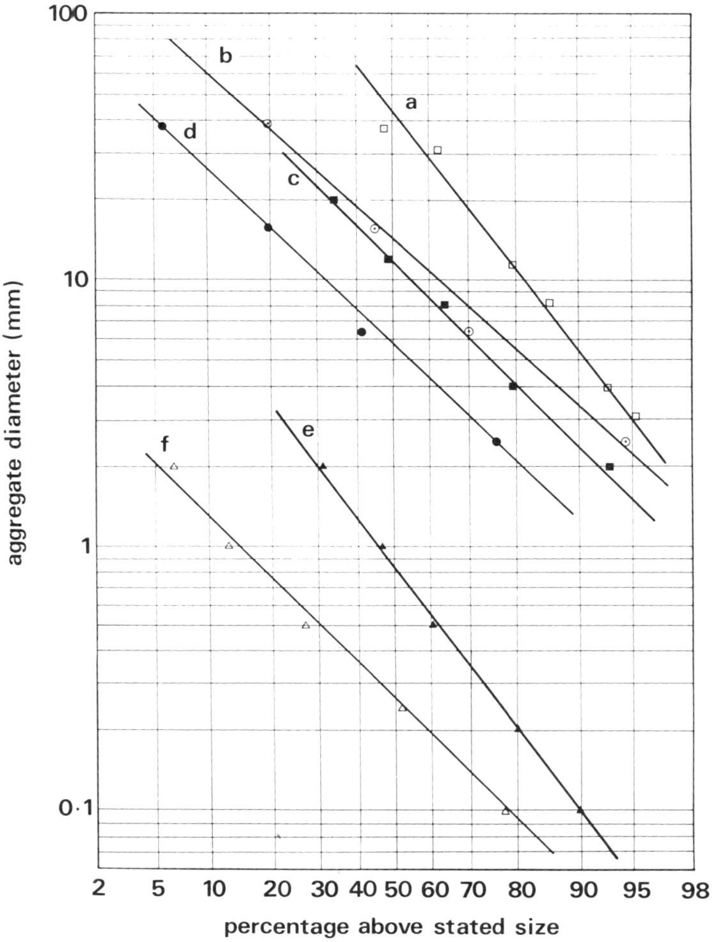

Fig. 2. Log-normal aggregate size distributions: (a), (c) Denchworth Series, England, arable and grassland, respectively (after Smith, 1980); (b), (d) clay loam, Rothamsted, England, arable and high-organic-matter horticultural soil, respectively (after Russell and Tamhane, 1940); (e), (f) Clarion silt loam, Iowa, arable and grassland, respectively (after Gardner, 1956).

of a sample by weight above a stated size (proportional to volume if a constant bulk density is assumed) on log-probability paper. A straight line signifies a log-normal distribution. Fig. 2 shows some examples; distributions a-d are for English clay loam/clay soils, while e and f are for a silt loam soil from Iowa, USA.

In addition to demonstrating the log-normal relationship very clearly, these plots also show interesting features relating to the extent of aggregation characteristic of soils of different climatic zones and under different systems of land use. The English soils all have mean aggregate sizes very much larger than those in the Iowa silt loam soil. Furthermore, whereas the Iowa soil has much larger aggregates when cropped with bluegrass (Fig. 2,e) than when under arable cultivation (Fig. 2,f), the Denchworth Series soil from England (the same series as the soil yielding the massive ped shown in Fig. 1) shows the opposite trend

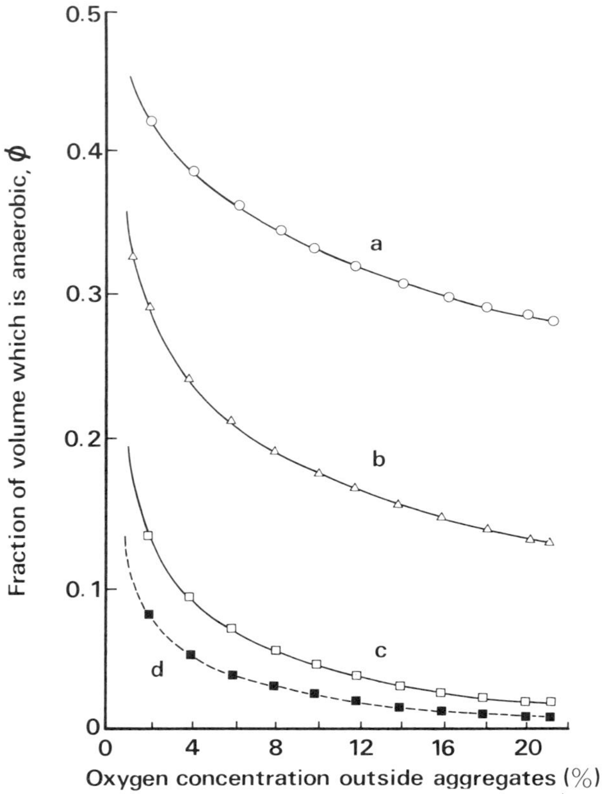

Fig. 3. Relationship between anaerobic volume fraction, ϕ, and oxygen concentration outside aggregates. (a) D = 10^{-6} cm^2 s^{-1}, μ = 0.7; (b) D = 10^{-5} cm^2 s^{-1}, μ = 0.7; (c) D = 10^{-4} cm^2 s^{-1}, μ = 0.7; (d) D = 10^{-4} cm^2 s^{-1}, μ = -0.3.

(Fig. 2 a,c). In both situations, the convergence towards an intermediate degree of aggregation under grassland is a move towards what is generally accepted as improved physical conditions for plant growth.

Smith (1977, 1980) applied the concepts of oxygen diffusion within aggregates developed by Currie (1961) and Greenwood and Berry (1962) for single aggregates to the log-normal population of aggregate sizes which exists in any bulk volume of soil, to calculate the fraction of the total volume which would be anaerobic, given particular values for the diffusion coefficients between and within the aggregates, the respiration rate, and the external oxygen concentration.

For any spherical aggregate of radius r, with an anaerobic zone at the centre of radius r_o, the fraction of the volume which is anaerobic is $(r_o/r)^3$, where

$$r_o/r = -\cos\theta/2\cos(\theta/3) \tag{2}$$

and

$$\theta = -\sin^{-1}(r_c/r) \tag{3}$$

For a log-normal radius distribution with mean μ and standard deviation σ, the fractional soil volume consisting of aggregates of radius between r exp(- ln r/2) and r exp(+ln r/2) is given by

$$\delta v_r = [1/\sigma(2\pi)^{1/2}] \exp[(\ln r - \mu)^2 / 2\sigma^2] \ \delta \ln r \tag{4}$$

The fractional volume of soil consisting of aggregates of radius r that is anaerobic is given by $(r_o/r)^3 v_r$. The fraction of the total volume which is anaerobic is obtained by substituting for v_r and integrating over r from r_c to a maximum radius r_{max}, taken as 10 cm.

$$\phi = [1/2.303\sigma(2\pi)^{1/2}] \int_{r_c}^{r_{max}} (r_o^3/r^4) \exp[(\log r - \mu)^2 / 2\sigma^2] \ dr \tag{5}$$

where, as above, $r_c = (6DC/Q)^{1/2}$ and logarithms are to the base 10.

Some typical relationships between the external oxygen concentration C, for different values of D, and different aggregate size distributions, are shown in Fig. 3.

Sexstone et al. (1985) measured the oxygen concentration profiles within individual silt loam soil aggregates with microelectrodes. They found that some of the aggregates had anaerobic zones within them (Fig. 4), the dimensions of which were in fair agreement with those predicted by the above model (Table 2).

Parkin and Tiedje (1984) and Parkin et al. (1984) developed a laboratory system for measuring denitrification in which acetylene and oxygen at the concentration measured in the soil pores was recycled through a soil core by means of a pump. When they plotted the denitrification rate as a percentage of the rate when completely anaerobic (in argon) against the external oxygen

Table 2. Comparison of denitrification rates and measured and calculated anaerobic radii of aggregates from cultivated and prairie soils. After Sexstone et al. (1985), with permission.

Aggregate	Denitrification rate radius (ng N g^{-1} h^{-1})	Aggregate (mm)	Anaerobic radius Measured[a]	Calculated
CA4[b]	6.3	7.0	3.3	0.5
CA1	13.0	8.0	1.3	2.4
CA2	9.2	12.0	5.5	6.1
PA1	0.03	9.0	0.3	0.0
PA2	1.7	18.0	14.0	13.6

[a] Mean radius measured on 4 to 10 profiles on each aggregate.
[b] obtained from a cultivated silt loam (CA) or from adjacent uncultivated native prairie (PA).

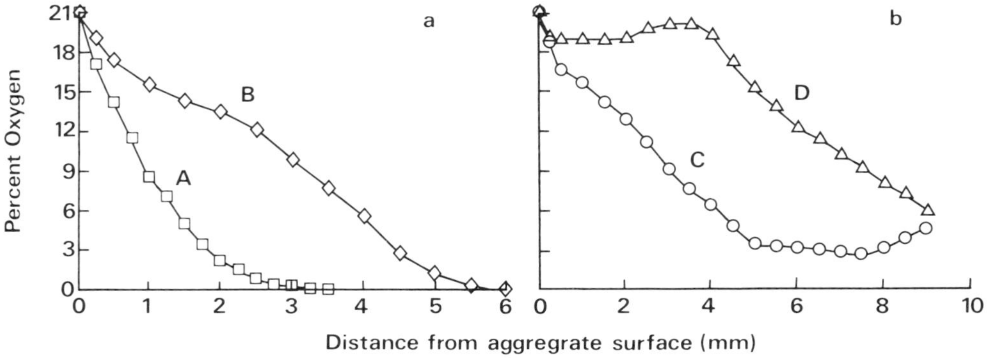

Fig. 4. Oxygen concentration profiles within individual soil aggregates, as measured by microelectrode. After Sexstone et al. (1985), with permission.

concentration, the shape of the relationship was in remarkably good agreement with the curve generated from the model outlined above (Fig. 5). This suggests that even with the gross simplification in the model that respiration rate is assumed to be constant throughout the soil mass, the physical conditions which determine the extent of anoxia also essentially determine the amount of denitrification activity. Recent development of the model to take into account spatial variations in respiration rate (Arah, 1988b; Arah and Smith, 1989) has shown that the close correlation between anaerobic fraction and denitrification rate is retained. If this relationship can ultimately be verified in a more or less natural situation it should provide a useful alternative approach to prediction of denitrification activity.

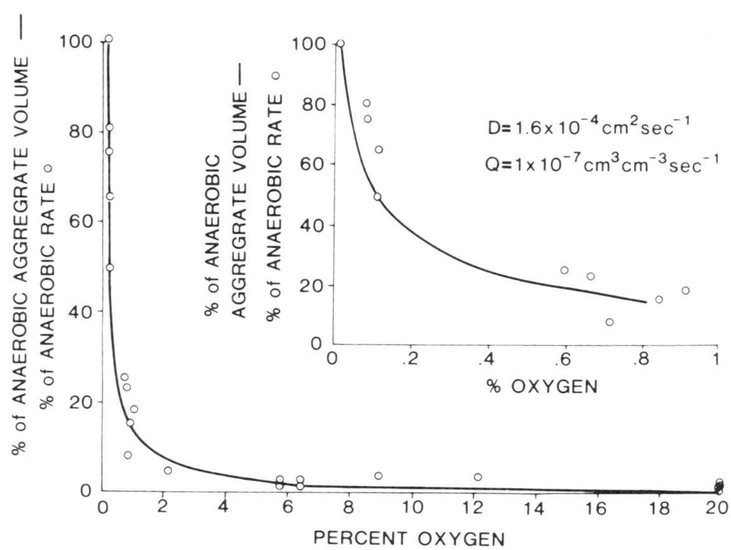

Fig. 5. Comparison of measured relationship between denitrification rate and external oxygen concentration (circles) with the corresponding theoretical relationship for anaerobic volume fraction (solid line) calculated from model of Smith (1980). From Parkin and Tiedje (1984), with permission.

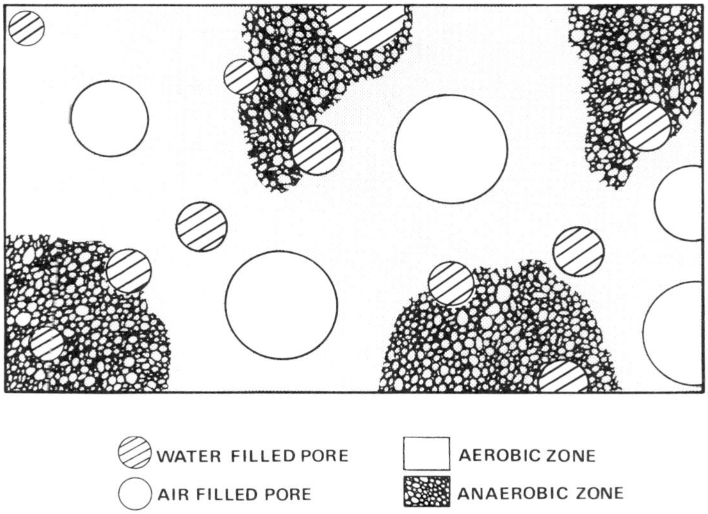

Fig. 6. Conceptual model of pores and anaerobic zones in a structureless soil. From Smith and Arah (1986), with permission.

STRUCTURELESS SOILS

In contrast with the situation just described, many soils, particularly those low in clay content and in organic matter, exhibit little or no aggregation. In effect, they are structureless. There is much evidence of anaerobic sites occurring in such soils (Burford and Millington, 1968; Rolston et al., 1978; Parkin, 1987; Arah, 1988a) and also direct measurement of low oxygen and the presence of another product of anaerobic conditions, ethylene, in a light sandy soil with no evident aggregation (Smith and Dowdell, 1974).

It may reasonably be assumed that the presence in these soils of the gaseous products of anaerobic activity is an indicator of anaerobic sites being present. Fig. 6 shows how these sites might occur because of the random distribution of air-filled pores. This concept is appropriate only for fairly high soil water contents, i.e. when all but the largest pores are water-filled, in situations in which the respiration rate is fairly uniform. However, as discussed above, if there are hotspots of microbial activity associated with localised concentrations of organic matter the range of water contents within which anoxia can occur will widen considerably.

A model of the development of anaerobic zones in essentially structureless soils has been developed (Smith and Arah, 1986), using the same concepts as outlined above for aggregated soils, but also incorporating a function for the probability of an air-filled pore occurring within a given distance from any point in the soil. In this model, the pores are considered to be circular, with a log-normal frequency distribution which may be derived from moisture characteristic curves. At any water tension the maximum radius of a water-filled pore is given by

$$r_a(\psi) = 2\gamma / \rho g \psi \tag{6}$$

where and are the surface tension and density of water, respectively, and g is the acceleration due to gravity.

The number of air-filled pores per unit area, $N(\psi)$, is given by

$$N(\psi) = \int_{r_a(\psi)}^{\infty} f(r)\, dr \tag{7}$$

and the air-filled porosity $\epsilon(\psi)$ is given by

$$\epsilon(\psi) = \int_{r_a(\psi)}^{\infty} \pi r^2 f(r)\, dr \tag{8}$$

Consider a point in any plane through the soil, at a distance x from the centre of the nearest air-filled pore. A term, p(x) is defined as p(x) = the probability of finding an air-filled pore at a distance x from the point multiplied by the probability of finding no air-filled pores within this distance. The former probability term is proportional to the area of the annulus of thickness δx at distance x. In the limit this is proportional to $2\pi x$. The latter term is given by the Poisson probability function $p(i) = (u^i/i!)e^{-u}$, where $i = 0$ and $u = N(\psi)\pi x^2$, the mean number of pores to be expected in the circle of radius x. Thus p(x) can be expressed as follows

$$p(x) = 2\pi x N \exp(-N\pi x^2) \tag{9}$$

and

$$x = \int_0^{\infty} x\, p(x)\, dx = 1/[2(N)^{1/2}] \tag{10}$$

where N is a function of ψ , i.e. the mean distance to the nearest air-filled pore in the plane is $1/[2(N)^{1/2}]$.

Since the soil is isotropic, this relationship must be true for all planes. Thus the point is surrounded by a shell of mean radius $\bar{x}$ containing its nearest-neighbour air-filled pores. This shell has a surface area $4\pi\bar{x}^2$ and a mean pore area equal to $\epsilon(\psi)4\pi\bar{x}^2$.

The supply of oxygen to the point by diffusion from the surrounding air-filled pores will depend on the pore area. In the "aggregate" model above, calculations are based on the concept of spheres of surface area $4\pi r^2$ surrounded by an external oxygen concentration C. By analogy, the spherical shell of radius x, with pore area $\epsilon(\psi)4\pi\bar{x}^2$, may be considered to correspond to a shell with all its surface exposed to an oxygen concentration of $\epsilon(\psi)C$. This value can then be substituted in Eqs. 1-5 and the anaerobic volume fraction calculated.

Again, as in the case of the "aggregate" model, an additional integration can in principle be performed to take account of spatial variability in the respiration rate. However, before such an integration can be usefully carried out, more information is required about the mean and variance of the respiration rate. Estimates of variance should be made on samples whose dimensions are, ideally, small enough to encompass only one microsite. Taking variability of respiration into account is particularly important in the context of structureless soil, where, as we have seen, the likelihood of anaerobic conditions occurring is more heavily dependent on this parameter than on, say, diffusion path length or very low diffusion coefficients.

EFFECT OF SOIL PHYSICAL CONDITIONS ON MEASUREMENT OF DENITRIFICATION

Denitrification flux rates and total losses from soils exhibit enormous variability over very short distances. The nature of the process makes this inevitable. In Chapters 13 and 15 of this book Arah and Parkin examine the spatial and temporal aspects of this variability; here, the question of whether existing data are also significantly affected by variability in the quality of measurement is considered.

Two principal techniques are available for assessments of denitrification fluxes and integrated total measurements of denitrification over a period. The first of these is the acetylene inhibition method, which relies on the prevention of N_2O reduction to N_2 by a concentration of acetylene of 0.1 % by volume (Yoshinari and Knowles, 1976), thus resulting in the entire gaseous flux being in the form of N_2O, which can be satisfactorily measured down to ambient concentrations (0.3 ppm) by electron capture gas chromatography (Smith, 1983). The second is the use of ^{15}N-labelled nitrate, and detection of any dinitrogen released by mass spectrometry. The ^{15}N/mass spectrometric method (Boast et al., 1988) has the advantage that reliance does not have to be placed on achieving satisfactory inhibition (and also avoids any problem due to the inhibition by acetylene of nitrification as well as denitrification). However, it has the drawback that in order to calculate the quantity of N coming from denitrification, it has to be assumed that there is only one equilibrium pool of nitrate in the soil. This is not always the case.

The acetylene/gas chromatographic method has the advantage of requiring cheaper and more easily used equipment, and has had widespread use over the last decade. The most widely used version of the technique for measurement of denitrification in the field is the "cover box" method, in which the soil is perfused by acetylene either from vertically installed tubes surrounding the microplot (Ryden et al., 1979), or from above by introduction of acetylene under the cover box (Egginton and Smith, 1986; Arah, 1988a). The air in the cover box may either be drawn through a trap to collect the evolved N_2O over a period, or sampled by syringe at intervals, and analysed in the laboratory.

The principal disadvantages of the acetylene inhibition method is that the acetylene may not penetrate to all the denitrifying sites. Other problems, in some circumstances, include the use of acetylene as a substrate by soil microorganisms and a reduction in its inhibitory effect by sulphide (Ryden and Dawson, 1982).

Examination of the data in the literature leads to the inescapable conclusion that the situations where this method has given the highest reported fluxes are on soils which are by no means the ones most likely to have anaerobic conditions at critical periods, e.g. after applications of substantial amounts of N fertiliser in the spring. For example, Ryden et al. (1979) found losses of up to 25 % of applied N in soils under horticultural production in California, whereas Colbourn et al. (1984a) found much smaller losses from a Denchworth clay under winter wheat in England.

Table 3. Observed rates of denitrification at sites on Edinburgh School of Agriculture farm, using cover box method.

	Site/Soil/Land Use	Denitrification rate ($g\ N\ ha^{-1}\ d^{-1}$)
1.	Macmerry SCL; grassland, reasonably structured	up to 2500
2.	Macmerry/Winton SCL/CL; continuous arable, bad structure	< 5
3.	Alluvial loam up to	600
4.	Darvel coarse sandy soilup to	180
5.	Winton CL; grassland, compacted;up to Measurement made within few days of surface application of NH_4NO_3	1800

Similar evidence comes from a series of experiments in Edinburgh on soils with widely varying texture and structure. The highest observed values (Table 3) were from soils whose texture/structure/drainage status were intermediate between the extremes represented by coarse sands and impermeable clay subsoils, respectively. It has been established by direct measurement that in the impermeable clay soil under arable cultivation no acetylene could be detected below 20 cm depth even after 9 hours of application to the surface (Arah, 1988a) (Fig. 7). It is evident that in such soils no satisfactory inhibition of N_2O reduction can be achieved. In the soils of intermediate physical properties, however, it seems reasonable to postulate that diffusion is sufficiently reduced to allow anaerobic sites to develop, but not reduced to the point where inhibitory concentrations of acetylene cannot be achieved. Thus the conditions are optimised for generating observable N_2O fluxes at the surface, as opposed to true total denitrification rates.

Further support for this hypothesis comes from an examination of data published by Colbourn et al. (1984b). When they used ^{15}N-labelled fertiliser on their site, maximum denitrification fluxes were an order of magnitude higher than those they reported earlier by the cover box method.

Many workers, of course, have recognised the problem of achieving satisfactory inhibition by acetylene, and this has led to the taking of soil cores and incubating them in a closed system, usually at field temperatures (either actually in the field (Ryden et al., 1987) or in an incubator set at field temperature. Usually, the entire outer surface of soil cores incubated in this way is exposed to the atmosphere in the incubation vessel (e.g. Parkin, 1987). This reduces the diffusion path length for acetylene, and ensures that it can reach any anaerobic microsite within the core. However, an additional consequence (which is only likely to be of major significance in soils with massive structures) is that it may also drastically alter the aeration conditions of the outer part of the core compared with those prevailing in situ in the field.

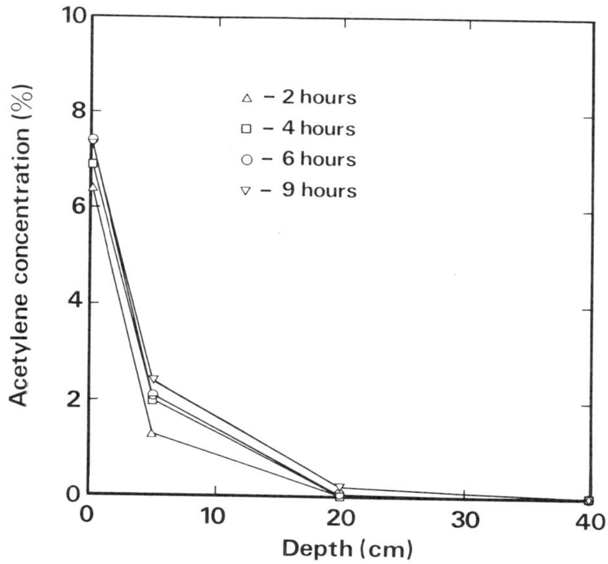

Fig. 7. Acetylene concentration in soil atmosphere at different times after the start of application to the surface of a Winton Series soil (Arah, 1988a).

Evidence in support of this hypothesis was obtained by incubating cores from Sites 2 and 3 in Table 3 at oxygen concentrations down to 5 %. (Measurements of the oxygen concentration in the soil profile at the time of taking the cores showed values down as far as 3 %.) For cores from Site 2 (Winton soil), even at an external oxygen concentration of 20 % the denitrification rate as measured by the acetylene method was more than an order of magnitude greater than with the cover box method, even though it was later in the growing season and less mineral N was likely to be present. The rate increased by a further order of magnitude when the oxygen was reduced to 5 % (Fig. 8). In contrast, the rates from the cores from Site 3 were of the same order as those measured by the cover box method.

It is worth noting that the shape of the denitrification rate-oxygen concentration relationship shown in Fig. 8 is similar to that of Parkin and Tiedje (1984) illustrated in Fig. 5. However, the steep increase in rate observed in the Edinburgh work took place at a much higher external oxygen concentration. This may be a consequence of the coarser soil structure prevailing in the Scottish soils.

In view of these observations it is concluded that the acetylene technique should not be used for field studies where the permeability of the soil is suspected to be low, and that it is highly desirable in any study to carry out direct measurements of acetylene down the soil profile to ensure that application times are adequate. Even then, the results should be treated with some scepticism: gas sampling probes give information about the content of large pores, not that of intra-aggregate pores.

Where it is necessary to resort to the core method in order to be able to make any measurement, it seems essential to match the external environment of the core at least approximately to that existing in the field, by lowering the

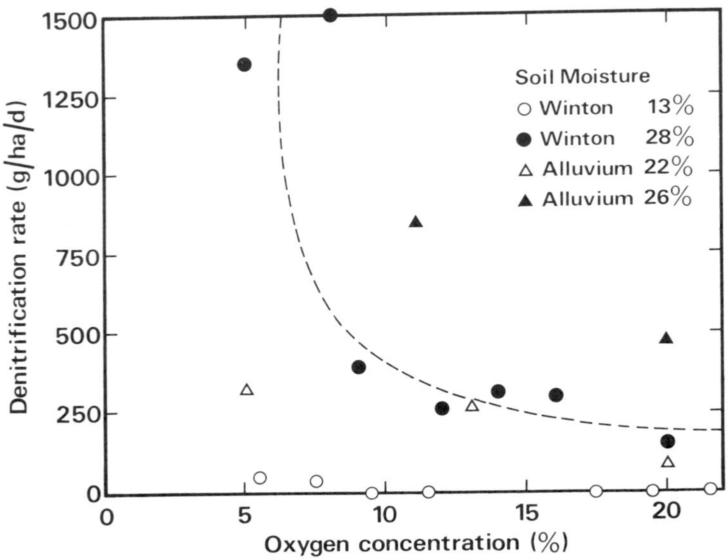

Fig. 8. Increase in denitrification rate in soil cores with decrease in external oxygen concentration and with increase in soil moisture content.

oxygen in the surrounding atmosphere to the concentration detectable in the macropores in the field. This may introduce an unwelcome complexity into the experimental procedures, but this is surely preferable to obtaining a potentially erroneous result.

SUMMARY

For denitrification to occur in soil, the essential requirements are the presence of nitrate and microbial activity and anaerobic or nearly anaerobic conditions. Soil water content is the principal factor controlling soil aeration, because of the barrier it presents to oxygen diffusion from the atmosphere. Anaerobic zones may occur as continuous layers below the water table or as localised regions within a generally aerated matrix. Examples of the latter are the centres of large aggregates of wet clay soil, and sites of high respiration rate, e.g. fragments of decomposing organic matter.

For structured soils, experimental evidence has been obtained showing good agreement with models of (a) the relationship between the fractional anaerobic volume of an individual aggregate and oxygen diffusion and respiration rate, and (b) the relationship between soil anaerobic volume and denitrification rate. Anoxia in structureless soils has also been modelled but not yet tested experimentally.

The acetylene inhibition method for measuring total denitrification cannot be applied in the field in coarsely structured clay soils, because the impermeability of these soils to gas diffusion, when wet, prevents the acetylene entering regions of anaerobic activity, and thus there is no inhibition of N_2O reduction. The popular alternative method of incubating soil cores in an atmosphere containing acetylene is also subject to major error unless the oxygen regime surrounding the cores approximates to that existing in the soil profile prior to disturbance.

REFERENCES

Arah, J.R.M., 1988a, "Measurement and modelling of denitrification in soil", Ph.D. Thesis, University of Edinburgh.

Arah, J.R.M., 1988b, Modelling denitrification in aggregated and structureless soils, in: "Nitrogen Efficiency in Agricultural Soils", D.S. Jenkinson and K.A. Smith, eds., Elsevier, London.

Arah, J.R.M. and Smith, K.A., 1989, Steady-state denitrification in aggregated soils: a mathematical model. J. Soil Sci., 40:139.

Armstrong, W., 1964, Oxygen diffusion from the roots of some British bog plants. London, 204:801.

Avery, B.W., 1964, "The soils and land use of the district around Aylesbury and Hemel Hempstead (Sheet 238)", Memoirs of Soil Survey of England and Wales, HMSO, London.

Boast, C.W., Mulvaney, R.L. and Baveye, P., 1988, Evaluation of nitrogen-15 tracer techniques for direct measurement of denitrification. I. Theory. Soil Sci. Soc. Am. J., 52:1317.

Burford, J.R. and Millington, R.J., 1968, Nitrous oxide in the atmosphere of a red-brown earth, Trans. 9th Int. Congr. Soil Sci., 2: 505.

Christensen, S., and Tiedje, J.M., 1988, Denitrification in the field, analysis of spatial and temporal variability, in: "Nitrogen Efficiency in Agricultural Soils", D.S.Jenkinson and K.A.Smith, eds., Elsevier, London.

Colbourn, P., Iqbal, M.M. and Harper, I.W., 1984a, Estimation of the total gaseous nitrogen losses from clay soils under laboratory and field conditions, J. Soil Sci., 35:11.

Colbourn, P., Harper, I.W. and Iqbal, M.M., 1984b, Denitrification losses from ^{15}N-labelled calcium nitrate fertilizer in a clay soil in the field, J. Soil Sci., 35:539.

Currie, J.A., 1961, Gaseous diffusion in the aeration of aggregated soils, Soil Sci., 92:40.

Currie, J.A., 1965, Diffusion within soil microstructure. A structural parameter for soils, J. Soil Sci., 16:279.

Dowdell, R.J. and Smith, K.A., 1974, Field studies of the soil atmosphere. II. Occurrence of nitrous oxide, J. Soil Sci., 25:231.

Drew, M.C. and Lynch, J.M., 1980, Soil anaerobiosis, microorganisms, and root function, Ann. Rev. Phytopath., 18:37.

Egginton, G.M. and Smith, K.A., 1986, Losses of nitrogen by denitrification from a grassland soil fertilized with cattle slurry and calcium nitrate, J. Soil Sci., 37:69.

Firestone, M.K., 1982, Biological denitrification, in: "Nitrogen in Agricultural Soils", F.J. Stevenson, ed., Am. Soc. Agron., Madison, Wisconsin.

Fluhler, H., Stolzy, L.H. and Ardakani, M.S., 1976, A statistical approach to define soil aeration in respect to denitrification. Soil Sci., 122:115.

Gardner, W.R., 1956, Representation of soil aggregate size distribution by a logarithmic-normal distribution, Soil Sci. Soc. Am. Proc., 20: 151.

Greenwood, D.J. and Berry, G., 1962, Aerobic respiration in soil crumbs, Nature, 195:161.

Jackson, M.B. and Campbell, D.J., 1976, Waterlogging and petiole epinasty in tomato: the role of ethylene and low oxygen. New Phytol., 76:21.

Marshall, T.J. and Holmes, J. W., 1988, "Soil Physics", 2nd ed., Cambridge University Press, Cambridge, England.

Parkin, T.B., 1987, Soil microsites as a source of denitrification variability, Soil Sci. Soc. Am. J., 51:1194.

Parkin, T.B. and Tiedje, J.M., 1984, Application of a soil core method to the investigation of oxygen effects on denitrification, Soil Biol. Biochem., 16:331.

Parkin, T.B., Kaspar, H.F., Sexstone, A.J. and Tiedje, J.M., 1984, A gas-flow soil core method to measure field denitrification rates, Soil Biol. Biochem., 16:323.

Patrick, W.H., 1982, Nitrogen transformations in submerged soil, in: "Nitrogen in Agricultural Soils", F.J.Stevenson, ed., Am. Soc. Agron., Madison, Wisconsin.

Payne, D. and Gregory, P.J., 1988, The soil atmosphere, in: "Russell's Soil Conditions and Plant Growth", 11th ed., A.Wild, ed., Longman, Harlow, England.

Rolston, D.E., Hoffman, D.L., and Toy, D.W., 1978, Field measurements of denitrification. I. Flux of N_2 and N_2O, Soil Sci. Soc. Am. J., 42:863.

Rowell, D.L., 1988, Flooded and poorly drained soils, in: "Russell's Soil Conditions and Plant Growth", 11th ed., A.Wild, ed., Longman, Harlow, England.

Russell, E.W. and Tamhame, R.V., 1940, The determination of the size distribution of soil clods and crumbs, J. Agric. Sci. Camb., 30:210.

Ryden, J.C. and Dawson, K.P., 1982, Evaluation of the acetylene inhibition technique for the measurement of denitrification in grassland soils, J. Sci. Food Agric., 33:1197.

Ryden, J.C., Land, L.J., Letey, J. and Focht, D.D., 1979, Direct measurement of denitrification loss from soils: II. Development and application of field methods, Soil Sci. Soc. Am. J., 43:110.

Ryden, J.C., Skinner, J.H. and Nixon, D.J., 1987, Soil core incubation system for the field measurement of denitrification using acetylene-inhibition, Soil Biol. Biochem., 19:753.

Sexstone, A.J., Revsbech, N.P., Parkin, T.B. and Tiedje, J.M., 1985, Direct measurement of oxygen profiles and denitrification rates in soil aggregates, Soil Sci. Soc. Am. J., 49:645.

Smith, K.A., 1977, Soil aeration, Soil Sci., 123:284.

Smith, K.A., 1980, A model of the extent of anaerobic zones in aggregated soils, and its potential application to estimates of denitrification, J. Soil Sci., 31:263.

Smith, K.A., 1983, Gas chromatographic analysis of the soil atmosphere, in: "Soil Analysis", K.A.Smith, ed., Marcel Dekker, New York.

Smith, K.A. and Arah, J.R.M., 1986, Anaerobic micro-environments in soil and the occurrence of anaerobic bacteria, in: "Microbial Communities in Soil", V. Jensen, A. Kjøller and L.H. Sørensen, eds., Elsevier, London.

Smith, K.A. and Dowdell, R.J., 1974, Field studies of the soil atmosphere. Relationship between ethylene, oxygen, soil moisture content and temperature, J. Soil Sci., 25:219.

Thomasson, A. and Robson, J.D., 1967, The moisture regime of soils developed on Keuper Marl, J. Soil Sci., 18:329.

Tiedje, J.M., Sexstone, A.J. Parkin, T.B. Revsbech, N.P. and Shelton, D.R., 1984, Anaerobic processes in soil, Plant Soil, 76:197.

Yoshinari, T. and Knowles, R., 1976, Acetylene inhibition of nitrous oxide reduction by denitrifying bacteria, Biochem. Biophys. Res. Commun., 69:705.

DIFFUSION-REACTION MODELS OF DENITRIFICATION IN SOIL MICROSITES

Jonathan R.M. Arah

The Edinburgh School of Agriculture
West Mains Road, Edinburgh EH9 3JG, UK

INTRODUCTION

Soil is a structured three-dimensional medium, exhibiting physical, chemical and biological heterogeneity over a wide range of scales. It is subject both to gradual and to rapid temporal change. Accommodating these features within a model of soil denitrification poses a number of problems.

Spatial variability

Denitrification rates recorded in the field are highly spatially variable (Folorunso and Rolston, 1984; Myrold, 1988): this observation has led to the development by several workers of the concept of the "hot-spot" (Parkin, 1987; Parkin et al., 1987; Christensen and Tiedje, 1988). A hot-spot is a rapidly denitrifying microsite, identified with a high local concentration of readily-oxidisable organic matter (dead beetles, dead leaves, etc.), and associated with a high rate of aerobic respiration. Spatial variation is explained in terms of the random distribution of such microsites in an essentially unreactive, homogeneous soil medium. Hot-spots are not conceived of as functionally identical, but rather are characterised by a log-normal probability distribution of denitrification intensity. Their internal structure, if any, generally seems to be ignored.

An alternative approach concentrates on structural factors (Smith, 1980; Arah, 1988). Soils are not generally homogeneous structures, and where any degree of aggregation exists it may favour the development of anaerobic microsites (at the centres of large saturated aggregates, for example) even where the supply of readily-oxidisable organic matter is relatively low. Thus in an aggregated soil even a "cold-spot" may denitrify.

The two approaches have points in common; it can be argued that structural effects are taken into account in the initial characterisation of the denitrification intensity distribution of the hot-spots, and conversely that different concentrations of readily-oxidisable organic matter may be introduced into the structural microsites model, thereby mimicking the hot-spot distribution. Where

Denitrification in Soil and Sediment, Edited by N. P. Revsbech and J. Sørensen, Plenum Press, New York, 1990

they diverge is in their concept of the internal structure of the active sites of denitrification.

Microsites may be categorised by the symmetry to which they most closely approximate. A one-dimensional, linear, model may be appropriate for describing sediments, biofilms and soils which are physically, chemically and biologically uniform (changes occurring only with depth). Cylindrical symmetry best describes the situation surrounding a root in a physically homogeneous soil. Spherical symmetry applies either within a physical aggregate or surrounding a localised hot-spot of activity in a structurally homogeneous medium. In the majority of cases a spherical microsite model is perhaps most appropriate.

Temporal variability

Most attempts to model soil denitrification, whether or not they take spatial variability into account, assume that the denitrifying system is in a state of pseudo-equilibrium - the current denitrification rate is determined by the current values of whatever parameters are taken to influence denitrification (Smith, 1980; Rolston et al., 1984; Myrold and Tiedje, 1985; Grundmann and Rolston, 1987; Arah, 1988; Arah and Smith, 1989a and b; Parkin and Robinson, 1989). It is assumed that, if we know the time-course of the controlling parameters, we can predict the time-course of a denitrification event simply by substituting the relevant numbers into our pseudo-equilibrium model. This may not always be the case; following a sufficiently rapid perturbation to one or more of the controlling variables there may in fact be an appreciable time-lag before the denitrifying system can adjust to its new situation. During this time the pseudo-equilibrium model does not apply. The response time of the denitrifying system determines the validity of the pseudo-equilibrium approximation - no pseudo-equilibrium model can cope with transient events occurring over times shorter than this.

McConnaughey and Bouldin (1985a) proposed a number of transient (i.e. non-equilibrium) models of denitrification in which oxygen, nitrate and nitrous oxide diffuse through a homogeneous soil medium and are simultaneously consumed by microbial reduction. The models are confined to one spatial dimension, and are therefore more appropriate to biofilms and sediments than to heterogeneous, structured soil. Arah (1989a and b) extended the method to cover transient processes in three dimensions (specifically in sites with spherical symmetry). By examining the processes of diffusion and reaction occurring within a generalised spherical microsite it is possible to set a lower limit to the microsite response time, and thus to assess the limits of applicability of any pseudo-equilibrium model.

TRANSIENT DIFFUSION/REACTION IN SPHERICAL MICROSITES

Consider a spherical microsite of radius R. A substrate diffuses into (or out of) the aggregate and is simultaneously consumed (or produced) by processes occurring within it. The equation governing such a diffusion/ reaction process in a spherically symmetric environment is:

$$\delta y(r,t)/\delta t = D(r,t)\ [\delta^2 y(r,t)/\delta r^2 + (2/r)\ \delta y(r,t)/\delta r] - Q(r,t) \tag{1}$$

range : $0 < r < R$ and $0 < t$
boundary conditions : $y(R,t) = Y(t)$ and $\delta y(0,t)/\delta r = 0$

r	:	radius within the microsite (or aggregate)
R	:	radius of the microsite (or aggregate)
t	:	time
y(r,t)	:	concentration of substrate (at r and t)
Y(t)	:	external concentration (at radius R) at time t
D(r,t)	:	diffusion constant of substrate (at r and t)
Q(r,t)	:	rate of reaction of substrate (at r and t)

As expressed in terms of Equation 1 a consumption rate Q(r,t) is positive; production is negative. The finite-difference approximation to Equation 1 may be solved for arbitrary values of Y, D, and Q using an iterative method (implemented in Pascal on a personal computer). It is possible, therefore, to simulate the evolution through time of the concentration profile of any substrate subject to any reaction within any spherically symmetric environment.

TRANSIENT DENITRIFICATION IN SPHERICAL MICROSITES

A simplified model of denitrification must consider the reactions of three mobile substrates - oxygen, nitrate, and nitrous oxide - and it must take account of the inhibition of nitrate reduction by oxygen, and of nitrous oxide reduction by oxygen and nitrate. Limiting the model to these three reactions involves a number of subsidiary assumptions: nitrite and nitric oxide are negligible intermediate products; oxidisable organic matter is relatively immobile; no other inhibitory effects are important. These simplifications are of course open to question, and where it can be shown that they are inappropriate the model must be altered to suit. In what follows it is assumed that the simple model described is adequate.

Biological reactions occurring in soil are enzyme-mediated. The reaction term Q(r,t) may be represented by Michaelis-Menten kinetics:

$$Q(r,t) = I(r,t)\ V(r,t)\ y(r,t)/[Km + y(r,t)] \qquad (2)$$

I(r,t)	:	inhibition function at radius r and time t
V(r,t)	:	reaction potential at radius r and time t
Km	:	Michaelis constant for reaction

The inhibition function I(r,t) reflects specific inhibitory effects (e.g., when modelling denitrification, those of oxygen on nitrate reduction, and oxygen and nitrate on nitrous oxide reduction). The reaction potential V(r,t) is the maximum rate at which the reaction would proceed in the absence of any inhibitory effects and if the concentration of substrate y(r,t) were not limiting (that is, if all enzymes present were saturated). It reflects the local concentration of readily-oxidisable organic matter, the microbial population, temperature, and any indirect effects of such factors as acidity. It is assumed in what follows that the local reaction potentials for the reduction of nitrate and nitrous oxide are proportional to that for oxygen.

Relevant microsite parameters, most of them adapted from Smith (1980), McConnaughey and Bouldin (1985b) and Myrold and Tiedje (1985), are given in Table 1. Inhibition of nitrate and nitrous oxide reduction by oxygen is linear up to a concentration of 10^{-4} mol/m^3, at which point both reactions are totally inhibited; nitrate inhibition of nitrous oxide reduction is complete at a concentra-

Table 1. Microsite parameters - transient diffusion/reaction model

Substrate	Michaelis constant Km (mol/m^3)	Reaction potential V ($mol/m^3/s$)	External concentration Y (mol/m^3)	Diffusion constant D (m^2/s)
oxygen	$1x10^{-3}$	$1x10^{-4}$	$1x10^{0}$	$5x10^{-9}$
nitrate	$2x10^{-1}$	$2x10^{-5}$	$1x10^{-1}$	$5x10^{-10}$
nitrous oxide	$1x10^{-2}$	$1x10^{-4}$	$1x10^{-5}$	$5x10^{-9}$

tion of $2x10^{-2}$ mol/m^3. The status of these numbers is uncertain; more work remains to be done to improve our knowledge in this area.

Intra-microsite concentration profiles

By substituting parameters from Table 1 into Equations 1 and 2, and solving the resulting finite-difference approximations by numerical methods, we can simulate the evolution through time (starting from arbitrary initial positions) of the intra-microsite concentration profiles of oxygen, nitrate and nitrous oxide. Figure 1 shows calculated oxygen concentration profiles in a 1 cm spherical microsite at various times following a perturbation. The microsite is internally homogeneous (intra-microsite diffusion constants and organic substrate concentrations are uniform) and at the start of the simulation it is in a state of pseudo-equilibrium (intra-microsite concentration profiles are static). The assumption of internal homogeneity is not essential to the model; any spherically symmetric distribution of D and V can be handled. The perturbation in question is an instantaneous tenfold reduction in intra-microsite gaseous diffusion constants; other parameters remain constant. This is a dramatic perturbation, roughly equivalent to instant saturation of the microsite from field capacity; the purpose is not to simulate a typical event in the soil but to establish likely limits to microsite response times.

Intra-microsite oxygen concentration profiles settle down to a new equilibrium position within 5 hours. Figure 2 shows the corresponding nitrate profiles; the adjustment in this case takes around 10 hours. Nitrous oxide profiles (not shown) take longer still to stabilise.

Microsite reaction rates

Intra-microsite concentration profiles may be employed to calculate microsite reaction rates by substitution in Equation 2 and integration. Figures 3, 4 and 5 show microsite reduction rates for oxygen, nitrate and nitrous oxide following the instantaneous perturbation described above. Rates for microsites 1 cm, 2 cm and 5 cm in radius are illustrated; microsites 5 mm in radius and smaller remain aerobic throughout. In all cases where denitrification occurs there is an initial peak in nitrate reduction as nitrate initially contained in the microsite is consumed, followed by a decline to a pseudo-equilibrium rate maintained by diffusive supply. Nitrous oxide reduction shows a similar pattern.

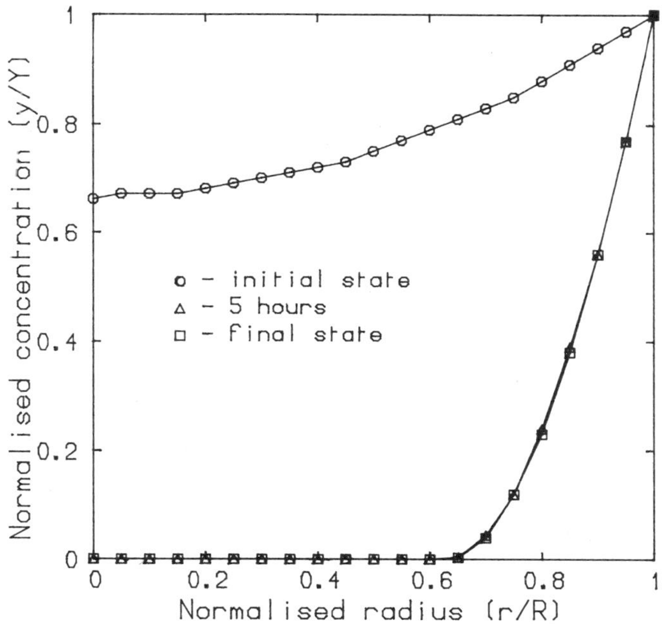

Fig. 1. Oxygen concentration profiles in a 1 cm microsite. Transient model.

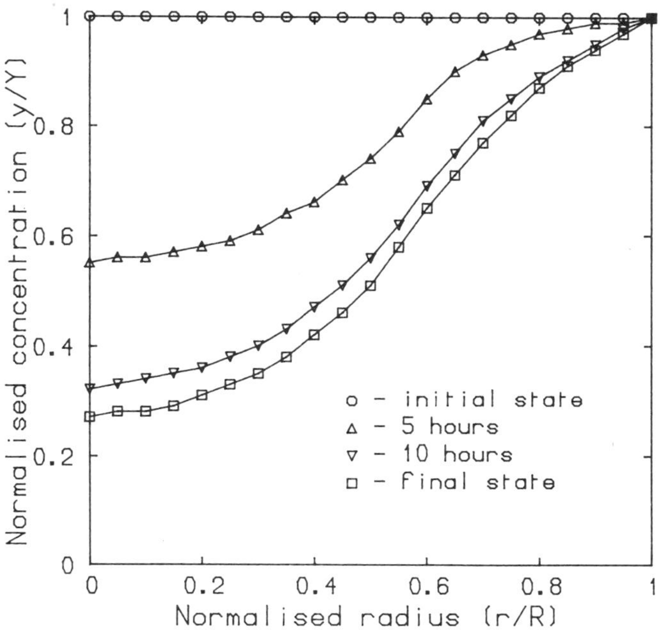

Fig. 2. Nitrate concentration profiles in a 1 cm microsite. Transient model.

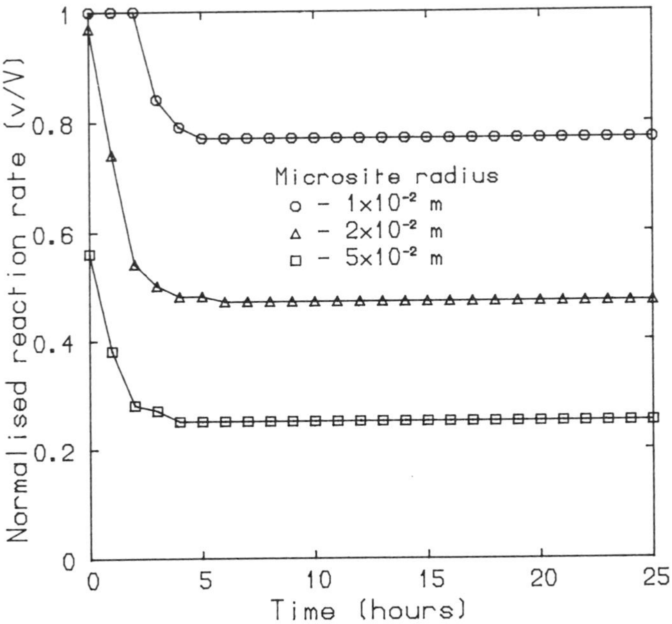

Fig. 3. Microsite respiration rates. Transient model.

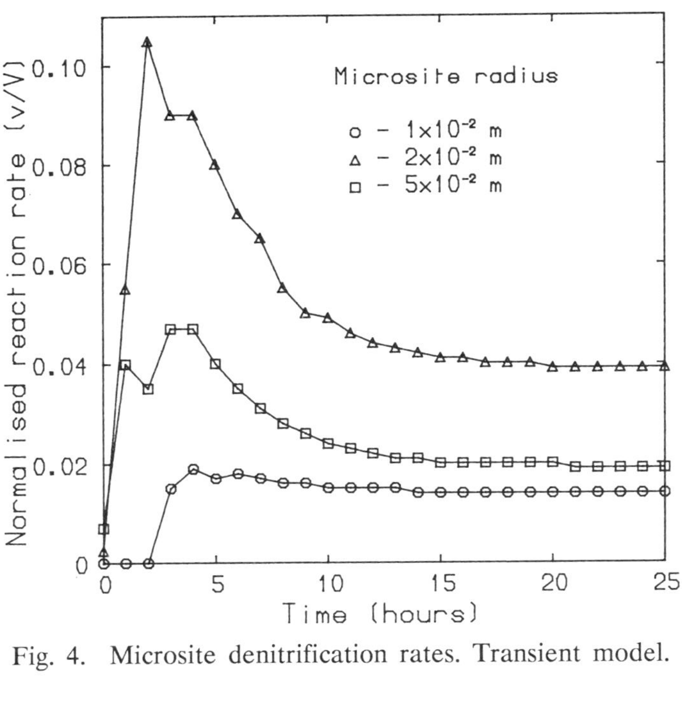

Fig. 4. Microsite denitrification rates. Transient model.

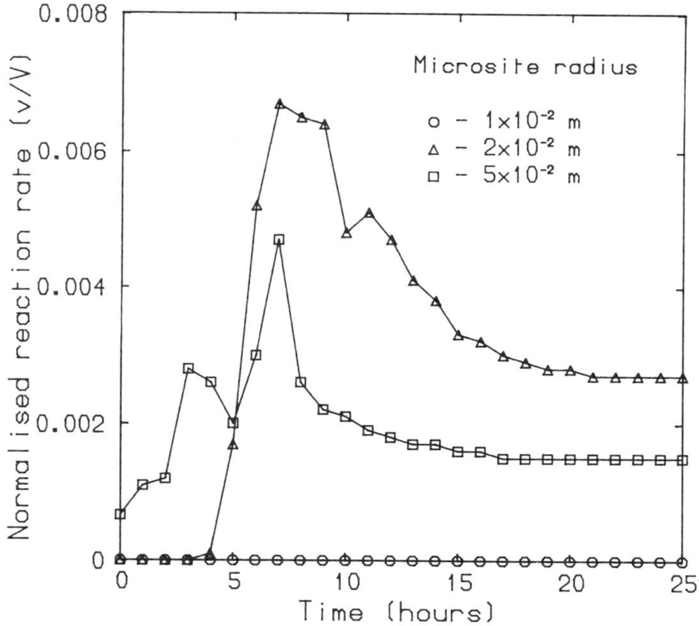

Fig. 5. Microsite nitrous oxide reduction rates. Transient model.

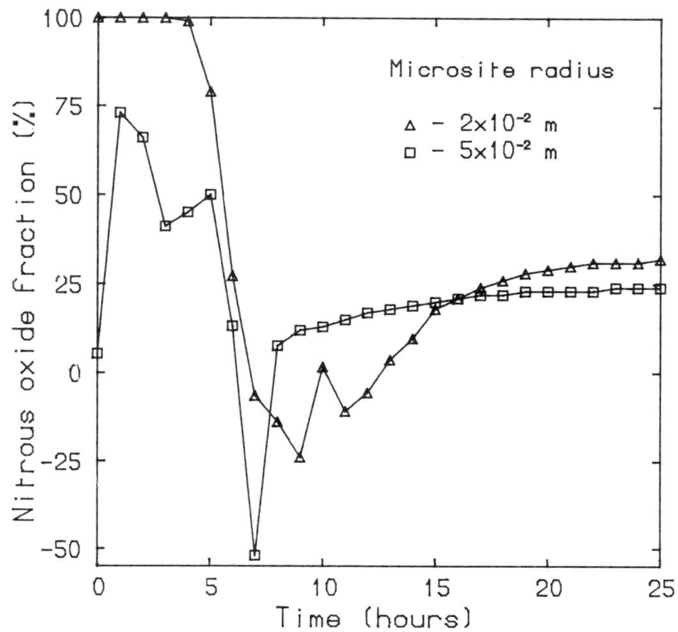

Fig. 6. Microsite nitrous oxide fraction. Transient model.

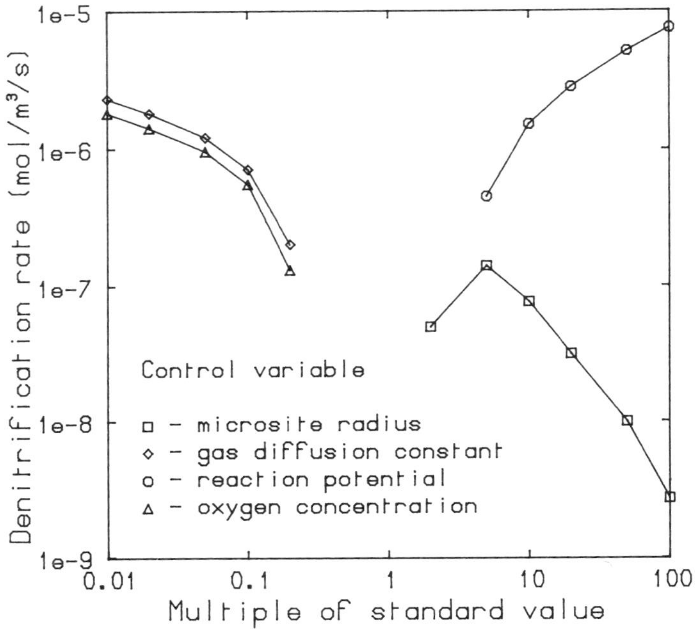

Fig. 7. Sensitivity analysis - unsaturated 1 cm microsite standard. Pseudo-equilibrium model.

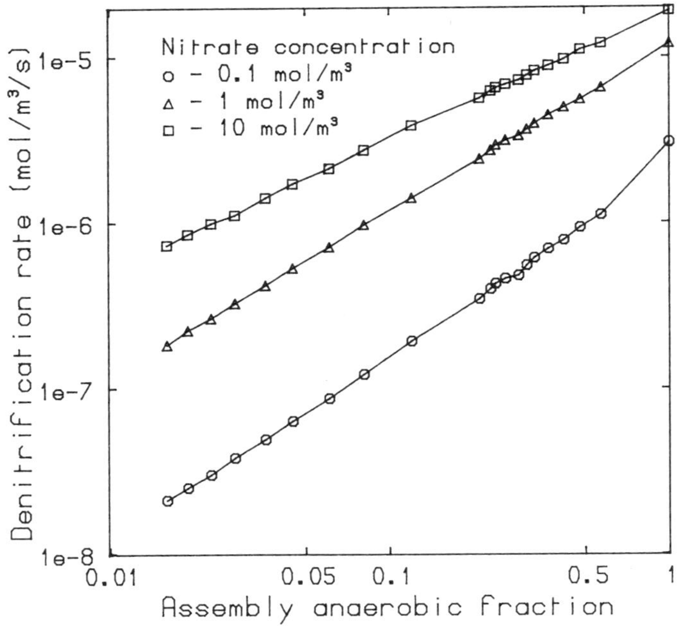

Fig. 8. Sensitivity analysis - saturated 1 cm microsite standard. Pseudo-equilibrium model.

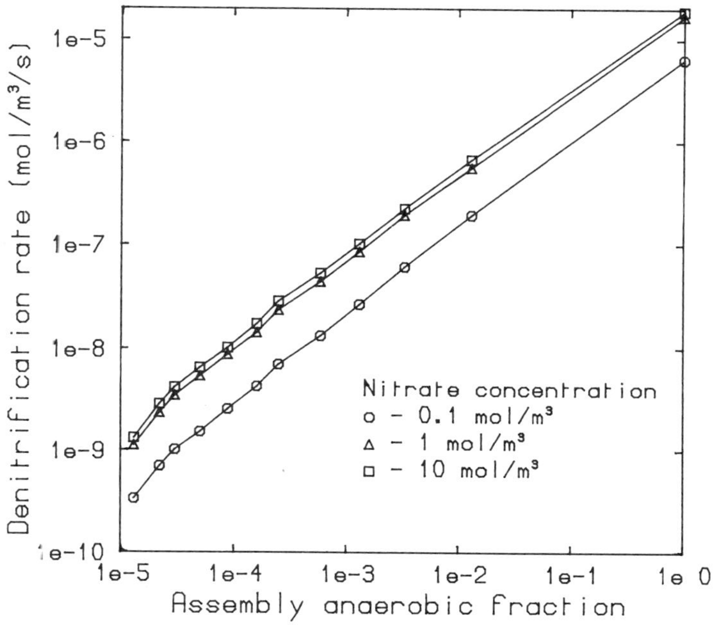

Fig. 9. Fine-structured assembly of microsites. Pseudo-equilibrium model.

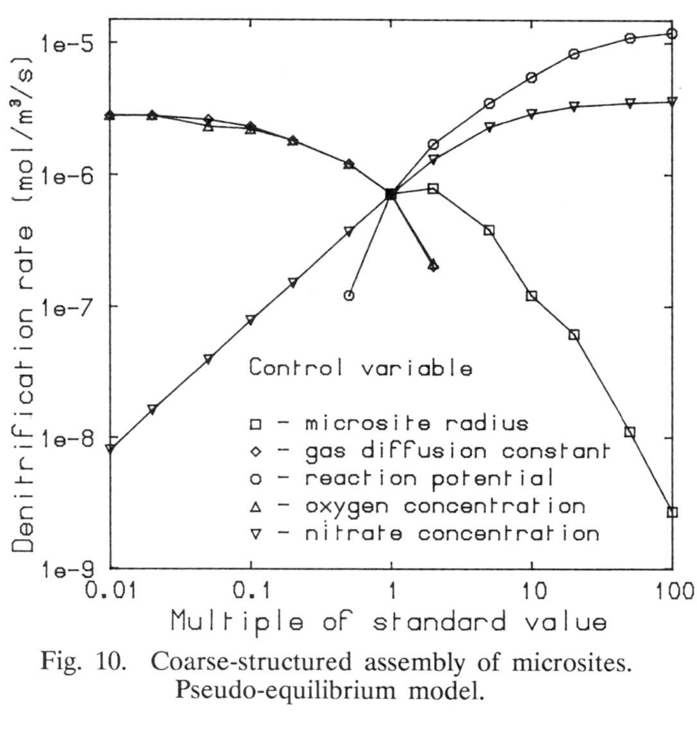

Fig. 10. Coarse-structured assembly of microsites. Pseudo-equilibrium model.

All microsite reaction rates stabilise within the 25 hours of the simulation.

Knowing microsite reaction rates for nitrate and nitrous oxide it is possible to calculate the "nitrous oxide fraction" - that fraction of the gaseous products of denitrification evolved as nitrous oxide. Figure 6 shows the time-course of this function for the 2 cm and 5 cm microsites, following the perturbation described above. A negative nitrous oxide fraction denotes a net consumption of nitrous oxide.

Microsite response times

The response time of the 1 cm microsite is about 16 hours, that of the 2 cm microsite is 21 hours, and that of the 5 cm microsite is 17 hours. Microsites 5 mm and smaller in radius have response times shorter than 10 hours. Thus, where nothing dramatic has happened in the soil for 24 hours or so, or where denitrification occurs predominantly in small microsites (as in the hot-spot model), a pseudo-equilibrium model of denitrification is appropriate. The pseudo-equilibrium (or "steady-state") analogue of Equation 1 is obtained by setting the left-hand side equal to zero. This is examined at length by Arah (1988) and by Arah and Smith (1989a and b), and a brief account is given below.

PSEUDO-EQUILIBRIUM DENITRIFICATION IN SPHERICAL MICROSITES

Figures 7 and 8 show sensitivity analyses for the pseudo-equilibrium model of microsite denitrification. The effect on microsite denitrification rate of varying the controlling parameters (one at a time, while all else remains constant) is examined. In Fig. 7 the reference standard is a 1 cm microsite with controlling parameters given by Table 1 (this "unsaturated" standard microsite does not actually denitrify). In Fig. 8 the standard is "saturated"; it has gaseous diffusion constants reduced by a factor of ten, and a denitrification rate of 7.1×10^{-7} $mol/m^3/s$. It is apparent that microsite denitrification rates are highly sensitive to alterations in microsite radius (small microsites do not denitrify at all, while very large ones are relatively inefficient) and reaction potentials; gas diffusion constants and external oxygen concentration are of similar importance; external nitrate concentration has a major effect only when the microsite is large and largely anaerobic.

Microsite variability

Not all microsites in the soil are characterised by the same controlling parameters (if microsites are to account for the observed spatial variability of denitrification then those parameters which determine microsite denitrification rate must vary from microsite to microsite). In particular, microsites differ in terms of their radius and their reaction potentials, both of which in this model are taken to follow log-normal probability distributions. Denitrification rates and anaerobic fractions of a "representative assembly" of microsites are calculated by integrating the relevant microsite parameters over independent probability distributions governing microsite radius and reaction potentials.

The denitrification rate of a representative assembly of denitrifying microsites is plotted against the assembly anaerobic fraction in Figs. 9 and 10. In Fig. 9 the mean microsite radius is 1 mm, in Fig. 10 it is 1 cm; in both cases the coefficient of variation is 50%. The soil represented in Fig. 9 is finely-structured; that in Fig. 10 is more heavily aggregated. In both model soils the mean oxygen and nitrous oxygen reaction potentials are 10^{-4} mol/m^3/s; mean nitrate reaction potential is 2x10^{-5} mol/m^3/s; the relevant coefficients of variation are all 50%. Michaelis constants and nitrate diffusion constants are as in Table 1, and results for saturated and unsaturated microsite assemblies are included.

In both Figs. there is an approximate proportionality between assembly denitrification rates and the corresponding anaerobic fractions. Increasing external nitrate concentration from 0.1 mol/m^3 to 1 mol/m^3 significantly increases assembly denitrification rates in both model soils; denitrification is constrained by diffusive limitation of nitrate at the lower concentration. A further increase to 10 mol/m^3 has less effect, particularly in the fine-structured model soil.

Steady-state concentration profiles are considerably easier to calculate than transient ones, and Smith (1980) included a useful formula for the radius of the steady-state anaerobic zone (if any) at the centre of a uniformly respiring homogeneous spherical microsite. This is particularly convenient in view of the relationships shown in Figs. 9 and 10; it appears to be possible to use the (relatively easy to calculate) assembly anaerobic fraction as an indicator of the assembly denitrification rate.

THEORETICAL CONSIDERATIONS

Spatial variability

Modelling spatial variability of denitrification necessarily involves the introduction of stochastic variables controlling the denitrification rate of soil microsites. The probability distributions governing these microsite parameters should have means and variances appropriate to the microsite scale; where they are actually measured using much larger volumes of soil the variances will almost certainly be underestimated. In order to be able to use microsites consistently it is important to have an estimate of their volume.

The spherical microsites of the model described above may be identified with actual physical aggregates in the soil. This has the advantage of making them easier to visualise; it also allows such parameters as gaseous diffusion constants to be calculated using relatively straightforward techniques (Nielson et al., 1984). Where the soil contains no visible aggregates microsites may be equated directly with localised high concentrations of oxidisable organic matter (hot-spots) in a structurally uniform medium. The difficulty then is to know exactly how big they are, and also how to determine their physical characteristics (moisture content, diffusion constants). These problems follow inevitably from the attempt to ascribe a structure to an entity previously conceived of as a point. An attempt to deal with them is described in Smith and Arah (1986) and Arah (1988), in which the Poisson distribution is employed to calculate the mean distance between a representative point and its nearest air-filled pore; in a randomly-structured medium this distance is the same in all planes, and thus effectively represents the radius of a notional spherical microsite.

Microsites are usually conceived of as independent, the evidence for this coming from core fractionation experiments (Parkin, 1987; Christensen and Tiedje, 1988) in which the total denitrification rate remains approximately constant whilst the core is divided and subdivided until the source of the activity can be identified. This is not always the case: Parkin (1987) also reports cores, exhibiting no discernible hot-spots, in which total denitrification activity decreases with each division. This observation, repeated in our lab (unpublished data), illustrates the effect of large-scale soil structure in maintaining the anaerobic conditions necessary for denitrification. It may be argued that this is a secondary phenomenon, that most denitrification occurs in hot-spots which are to all intents and purposes unaffected by their position in the soil, and that it is therefore fair to treat microsites as independent. If this is the case it should be borne out by the statistics: were the soil really to consist of independently denitrifying point-microsites, a reduction by a factor n in sample size would result in an increase by the same factor in the variance of sample denitrification rate. This is not observed (Parkin et al., 1987; unpublished data), possibly due to undersampling of the soil. To what extent can we picture microsites as independent?

Temporal variability

Most published models of denitrification assume a state of pseudo-equilibrium in the soil, for the very good reason that pseudo-equilibrium relationships are easier to handle than transient ones. Departure from pseudo-equilibrium, even at the microsite scale, may be caused by two distinct phenomena, discussed below.

Microbiological adjustments are not instantaneous. More work remains to be done to evaluate how rapidly soil microbes respond to changes in their environment. Evidence from laboratory incubations suggests that in many cases the transition from oxygen to nitrate reduction is rapid - the relevant enzymes are already present when the transition to anaerobic conditions occurs. Subsequent alterations in enzyme synthesis and microbial populations may be more gradual. In the models presented here it is assumed that microsite reaction potentials remain constant; any changes could of course be incorporated into the transient model.

The physical structure of the microsites in which denitrification occurs may itself cause a significant delay in the response of a microsite to altered conditions in the soil. The work presented here suggests that the response of a spherical microsite smaller than 1 cm in radius is in fact fairly rapid (under 16 hours, even where the disturbance is as severe as an instantaneous tenfold reduction in intra-microsite gaseous diffusion).

It is therefore concluded that a pseudo-equilibrium model of microsite denitrification is adequate provided no major microbiological changes occur during the course of the simulation, and that the soil contains few large aggregates. Where soil structure is a major factor a transient model of denitrification is more appropriate.

Model evaluation

We are currently conducting incubations of artificial soil aggregates under controlled conditions in an attempt to evaluate the spherical microsite model described above. Other workers have independently obtained encouraging results. Figures 9 and 10 suggest an approximate proportionality between the denitrification rate and the calculated anaerobic fraction of a representative assembly of microsites. Parkin and Tiedje (1984), incubating intact soil cores under controlled oxygen concentrations, found just such a proportionality. de Beer and Sweerts (1989) used an ion-specific micro-electrode to measure nitrate concentration profiles inside spherical agar beads containing immobilised denitrifying bacteria, incubated anaerobically in the presence of constant external nitrate concentrations. Measured concentration profiles agreed extremely well with those calculated using a model essentially identical to the pseudo-equilibrium model described above, without the complication of oxygen.

It seems probable that mathematical models of simultaneous diffusion and reaction in spherically symmetric environments can indeed describe what they set out to describe, and that experiments on sufficiently well-defined model systems will confirm this. What is more in doubt is whether the situations described by these models actually occur in the soil. How common is (approximate) spherical symmetry? This question is related to that of the independence of microsites; if microsites are not in fact independent this is presumably because their spheres of influence overlap. Should reaction potentials decline or increase towards the centre of the microsite? Does the mixed population of microorganisms in the soil behave "rationally", predominantly reducing only that substrate which is energetically most favourable? Is it reasonable to neglect nitrite as an intermediate?

It is unlikely that an answer applicable to all soils will be found for all, or indeed any, of these questions. Rather it will be necessary to delineate circumstances in which one or another of a range of approximations is least objectionable. This task is onerous, but not impossible.

SUMMARY

Spatial variability is accommodated within a model of soil denitrification by postulating the existence of independent spherical microsites characterised by a number of local parameters: reaction potentials (related to the local supply of readily-oxidisable organic matter), radii, internal diffusion constants (determined by structure and moisture content) and the external concentrations of oxygen and nitrate. Organic matter "hot-spots" (Parkin, Chapter 13 in this book) and large saturated aggregates (Smith, Chapter 14 in this book) can be viewed as subsets of this species of microsite. Intra-microsite processes involving simultaneous diffusion and reaction of oxygen, nitrate and nitrous oxide can be modelled mathematically for any combination of microsite parameters. An examination of microsite response to sudden perturbations suggests that, except where large microsites predominate, a pseudo-equilibrium model of denitrification may be adequate. This greatly simplifies the model. The net denitrification rate of a large assembly of microsites differing in some or all of their controlling parameters may be calculated by integrating over the probability distributions (usually log-normal) governing these parameters. Such an integration performed on two model soils suggests a simple relationship between assembly denitrification rate and anaerobic fraction. More information is required on the spatial and temporal variability of microsite parameters in real soils.

ACKNOWLEDGEMENT

I would like to thank the Agricultural and Food Research Council for financial support.

REFERENCES

Arah, J.R.M., 1988, Modelling denitrification in aggregated and structureless soils, in: "Nitrogen Efficiency in Agricultural Soils", D.S. Jenkinson and K.A. Smith, eds., Elsevier, London.

Arah, J.R.M., 1989a, Steady-state denitrification: an adequate approximation?, Mitt. Deutsch. Bodenkund. Gesell., (in press).

Arah, J.R.M., 1989b, Modelling spatial and temporal variability of denitrification, Biol. Fert. Soils., (in press).

Arah, J.R.M., and Smith, K.A., 1989a, Steady-state denitrification: a mathematical model, J. Soil Sci., 40:139.

Arah, J.R.M., and Smith, K.A., 1989b, Modelling denitrification in aggregated soils: relative importance of moisture tension, soil structure, and oxidisable organic matter, in: "Nitrogen in Organic Wastes Applied to Soils", J.Aa. Hansen and K. Henriksen, eds., Academic Press, London.

Christensen, S., and Tiedje, J.M., 1988, Denitrification in the field, analysis of spatial and temporal variability, in: "Nitrogen Efficiency in Agricultural Soils", D.S. Jenkinson and K.A. Smith, eds., Elsevier, London.

de Beer, D, and Sweerts, J-P.R.A., 1989, Measurement of nitrate gradients with an ion-selective microelectrode, Anal. Chim. Acta., 219:351.

Folorunso, O.A., and Rolston, D.E., 1984, Spatial variabilty of field-measured denitrification gas fluxes. Soil Sci. Soc. Am. J., 48:1214.

Grundmann, G.L., and Rolston, D.E., 1987, A water function approximation to degree of anaerobiosis associated with denitrification, Soil Sci., 144:437.

McConnaughey, P.K., and Bouldin, D.R., 1985a, Transient microsite models of denitrification: I. Model development, Soil Sci. Soc. Am. J., 49:886.

McConnaughey, P.K., and Bouldin, D.R., 1985b, Transient microsite models of denitrification: I. Model results, Soil Sci. Soc. Am. J., 49:891.

Myrold, D.D., 1988, Denitrification in ryegrass and winter wheat cropping systems in western Oregon, Soil Sci. Soc. Am. J., 52:412.

Myrold, D.D., and Tiedje, J.M., 1985, Diffusional constraints on denitrification in soil, Soil Sci. Soc. Am. J., 49:651.

Nielson, K.K., Rogers, V.C., and Gee, G.W., 1984, Diffusion of radon through soils: a pore distribution model, Soil Sci. Soc. Am. J., 48:482.

Parkin, T.B., 1987, Soil microsites as a source of denitrification variability. Soil Sci. Soc. Am. J., 51:1194.

Parkin, T.B. and Robinson, J.A., 1989, Stochastic models of soil denitrification, Appl. Environ. Microbiol., 55:72.

Parkin, T.B., Starr, J.L., and Meisinger, J.J., 1987, Influence of sample size on measurement of soil denitrification, Soil Sci. Soc. Am. J., 51:1492.

Parkin, T.B., and Tiedje, J.M., 1984, Application of a soil core method to investigate the effect of oxygen concentration on denitrification, Soil Biol. Biochem., 16:331.

Rolston. D.E., Rao, P.S., Davidson, J.M., and Jessup, R.E., 1984, Simulation of denitrification losses of nitrate fertilizer applied to uncropped, cropped, and manure-amended field plots, Soil Sci., 137:270.

Smith, K.A., 1980, A model of the extent of anaerobic zones in aggregated soils, and its potential application to estimates of denitrification, J. Soil Sci., 31:263.

Smith, K.A., and Arah, J.R.M., 1986, Anaerobic micro-environments in soil and the occurrence of anaerobic bacteria, in: "Microbial Communities in Soil", V. Jensen, A. Kjøller and L.H. Sørensen, eds., Elsevier, London.

COMBINED USE OF THE ACETYLENE INHIBITION TECHNIQUE AND MICROSENSORS FOR QUANTIFICATION OF DENITRIFICATION IN SEDIMENTS AND BIOFILMS

Niels Peter Revsbech and Jan Sørensen*

Department of Ecology and Genetics
University of Aarhus, Ny Munkegade
DK-8000 Aarhus C, and
*Department of Ecology and Molecular Biology
Section of Microbiology
Royal Veterinary and Agricultural University
Rolighedsvej 21, DK-1958 Frederiksberg C
Denmark

INTRODUCTION

The introduction of the acetylene inhibition technique for quantification of denitrification (Balderston et al. 1976; Yoshinari and Knowles, 1976) supplied researchers of this reaction with a highly sensitive and relatively simple assay. Nitrous oxide is a free intermediary in the bacterial reduction of NO_3^- to N_2, and the reduction of N_2O is inhibited by acetylene. The accumulation rate of N_2O in the inhibited sample is thus a measure of the denitrification activity. Gas chromatographs equipped with ECDs (Electron Capture Detectors) can be used to quantify the evolved N_2O with good accuracy and extremely high detection limits. A problem by use of gas chromatography is, however, the need to extract nitrous oxide from the sample before analysis. The procedures for extraction of sample segments may be elaborated to yield a reasonable spatial resolution of the assay (e.g., Sørensen et al, 1979), but the highest resolution by such an approach is still rather on a centimeter than on a millimeter scale while denitrifying microenvironments such as biofilms often have dimensions of less than 1 mm. The recent introduction of a microsensor for simultaneous detection of O_2 and N_2O has now made it possible to determine the activity within denitrifying microenvironments with a spatial resolution of less than 0.1 mm.

In this chapter we describe various experimental approaches used to determine rates and distributions of denitrification activity by use of the acetylene inhibition technique with particular emphasis on application of the O_2/N_2O microsensor. A very recently developed microsensor for NO_3^- is also described, and the capabilities and limitations associated with the use of these two microsensors in ecological research are discussed.

METHODS

Denitrification estimates based on N_2O extraction

Most applications of the acetylene inhibition technique for the study of denitrification in sediments have been based on incubation of sediment cores with acetylene and subsequent extraction of the evolved N_2O after a relatively short incubation period (a few hours). A universally applicable procedure for introduction of acetylene to the denitrification zones within the sediment was developed by Sørensen (1978), who injected acetylene-saturated water into sediment cores until a partial pressure of about 3 kPa was obtained within both sediment and overlying water. Fig. 1 shows how such injection may be performed through silicone-rubber injection ports. The denitrification zone within shallow sediments is generally found so close to the surface or to ventilated infaunal burrows that diffusional supply of acetylene from the overlying water should suffice to stop N_2O reduction, and it may thus not be necessary to perform the injection of acetylene in such sediments. The injection of relatively large volumes of water does disturb the natural chemical gradients in the sediment core to some extent, and it would therefore be an advantage to avoid this step (Christensen et al., 1989). However, experiments comparing the two approaches have not been conducted. Nitrate may be found in deep sediment layers due to nitrifying activity around plant roots (Christensen and Sørensen, 1986) or due to supply from ground water (Capone and Bautista, 1985), and acetylene would have to be supplied to such deep sediment zones by injection. It is essential that the gas headspace shown in Fig. 1 and the overlying water are saturated with a gas mixture containing identical partial pressures of acetylene as diffusional equilibration of gases across the gas-water interface will otherwise cause a buildup of over- or underpressure within the core. The easiest way to get around this problem is to fill the cores completely with water to avoid a gas headspace.

A significant bacterial acetylene oxidation has not been reported during the short incubations used by determination of denitrification in sediments whereas the growth of acetylene-oxidizing microbial populations may be a problem in soils (Chapter 9, this volume) where long incubations with acetylene may be necessary to get acetylene into the central parts of soil aggregates.

The N_2O is usually extracted by a headspace extraction where the sample to be analyzed is brought in diffusional equilibrium with a known volume of N_2O-free gas (Andersen et al., 1984). A sample of this gas is then analyzed on a gas chromatograph, usually stored in gastight Venoject (Terumo Europe N.V., Leuven, Belgium) tubes until analysis is convenient. Acetylene inhibited cores may be sectioned before analysis of N_2O, and a depth distribution of N_2O accumulation can thus be obtained. The accumulation rate of N_2O in each depth interval is often considered as the rate of denitrification. The N_2O found in the overlying water has usually been added to the pool found in the uppermost analyzed segment. Such a simple attempt to describe the depth distribution works quite well if the denitrification activity is changing slowly over a scale of centimeters, but when the scale is in millimeters, it gives very crude pictures of the depth distribution due to the high diffusion coefficient of a small uncharged molecule as N_2O (Broecker and Peng, 1974). Mathematical modelling of the N_2O profiles obtained from sectioned sediment cores have never been used to improve the spatial resolution of the denitrification assay, and in most sediments such modelling would also be complicated by the need to take denitrification around infaunal burrows into account.

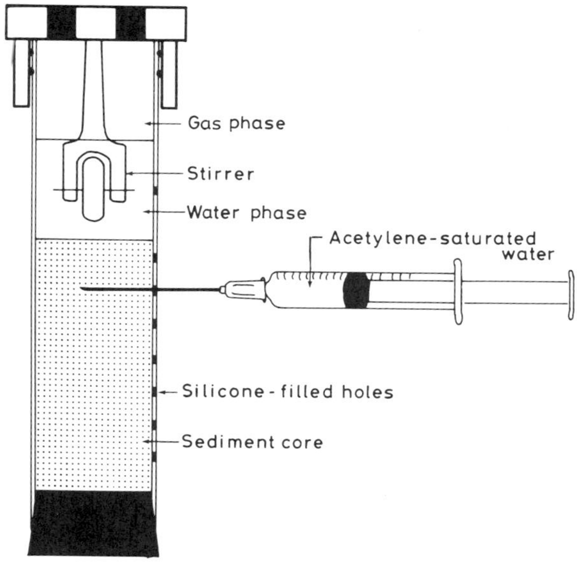

Fig. 1. Injection of acetylene saturated water into sediment core. The injection is done through silicone rubber filled injection ports in the Plexiglas cylinder. Injections in three directions are performed through each port to ensure a relatively homogenous distribution of acetylene. From Koike and Sørensen (1988).

A major potential problem with the acetylene-inhibition method is concurrent inhibition of nitrification activity by the acetylene. The effect of this may be small in sediments or biofilms overlaid by water with high nitrate concentration. However, microenvironments with tightly coupled nitrification-denitrification reactions are common. A typical example is the zone around an oxygen-excreting root in an otherwise anoxic and ammonia-rich sediment. The only source of NO_3^- in the rhizosphere is local nitrification. The turnover rate of the nitrate pool can be so rapid that the acetylene inhibition method will underestimate the actual activity because of decrease in NO_3^- supply rate. Methods based on ^{15}N should generally be applied in such environments.

An additional problem with the acetylene inhibition technique for assay of denitrification activity is the production of a more oxidized end product of the denitrification reaction after acetylene is added. Nitrate will accept 5 electrons when reduced to N_2, but only 4 when reduced to N_2O. This difference may cause an overestimation of the natural rate of denitrification in environments where the activity is regulated by the availability of organic carbon. The difference may be compensated for in stirred media where NO_3^- and organic carbon are found homogeneously distributed. The actual denitrification activity after correction for the lower electron accepting efficiency in the presence of acetylene would correspond to 80% of the denitrification activity determined by N_2O evolution. It is more difficult to evaluate the effect in stratified microbial communities such as sediments where nitrate is supplied to the denitrification zone by diffusion. The addition of acetylene elevates the demand for NO_3^- in each layer by about 25%, which eventually results in decreased penetration depth of NO_3^-. The exact elevation of the total rate of NO_3^- reduction by addition of acetylene then depends on the vertical microdistribution of available electron donors.

The analysis of N_2O accumulation within acetylene-inhibited samples may be combined with a determination of the NO_3^- uptake during the incubation to determine the difference between the total disappearance of NO_3^- and the rate of denitrification. The difference may be either due to assimilation or to dissimilatory reduction to ammonia. Assimilation can often be inhibited by adding relatively high concentrations of ammonium to the overlying water. The quantification of dissimilatory reduction of nitrate to ammonia by this approach is complicated by a the analytical difficulties in determining the NO_3^- concentrations before and after the incubation with sufficient accuracy and also by the often large variations in both denitrification and total NO_3^- uptake among individual sediment cores. A more satisfactory approach to quantify dissimilatory NO_3^- reduction is offered by continuous flow incubations of sediment cores where the overlying water is renewed continuously from a reservoir with well defined concentrations of both electron acceptors, ammonium, and dissolved N_2 (Nishio, 1982). Such continuous flow systems where the uptake or production of individual nitrogen compounds can be measured accurately are described in Chapter 18, this volume. Combined use of continuous flow and ^{15}N techniques (Chapter 18, this volume) should provide a great amount of information about the nitrogen transformations in the investigated sediment cores, and such systems are therefore also very well suited for calibration of other less labor-intensive methods such as the conventional acetylene inhibition technique for determination of denitrification. There is increasing evidence that the acetylene inhibition technique does not operate satisfactorily in environments characterized by low concentration of NO_3^- (Kaspar et al., 1981; Oremland et al., 1984; Slater and Capone, 1989) or presence of sulfide (Tam and Knowles, 1979; Sørensen et al., 1987), and it may therefore be necessary to calibrate the acetylene inhibition technique against more reliable methods when it is used in such environments.

The combined O_2 and N_2O microsensor

Microprofiles of oxygen have been used to characterize the chemical micro-environment within denitrifying sediments (Sørensen et al., 1979; Andersen et al., 1984) and soils (Sexstone et al., 1985), but a quantification of denitrification based on microsensors was not possible until the combined O_2 and N_2O microsensor (Fig. 2) was developed (Revsbech et al., 1988).

A procedure for the construction of this microsensor was described in an earlier publication (Revsbech et al., 1988). Since then we have tried to simplify the rather complicated construction procedure, but with poor success. The microsensor contains 3 cathodes and an internal Ag/AgCl anode enclosed in an outer glass casing containing an alkaline electrolyte. The silicone rubber membrane is highly permeable to small uncharged molecules, such as O_2 and N_2O (Chang and Mayerhof, 1986), which both reach the foremost gold cathode where oxygen is reduced. The current in the measuring circuit originating from this reduction is linearly proportional to the concentration of O_2 in the analyzed medium. The potential at which this cathode is charged is a compromise between a sufficiently negative potential to get a complete reduction of O_2 to OH^- and a potential which is not so negative that N_2O is reduced to a major extent. This compromise usually results in a potential of about -1.0 V relative to the anode. It is, however, essential that the porous gold matrix of the cathode has suitable characteristics. It should completely fill the lumen within the outer casing over a relatively short distance, preferably 20-50 μm, as longer gold cathodes may reduce a significant fraction of the N_2O which enters through the silicone membrane. Such an interference from N_2O would result in an inaccurate determination of O_2,

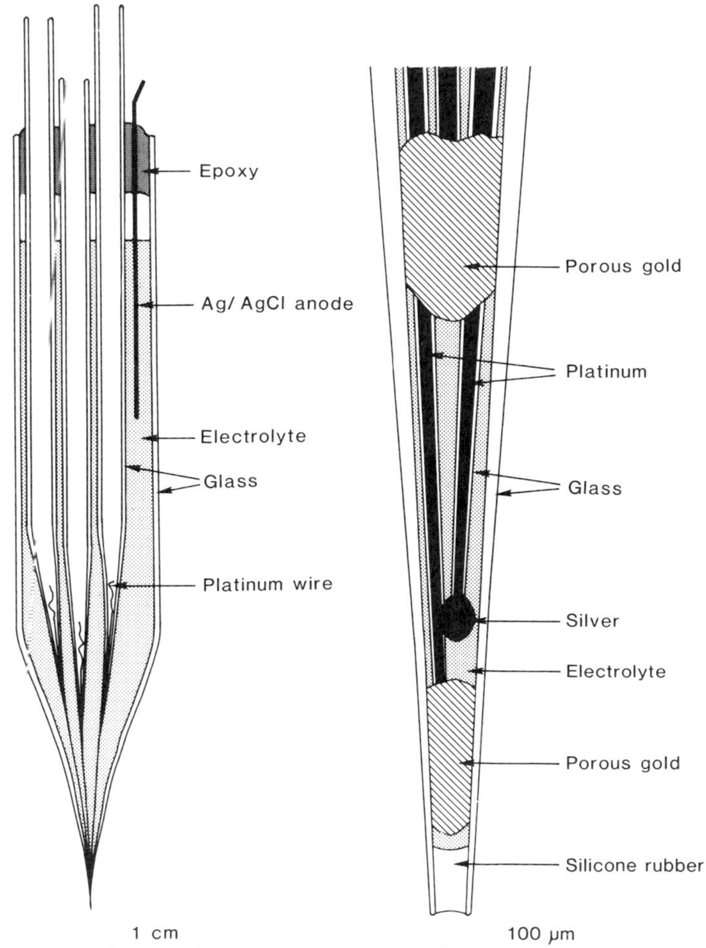

Fig. 2. A microsensor for simultaneous analysis of O_2 and N_2O. An enlarged view of the microsensor tip is given to the right. After Revsbech et al. (1988).

but an even more negative effect would be a reduction in the sensitivity towards N_2O. A N_2O sensing silver cathode is situated just behind the O_2 sensing cathode. The best compromise between sensitivity towards N_2O and a low, stable current for zero N_2O is usually obtained at a potential of about -1.1 V. The backmost cathode consists of porous gold, and it is charged at -1.1 V. It prevents an interfering diffusive flux of reducible substances such as O_2 from the bulk of internal electrolyte towards the sensing cathodes.

The measuring circuit for analysis with amperometric microsensors has been previously described (Revsbech and Jørgensen, 1986). It is essential to use a high quality ammeter which can resolve at least 1 x 10^{-13} A. The microsensors must be introduced into the analyzed substrates by use of a micromanipulator. Both the position of the micromanipulator (i.e., microsensor tip) and currents in the measuring circuits may be transferred directly to a personal computer (Revsbech and Jørgensen, 1986).

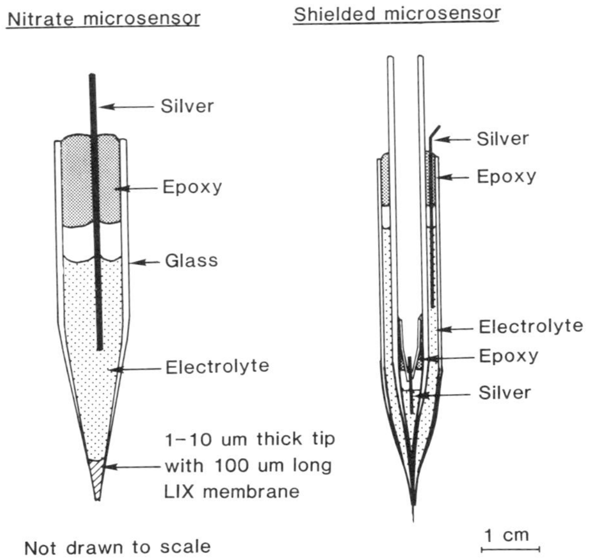

Fig. 3. A microsensor for NO_3^-. The microsensor itself is shown to the left, and to the right is shown how the microsensor can be protected from electrical noise by an outer electrolyte-containing casing.

The detection limits of the microsensor can be less than 1 μM for both O_2 and N_2O. The calibration curves are linear with typical currents of about 10 pA for zero N_2O and 35 pA for 100 μM N_2O. The currents for oxygen are typically 50 pA for zero and 2000 pA for 250 μM O_2. The N_2O signal is virtually unaffected by the diffusional characteristics of the analyzed medium whereas the O_2 signal is affected by changes in flow or diffusional characteristics. Usually the oxygen signal exhibits a 10-20% higher current for vigorously stirred water than stagnant water. It has previously been described how such an non-ideal behavior of the sensor affects the measured concentration profiles near a sediment-water interface (Gust et al., 1987; Revsbech, 1989a). It is, however, relatively easy to construct oxygen microsensors with inner guard cathodes where there is less than 2% or even 1% difference in signal between stirred and stagnant medium (Revsbech 1989a).

Interferences on the N_2O signal may occur by some sensors; others are unaffected by all the possible interfering agent which were tested (excluding sulfide). Changes in the pH of the electrolyte around the N_2O cathode may result in erratic signals. Such changes may be induced by changes in ambient O_2 which is reduced to OH^- within the sensor, or they may be caused by changes in the concentration of acid-reacting CO_2. For these reasons, it is necessary to stabilize the pH in the tip region of the microsensor by constructing a very conical outer casing where the diffusional supply of OH^- from the electrolyte behind stabilizes the pH. Correctly constructed microsensors do not have pH-related interference. Sulfide destroys the catalytic surface of the N_2O cathode and should therefore not be present in analyzed media. The presence of sulfide results in a continuous negative drift in N_2O signal. The interference can be avoided by addition of molybdate (Dalsgaard, 1988) which inhibits sulfate reduction and binds dissolved sulfide chemically. However, the removal of sulfide may change the rate of denitrification by organisms such as Thiobacillus denitrificans. An alternative would be to put a sulfide oxidizing anode in front of the 2 sensing cathodes within the sensor, but attempts to do this have so far not been successful. The most disturbing interference when working with the combined O_2 and N_2O microsensor is, however, the effect of acetylene on the signal of the N_2O cathode. When the microsensors are made as described originally (Revsbech et al., 1988), the interference is often very low and even negligible, but some sensors are highly sensitive to acetylene. All modifications of the procedure used to make the catalytic silver cathode we have tried so far have produced microsensors which were very sensitive to acetylene, and usually they also exhibited a downward drift in the sensitivity to N_2O. A slight interference from acetylene does not present a problem in thin biofilms and similar substrates which equilibrate to a homogeneous distribution of acetylene within a few minutes. However, only a relatively small proportion of the constructed microsensors may be satisfactory for use in moderately active soils and sediments where acetylene concentrations gradients may result in differences in signal which are higher than those caused by N_2O itself. The sensitivity towards acetylene may, however, change with time, and an unacceptable sensor may be acceptable the next day. Due to the corrosive effects of the highly alkaline electrolyte, the performance of most microsensors begin to degrade within a few days, and only a few have acceptable characteristics after more than 1 week. The lifetime may be improved by substituting the alkaline electrolyte with KCl solution in periods when the electrode in not in use.

NO_3^- microsensor

Microprofiles of N_2O can be used to calculate depth profiles of denitrification. To investigate interactions between denitrification and other NO_3^--consuming processes within sediments, soils and biofilms, it is also necessary to obtain microprofiles of NO_3^-. A microsensor for NO_3^- (Beer and Sweerts, 1989) is shown in Fig. 3A. The construction of the NO_3^- microsensor is quite simple. It consists of a glass micropipette which is partially or totally silanized (Thomas, 1978) to make the glass hydrophobic. The tip region of the micropipette is then filled with a liquid ion exchanger (e.g., Orion 92-07-01), and the rest of the lumen within the micropipette is filled with an inner electrolyte usually containing both NO_3^- and Cl^-. Due to the ion selectivity of the liquid ion exchanger, a potential is formed between an external reference electrode and the internal Ag/AgCl electrode, and this potential is read on a high impedance voltmeter (e.g., Keithley 617). The signal from the microsensor to the voltmeter may be conducted through a triaxial cable (e.g., Keithley 6011) which decreases the response time of small microsensors with high electrical resistance through an electronic dissipation of capacitance phenomena. The unshielded microsensor is difficult to use in an ordinary laboratory environment due to the high impedance of the liquid ion exchanger which makes the microsensor extremely sensitive to electrical noise. Usually such noise is reduced by use of a Faraday cage surrounding the whole experimental setup. The noise can, however, be more efficiently eliminated by using a coaxial design in which the microsensor is surrounded by an electrolyte kept on the potential of the external reference (Fig. 3B). In case of "guarded mode" triaxial signal cunductance, it may also be possible to apply the potential of the innermost shielding in the cable to the Ag/AgCl electrode immersed into the shielding electrolyte, and the sensor should then both have a very rapid response and be unaffected by electrical noise. We have not yet tried to use such liquid shielding with guarded mode signal conductance which may be necessary by use of very small, high resistance microelectrodes. The construction of the shielded sensor shown in Fig. 3B makes it possible to use the same desiccant-containing electrode holder as used for oxygen microsensors (Revsbech, 1983) when a normal low-noise coaxial cable is used. It is highly essential always to use clean glass surfaces enclosed in desiccant-containing electrode holders around the signal-conducting parts of microsensors, as shortcircuits caused by humidity can be totally detrimental to any otherwise good microsensor or measuring circuit. It does not help you to use a good voltmeter with an internal impedance of $>10^{14}$ Ohms if the impedance along a critical glass surface is only 10^{10} ohms!

The lowest possible detection limit for NO_3^- microsensors is about 1 μM. The sensitivity varies somewhat among microsensors, but in demineralized water most have an almost ideal response at concentrations above 10 μM with a slope of the calibration curve which approaches the theoretical 59 mV per decade (usually about 55 mV). Use of such microsensors is, however, restricted to environments with low concentrations of monovalent anions. Especially Cl^- interferes with the measurement of NO_3^-, although the selectivity coefficient for most ion exchangers used for NO_3^- sensors are highly in favor of NO_3^-. Due to high concentrations of Cl^-, NO_3^- microsensors cannot be used in marine environments. Even for use in environments with relatively low electrolyte concentrations as found in freshwater, it is essential that the calibration curves are made in a medium containing identical concentrations of the dominating anions. Bicarbonate does, however, also interfere, and this interference is difficult to compensate for by the calibration, as all stratified substrates exhibiting denitrification activity and NO_3^- gradients are also characterized by concentration gradients of bicarbonate. The microsensors can only be used for 1 day, but the inconvenience caused by their short lifetime is largely balanced by their ease of construction.

RESULTS OBTAINED BY USE OF MICROSENSORS

The combined O_2 and N_2O microsensor

The data obtained by use of the combined O_2 and N_2O microsensor may be treated as the data obtained by conventional extraction techniques followed by GC analysis. However, the real advantage of using such a technique with a high spatial resolution only emerges when diffusion-reaction models are used to deduce microprofiles of denitrification from the analyzed N_2O profiles. The following equations are only valid for substrates where the concentration changes occur in one dimension only. Relatively simple systems where the diffusion must be described in 3 dimensions may, however, also be described mathematically (Aller, 1980). The simplest diffusion equation is Fick's first law (Berner 1980):

$$J = -\phi D_s \frac{\partial C}{\partial x} \qquad (1)$$

The equation expresses that the transport (diffusive flux) rate, J, of a solute through a unit area perpendicular to the concentration gradient is equal to the concentration (C) gradient with depth (x), multiplied with the apparent diffusion coefficient within the substrate, D_s, and the porosity, ø. Equation 1 may, for example, be used to calculate the flux of N_2O out of a sediment which has been incubated with acetylene until a steady state N_2O emission is approached. Most sediments must, however, be incubated for a very long time before a steady state is reached, and the change is concentration (C) with time (t) therefore has to be taken into account (Berner, 1980):

$$\frac{\partial C}{\partial t} = D_s \frac{\partial^2 C}{\partial x^2} + P - R \tag{2}$$

P is the production rate of the investigated chemical species (at depth x), and R is the corresponding consumption rate. The porosity and diffusion coefficient usually varies with depth, and taking this into account, Eq. 2 gets somewhat more complicated (Berner, 1980):

$$\frac{\partial C}{\partial t} = D_s \frac{\partial^2 C}{\partial x^2} + \left[\frac{\partial D_s}{\partial x} + \frac{D_s}{\phi} \cdot \frac{\partial \phi}{\partial x} \right] \cdot \frac{\partial C}{\partial x} + (P - R) \tag{3}$$

Equation 3 can be transformed into a discrete form which can be implemented on a computer (Revsbech et al., 1986). Details about computer programming using Eq. 3 have been published previously (Revsbech et al., 1986; Revsbech, 1989b, Revsbech et al., 1989). In the following it will be shown how computer-simulation of measured N_2O profiles can be used to calculate the distribution of denitrification within the substrate. To perform such a simulation it is necessary to know the vertical distribution of diffusion coefficient and porosity within the substrate, and microtechniques for determination of these parameters will be described in the last section of this chapter.

Microprofiles of O_2 and N_2O were measured 9, 28, and 55 min. after the addition of acetylene to a sediment core from a small Danish river (Fig. 4). The acetylene was introduced by injection as shown in Fig. 1. The measured profiles of N_2O are shown with symbols (Fig. 4A), whereas the continuous depth profiles are simulated curves obtained by insertion of the denitrification profile shown in Fig. 4B into the computerized, discrete version of the diffusion-reaction model (Eq. 4). Denitrification occurred in a zone extending from just above the oxic-anoxic interface at 1 mm down to a depth of 5 mm. Insertion of a production rate of N_2O (denitrification) into the model could, however, not reproduce the measured N_2O profiles completely, and it was necessary to insert a consumption of N_2O below the denitrification zone. Along with the production rates shown in Fig. 4B, a consumption rate of 22 nmol $cm^{-3}h^{-1}$ in all N_2O-containing layers below 5 mm depth had to be inserted into the model to give the almost perfect curve fit shown in Fig. 4A. Analysis of sediment from another river showed that also in this sediment there was a significant reduction of N_2O in layers below the denitrification zone. It has been mentioned above that the inhibition of N_2O reductase by acetylene may be contingent upon the simultaneous presence of NO_3^-, and the reduction of N_2O in the deeper layers of the sediments could thus be due to the absence of NO_3^- in these layers. Microsensor analysis of N_2O microprofiles in acetylene inhibited epilithic biofilms shows, however, that some biofilms do not reduce N_2O in deeper layers even if NO_3^- is totally depleted (Revsbech et al., 1989). Other biofilms exhibit a very high rate of N_2O consumption in deeper, reduced layers (T. Dalsgaard, 1988), and it seemed likely that sulfide, which alleviates the inhibition of N_2O reductase by acetylene (Sørensen et al., 1987), caused this consumption of N_2O.

The ability to distinguish between zones of N_2O production and N_2O consumption by the use of N_2O microsensors is a clear advantage in comparison with gas chromatographic techniques, as such a separation makes it possible to compensate denitrification estimates for the simultaneous consumption of N_2O. In the sediment analyzed in Fig. 4, only 7% of the N_2O formed during the 55-min incubation was reduced, but this percentage has been much higher in other

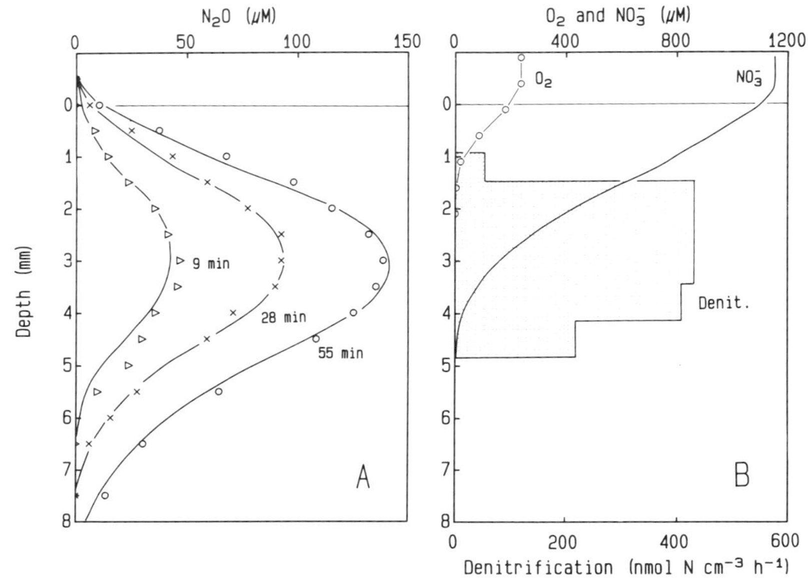

Fig. 4. A: Microprofiles of N_2O in a riverine sediment at various times (in minutes) after adding acetylene to both sediment and overlying water. B: Measured microprofile of O_2 and simulated microprofiles of NO_3^- and denitrification in the sediment. The simulated values were obtained from analysis of the N_2O data shown in Fig. 4A. From Christensen et al. (1989).

sediments and biofilms. It cannot be excluded that some N_2O reduction took place within the denitrification zone, as profile analysis only gave information about net rates. The denitrification zone was actually defined above as the layer where the net rate of N_2O evolution was positive. The specific rates of N_2O reduction estimated for the deeper layers were, however, only about 5% of the specific rates of denitrification within the denitrification zone, and the presence of nitrate combined with an absence of sulfide within the denitrification zone should have resulted in an even lower N_2O reduction activity within this layer. The simulated NO_3^- profile shown in Fig. 4B also supports the assumption of a negligible N_2O reduction within the denitrification zone.

The NO_3^- profile (Fig. 4B) was simulated using the same diffusion-reaction model as used to model the N_2O profiles, but the reaction rate in each layer was now assumed to be a consumption of NO_3^- which on a molar basis equalled twice the production rates of N_2O illustrated in Fig. 4B. The resulting NO_3^- profile exhibited a concentration in the overlying water of 1,170 μM. Denitrification alone could thus explain that NO_3^- decreased from 1,170 μM in the overlying water to 0 at the bottom of the denitrification zone. The NO_3^- concentration actually present in the overlying water was 1,250 μM, and the concentration gradient upon which other NO_3^- consuming processes would depend should then be based on an "available" concentration in the water of (1,250-1,170) μM = 80 μM. This fraction constitutes only 6% of the NO_3^- concentration, and can also with some caution be considered as the percentage of the NO_3^- which was reduced to ammonia (assimilitative or dissimilative reduction). The basic assumption for such an estimate of the relative importance of denitrification and other nitrate reduction pathways is that the percentage going to denitrification and to the combined assimilatory and dissimilatory reduction to ammonium was

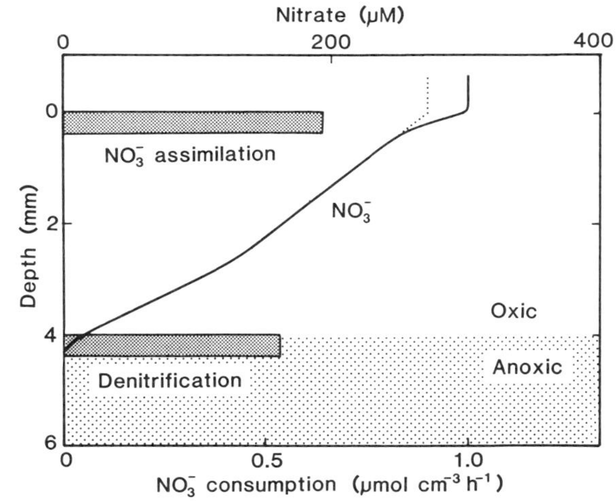

Fig. 5. Effect of a NO_3^- assimilation zone and a denitrification zone on a profile of NO_3^-. The solid line is the NO_3^- profile when both zones are active. The dotted line represents a NO_3^- profile which could support the same rate of denitrification in the absence of a NO_3^- assimilation zone. It is evident that the NO_3^- assimilation only had a small effect on the NO_3^- profile and hence on the rate of denitrification.

constant throughout the NO_3^- -containing zone. We do not yet know the depth distribution of dissimilative NO_3^- reducers relative to the denitrifiers, so the assumption of unchanged relative rates of these two reactions throughout the denitrification zone cannot be supported. Assimilative reduction of NO_3^- to ammonium can occur at very high rates in illuminated benthic communities of microalgae, and the processes of denitrification and reduction to ammonium may in such cases be widely separated. To illustrate this, microprofiles of oxygen, oxygenic photosynthesis, and N_2O were measured in a sediment with a dense benthic community of microalgae (L.P. Nielsen et al., submitted to Limnol. Oceanogr.). The microprofiles were measured both with virtually NH_4^+ free overlying water and with 27 μM NH_4^+. The microprofiles were virtually identical with and without NH_4^+. A 0.5-mm thick photosynthetically active surface layer produced sufficient oxygen to extend the oxic layer down to the 4-mm depth, below which was a denitrification zone from 4.0 to 4.5 mm. The microprofile of NO_3^- was not measured, but it must have decreased from 300 μM in the overlying water to 0 at 4.5 mm depth. The exchange of NO_3^- across the sediment-water interface in the experiments with and without NO_3^- were also measured. The uptake of NO_3^- in the experiment without NH_4^+ turned out to be 2.4 times higher than when NH_4^+ was added to inhibit assimilative NO_3^- reduction.

At first glance the data from microprofile and NO_3^- exchange experiments may seem to conflict. The N_2O profile was unaffected by addition of NH_4^+, and the overall NO_3^- profile must consequently also have been virtually unaffected by the addition. Yet the NO_3^- exchange data indicated that the gradient at the sediment-water interface must have been 2.4 times steeper without NH_4^+ as compared to the experiment with NH_4^+. The explanation for these apparently conflicting observations is that NO_3^- assimilation took place in the uppermost 0.5-mm layer where the microalgae were active, and the proximity of the assimilation zone to the NO_3^- source (overlying water) only required a steep concentration

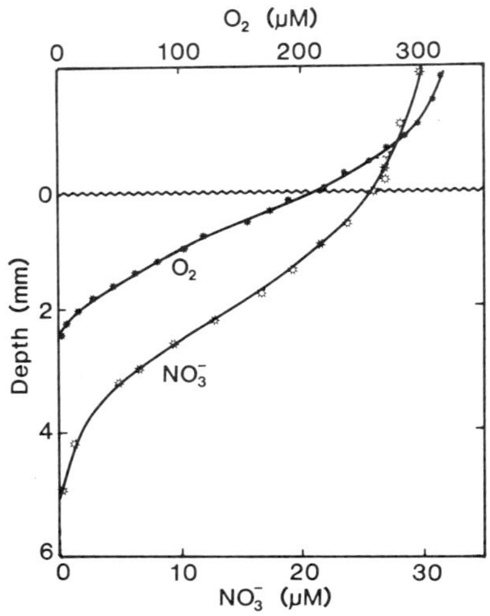

Fig. 6. Microprofiles of O_2 and NO_3^- in a lake sediment. The analysis was performed with one microsensor for O_2 and one for NO_3^-. Redrawn from data published by Sweerts and de Beer (1989).

gradient over a very short depth interval, thus not changing the overall NO_3^- profile significantly. The denitrification profile in illuminated sediment without NH_4^+ in the overlying water is shown in Fig. 5 together with modeled profiles of NO_3^- and NO_3^- assimilation. The dotted line in Fig. 5 illustrates that the effect of NO_3^- assimilation on denitrification corresponds to the effect of a decrease in the NO_3^- concentration in the overlying water from 302 to 271 μM, or 10%. The percent decrease in denitrification would then also be roughly 10%.

The NO_3^- microsensor

Only a few data obtained by use of NO_3^- microsensors have been published (Beer and Sweerts, 1989; Sweerts and Beer, 1989). Profiles of oxygen (obtained by use of an oxygen microelectrode) and NO_3^- from a Dutch lake are shown in Fig. 6.

Like the data measured by use of the combined O_2 and N_2O microsensor, the profiles shown in Fig. 6 show that (dissimilative) nitrate reduction takes place in a narrow layer immediately below the oxic layer. Detailed microprofiles of total NO_3^- reduction may be obtained from such microprofiles of NO_3^- if it is possible to measure NO_3^- with sufficient accuracy. It would be interesting to observe how such microprofiles of NO_3^- reduction are affected by acetylene, and it would also be highly intriguing to see how such microprofiles compare to denitrification profiles obtained from N_2O microsensor data. In principle, it should be possible to calculate the depth distribution of both denitrification and reduction to ammonium from such coupled data. If sufficiently accurate and sensitive, the NO_3^- microsensor will also enhance our understanding of nitrification and the coupling between nitrification and nitrate reduction in various microenvironments (e.g., the surroundings of an oxygen excreting root in an otherwise anoxic sediment or soil).

DIFFUSION COEFFICIENTS

In the description of results obtained by use of microsensors we have probably quite convincingly shown that very detailed microprofiles of O_2, N_2O, and NO_3^- can be obtained. However, to calculate metabolic rates from these microprofiles it is necessary to know how these chemical species diffuse in the analyzed media; i.e., we need to know depth profiles of the apparent diffusion coefficient (D_s) and the porosity (ø). Methods have been developed which can give quite satisfactory microprofiles of D_s for gases and to some extent also of ø (Revsbech, 1989b). It is simplest to perform calculations on steady-state concentration profiles, as these can be analyzed by use of Fick's first law of diffusion (Eq. 1). To calculate a diffusive flux, it is only necessary to know the slope of the concentration profile, as obtained by microsensor analysis, and the value of $øD_s$. Microprofiles of $øD_s$ are quite easily obtained for gases by use of the diffusion chamber shown in Fig. 7. The sample to be analyzed should be placed on top of a silicone rubber membrane where it is situated in a steep steady-state diffusion gradient of gases established between the surrounding atmosphere and the gas in the reservoir below the membrane. It is essential to either place a 0.5-1.0 mm layer of water above the sample or to put a layer of 1% agar just above the silicone membrane. The diffusion flux through the sample and the thin layer of water or agar is constant through all layers at steady state, and according the Eq. 1, the change in slope between layers can thus only be due to a change in the value of $øD_s$. The water or agar layer serves as a reference through which the value for $øD_s$ is known, and the difference in slope between the reference and any layer in the sample is thus proportional to the difference in $øD_s$. There should be no transformations of the gasses to be analyzed, so in the case of

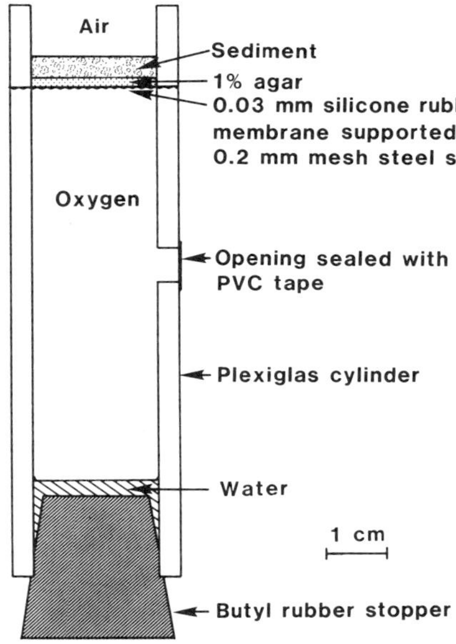

Fig. 7. Diffusion chamber for establishment of a concentration gradient of gases through a sample. The diffusional characteristics of the sample are then determined from the difference in concentration gradient between the sample and a reference layer. From Revsbech (1989b).

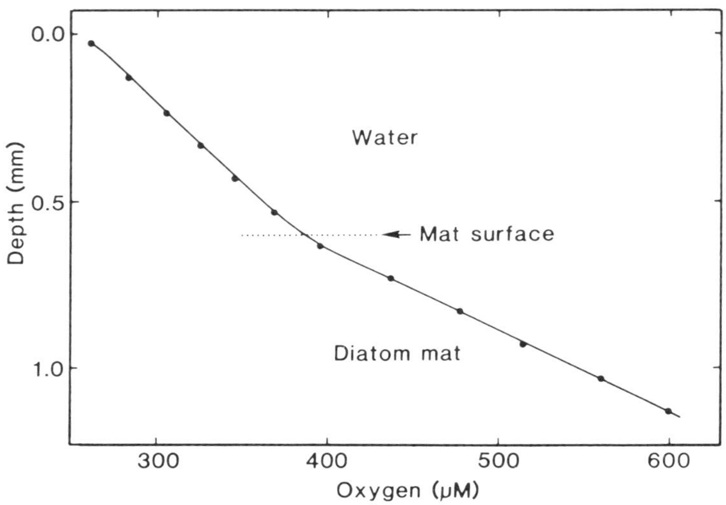

Fig. 8. Concentration profile of O_2 through a diatom biofilm and a 0.6 mm layer of overlying water. The change in slope is due to a difference in the diffusional characteristics between the two layers. From Revsbech (1989b).

oxygen, the sample has to be biologically inactivated with $HgCl_2$ before the analysis. The transformation of N_2O can also be prevented by either creating high oxygen concentrations within the sample or by including high concentrations of acetylene in the gas reservoir. An example of a concentration profile through a Hg-killed diatom biofilm is shown in Fig. 8. The øDs value in this biofilm was about 51% of the value in water corresponding to a value of 1.05 cm^2s^{-1} at 20 °C.

By use of experimentally obtained N_2O microprofiles and also experimentally obtained microprofiles of $øD_s$ it is thus possible to obtain estimates for depth profiles of denitrification. These profiles are in units of concentration change per volume of porewater, and to integrate such microprofiles and to obtain the total activity it is thus necessary to multiply the rates in individual layers with the porosities in these layers.

It is much more difficult to obtain good estimates for the diffusion coefficient of NO_3^-. A first approximation may be obtained from literature values of the NO_3^- diffusion coefficient in water (Li and Gregory, 1974) and then assume a similar decrease in the biofilm as measured for O_2 or N_2O. However, the diffusion of a charged species as NO_3^- within a matrix containing fixed positive and negative charges may be quite different from that of an uncharged gas. It is possible that part of the porewater space is made unavailable by electric repulsion from fixed negative charges (Riemer and Harremoes, 1978). It is also possible that the diffusion and reaction of charged species create counter-diffusion phenomena (e.g., Li and Gregory, 1974), where the diffusion of NO_3^- is either facilitated or slowed down because of a build-up of an electrical potential between substrate and overlying water. Such counter-diffusion phenomena would only occur in metabolically active biofilms, and a determination of a diffusion coefficient in an inactivated biofilm would thus be of limited value. Calculations were performed on the N_2O data shown in Fig. 4 to obtain profiles of NO_3^-, but these calculations were based on an assumed value for the diffusion coefficient for NO_3^-. Incorrectly estimated diffusion coefficients will invariably lead to erroneous microprofiles of NO_3^-. The fractions of NO_3^- being consumed by denitrification and dissimilatory reduction to NH_4^+ were estimated from the calculated NO_3^- profiles and these estimates may thus also be far from reality. Also calculations based on NO_3^- microprofiles measured with NO_3^- microelectrodes would be invalidated by erroneously estimated diffusion coefficient for NO_3^-. The conclusion is, that by use of the O_2/N_2O microsensor it is quite simple to obtain microprofiles of denitrifying activity, but it may be difficult to get a similar microprofile of NO_3^- reduction to ammonia (assimilative or dissimilative).

SUMMARY

The acetylene inhibition technique is still the most convenient method for the quantification of denitrification activity. When used in combination with the microsensor for O_2/N_2O, the method can yield determinations of denitrification with a spatial resolution of less than 0.1 mm. By use of this microsensor it became evident that N_2O may be reduced in deeper, reduced layers of many denitrifying environments, and the rates of both N_2O production and consumption were calculated from the microprofiles. Denitrification may also be studied by use of a NO_3^- microsensor, but calculations on NO_3^- profiles must be based on diffusion coefficients for NO_3^- which are difficult to measure in biologically active substrates.

RERERENCES

Aller, R.C., 1980, Quantifying solute distributions in the bioturbated zone of marine sediments by defining an average microenvironment, Geochim. Cosmochim. Acta, 44:1955.

Andersen, T.K., Jensen, M.H., and J. Sørensen, 1984, Diurnal variation of nitrogen cycling in coastal, marine sediments. I. Denitrification, Mar. Biol., 83:171. Broecker, W.S., and Peng, T.-H., 1974, Gas exchange rates between air and sea, Tellus, 26:21.

Balderston, W.L., Sherr, B., and Payne, W.J., 1976, Blockage by acetylene of nitrous oxide reduction in Pseudomonas perfectomarinus, Appl. Environ. Microbiol., 31:504.

Berner, R.A., 1980, "Early diagenesis. A theoretical approach," Princeton University Press, Princeton.

Broecker, W.S., and Peng, T.-H., 1974, Gas exchange rates between air and sea, Tellus, 26:21.

Capone, D.G., and Bautista, M., 1985, Direct evidence for a groundwater source for nitrate in nearshore marine sediments, Nature, 313:143.

Chang, Q., and Mayerhof, E., 1986, Membrane-dialyzer injection loop for enhancing the selectivity of anion-responsive liquid-membrane electrodes in flow systems, Anal. Chim. Acta., 186:81.

Christensen, P.B., Nielsen, L.P., Revsbech, N.P., and Sørensen, J., 1989, Microzonation of denitrification activity in stream sediments as studied with a combined oxygen and nitrius oxide microsensor, Appl. Environ. Microbiol., 55:1234.

Christensen, P.B., and Sørensen, J., 1986, Temporal variation of denitrification activity in plant covered littoral sediment (Lake Hampen, Denmark). Appl. Environ. Microbiol., 51:1174.

Dalsgaard, T., 1988, "Denitrification in biofilms," M.Sc. thesis, University of Aarhus, Aarhus.

de Beer, D., and Sweerts, J.-P.R.A., 1989, Measurement of nitrate gradients with an ion-selective microelectrode. Anal. Chim. Acta, 219:351.

Gust, G., Booij, K., Helder, W., and Sundby, B., 1987, On the velocity sensitivity (stirring effect) of polarographic oxygen microelectrodes, Neth. J. Sea Res., 21:255.

Kaspar, H.F., Tiedje, J.M., and Firestone, R.B., 1981, Denitrification and dissimilatory nitrate reduction to ammonium by digested sludge, Can. J. Microbiol., 38:486.

Koike, I., and Sørensen, J., 1988, Nitrate reduction and denitrification in marine sediments, in: "Nitrogen cycling in coastal marine environments," T.H. Blackburn and J. Sørensen, eds., SCOPE 33, John Wiley & Sons, Chichester.

Li, Y.-H., and Gregory, S., 1974, Diffusion of ions in sea water and in deep sea sediments, Geochim. Cosmochim. Acta., 38:703.

Nishio, T., 1982, "Nitrogen cycling in coastal and estuarine sediments with special reference to nitrate reduction, denitrification and nitrification," Doctoral dissertation, University of Tokyo, Tokyo.

Oremland, R.S., Umberger, C., Culbertson, C.W., and Smith, R.L., 1984, Denitrification in San Fransisco intertidal sediments, Appl. Environ. Microbiol., 47:1106.

Revsbech, N.P., 1983, In situ measurement of oxygen profiles of sediments by use of oxygen microelectrodes, in: "Polarographic oxygen sensors: Aquatic and physiological applications," E. Gnaiger and H. Forster, eds., Springer, Heidelberg.

Revsbech, N.P., 1989a, An oxygen microsensor with a guard cathode, Limnol. Oceanogr., 34:474.

Revsbech, N.P., 1989b, Diffusion characteristics of microbial communities determined by use of oxygen microsensors. J. Microbiol. Meth., 9:111.

Revsbech, N.P., Christensen, P.B., Nielsen, L.P., and Sørensen, J., 1988, A combined oxygen and nitrous oxide microsensor for denitrification studies. Appl. Environ. Microbiol., 54:2245.

Revsbech, N.P., Christensen, P.B., Nielsen, L.P., and J. Sørensen, 1989, Denitrification in a trickling filter biofilm studied by a microsensor for oxygen and nitrous oxide, Water Res., 23:867.

Revsbech, N.P., and Jørgensen, B.B., 1986, Microelectrodes: Their use in microbial ecology, in: "Advances in Microbial Ecology, vol. 9," K.C. Marshall, ed., Plenum, New York.

Revsbech, N.P., Madsen, B., and Jørgensen, B.B., 1986, Oxygen production and consumption in sediments determined at high spatial resolution by computer simulation of oxygen microelectrode data, Limnol. Oceanogr., 31:293.

Riemer, M., and Harremoes, P., 1978, Multi-component diffusion in denitrifying biofilms, Prog. Wat. Res., 10:149.

Sextone, A.J., Revsbech, N.P., Parkin, T.B., and Tiedje, J.M., 1985, Direct measurement of oxygen profiles and denitrification rates in soil aggregates, Soil. Sci. Soc. Am. J., 49:645.

Slater, J.M., and Capone, D.G., 1989, Nitrate requirement for acetylene inhibition of nitrous oxide reduction in marine sediments, Microb. Ecol., 17:143.

Sørensen, J, 1978, Denitrification rates in a marine sediment as measured the acetylene inhibition technique, Appl. Environ. Microbiol., 36:139.

Sørensen, J., Jørgensen, B.B., and Revsbech, N.P., 1979, A comparison of oxygen, nitrate, and sulfate respiration in coastal marine sediments, Microb. Ecol., 5:105.

Sørensen, J., Rasmussen, L.K., and Koike, I., 1987, Micromolar sulfide concentration alleviate acetylene blockage of nitrous oxide reduction by denitrifying Pseudomonas fluorescens. Can. J. Microbiol., 33:1001.

Sweerts, J.-P.R.A., and de Beer, D., 1989, Microelectrode measurements of nitrate gradients in the littoral and profundal sediments of a meso-eutrophic lake (Lake Vechten, The Netherlands), Appl. Environ. Microbiol., 55:754.

Tam, T.-Y., and Knowles, R., 1979. Effects of sulfide and acetylene on nitrous oxide reduction by soil and by Pseudomonas aeruginosa, Can. J. Microbiol., 25:1133.

Thomas, R.C., 1978, "Ion-sensitive intracellular microelectrodes, how to make and use them". Academic Press, London.

Yoshinari, T., and Knowles, R., 1976, Acetylene inhibition of nitrous oxide reduction by denitrifying bacteria. Biochem. Biophys. Res. Commun., 69: 705.

DENITRIFICATION IN STREAM BIOFILM AND SEDIMENT: IN SITU VARIATION AND CONTROL FACTORS

Jan Sørensen and Niels Peter Revsbech*

Department of Ecology and Molecular Biology
Section of Microbiology
Royal Veterinary and Agricultural University
Rolighedsvej 21, DK-1958 Frederiksberg C, and
*Department of Ecology and Genetics
University of Aarhus, Ny Munkegade
DK-8000 Aarhus C, Denmark

INTRODUCTION

Seasonal and diurnal variations of denitrification activity have been studied in sediments from lakes (Christensen and Sørensen, 1986), estuaries (Jørgensen and Sørensen, 1988) and other near-coastal areas (Andersen et al., 1984). Only recently, denitrification has also been determined in riverine systems including sediment and epilithic biofilms (Christensen et al., 1989; Christensen et al., in press; Duff et al., 1984; Nielsen et al., in press; Sørensen et al., 1988). Direct measurements of the process is essential to provide detailed information about temporal and spatial variation and to evaluate the important factors of environmental control. It is often possible to judge the relative influence of primary control parameters such as temperature, substrate availability (nitrate and organic matter) and O_2 inhibition. Sometimes, even the effects of changes in the biological community, e.g. shifts in benthic algae metabolism, bioturbation and grazing effects by animals, effects of plant metabolism in the rhizosphere, etc. can be observed.

After the observation that active microbial communities such as sediments generally contain a thin oxic surface zone (e.g. Revsbech et al., 1986), it may be expected that denitrification is also confined to a narrow suboxic zone. This is due to the requirement of a NO_3^- supply by diffusion transport, either from the overlying water phase or from NO_3^- production (nitrification) within the thin oxic surface layer. By comparison with the excessive amount of data, which has recently been obtained by O_2 microsensor studies of both photosynthesis and O_2 respiration in narrow surface films, detailed microzonation studies of anaerobic respiration such as denitrification have long been impeded by the low spatial resolution of available techniques. Where the acetylene inhibition technique has been used to measure denitrification, a critical step has been to detect N_2O accumulation at its actual production site in the sediment, particularly when the process occurs in a few mm deep layer below the oxic surface zone. Recent attempts to determine microprofiles of both NO_3^- concentrations and denitrification activity were made by sectioning sediment cores in thin, 1- to 5-mm segments (Lomstein et al., in press). This is only possible, however, where the sediment is a soft, but relatively well consolidated mud. Great progress is therefore to be expected from a new N_2O microelectrode, which was employed to demonstrate microprofiles of denitrification activity in both biofilms (Nielsen et al., in press; Revsbech et al., 1989) and sediments (Christensen et al., 1989; Nielsen et al., in press).

The present review describes a number of observations made by the new microelectrodes, e.g. on depth distribution and environmental control of microbial photosynthesis, O_2 respiration and denitrification in biofilms and sediments from freshwater environments. Both the biofilms from trickling filters or streams and the sediment from shallow lowland streams provide useful environments for laboratory and field studies of the diurnal and seasonal variations of denitrification and have allowed us to reveal some of its important mechanisms of control in situ.

THE EPILITHON-MICROBENTHOS COMMUNITY

Biological components and processes

In order to explain seasonal or even regional variations of the denitrification, parameters like temperature and overall availability of organic substrate affecting bacterial growth must be important. Benthic microalgae comprises an important autotrophic component in most eutrophic, lowland streams of temperate regions (Sand-Jensen et al., 1988). The microalgal community, which developes rapidly in early spring, in turn provides an important source of organic substrate for the heterotrophic (Haack and McFeters, 1982), and thus the denitrifying bacteria. The early development of algal biomass in spring is strongly coupled to microbial degradation so long as the abundance of grazing invertebrates is relatively low. This predominantly microbial community, discerned in the spring to early summer, is therefore characterized by direct interactions of algal - bacterial metabolism and well suited for direct studies of denitrification supported by algal growth. After the spring, photoautotrophic activity may decrease in the stream (Christensen and Sørensen, 1988) and the algal community is often subject to increasing grazing pressure and sediment reworking (bioturbation) by the bottom invertebrates.

Diel cycles resulting from short-term interactions between algal metabolism and denitrification may be simplest to predict in the spring when the microalgae and bacteria are predominant in the benthic community. The previous studies have all demonstrated the inhibitory effect of light, photosynthesis and O_2 production on denitrification activity in sediments of shallow waters (Andersen et al., 1984; Christensen and Sørensen, 1986; Jørgensen and Sørensen, 1988; Sørensen et al., 1988): photosynthetic O_2 production results in a deeper extension of the oxic surface zone and denitrification is inhibited during the day. However, even if photosynthesis is expected to limit denitrification activity to deeper layers in the sediment, the actual change in denitrification rate must depend on depth distributions of both organic substrate, NO_3^- and the denitrifying population. The new N_2O microsensor technique has allowed, for the first time, a determination of specific activities of denitrification (activity per unit of volume) at a given depth. A much more detailed understanding of the limiting factor(s) of denitrification, e.g. under dark and light conditions, has therefore become available (Christensen et al., 1989; Nielsen et al., in press; Nielsen et al., in press).

New methodology in the microscale: Diffusion - reaction model to determine denitrification from N_2O concentration profiles

As for the previous studies of photosynthesis and O_2 respiration based on determination of O_2 microprofiles (Revsbech et al., 1986), denitrification activity may be approached by determination of N_2O microprofiles and application of Fick's 2. law of diffusion (Christensen et al., 1989; Nielsen et al., in press, Revsbech et al., 1989; Revsbech and Sørensen, this book). After incubation of the sediment or biofilm with acetylene to inhibit N_2O reduction by the denitrifying bacteria, the N_2O accumulation is followed by the microelectrode technique. A diffusion-reaction model containing only the parameters of N_2O diffusion and accumulation should adequately describe the depth distribution of denitrification. Thus, if N_2O consumption is absent (as in the presence of acetylene), the denitrification rate at a given depth relates to the steady state profile of N_2O as follows:

$$P(z) = D\, d^2C/dz^2 \qquad (1)$$

where P is the specific activity of N_2O production (denitrification), D is diffusivity, C is concentration and z is depth.

Assuming that production of N_2O equals consumption of NO_3^- (denitrification accounting for all NO_3^- reduction), the NO_3^- profile can be calculated directly from the steady state profile of N_2O. The former should be a mirror image of the N_2O profile, corrected for the difference in diffusion coefficient and aligned to a known NO_3^- concentration in the overlying water (Nielsen et al., in press).

CONTROL OF DENITRIFICATION IN SURFACE FILMS

The microsensor studies presented here were made in Danish lowland streams characterized by relatively high NO_3^- levels in the water (about 0.1 - 1 mM NO_3^-), depending on season and study site. The high nutrient level in the streams is primarily due to run-off from agricultural areas.

Figure 1 shows an example of a depth distribution of O_2 respiration and denitrification in a biofilm, measured by microelectrode technique. Oxygen and NO_3^- concentrations in the water phase were about 0.2 mM and 1.2 mM, respectively, and the recorded steady state profiles of O_2 and N_2O were obtained after about 1 h of incubation with 10 kPa acetylene added to the water phase. As judged from the non-linear part of the N_2O microprofile, denitrification activity was limited to a zone beginning ing at the oxic - anoxic interface at 0.3 mm depth and extending to almost 0.9 mm depth. The biofilm surface (zero depth) is here and in the following paragraphs defined by the second inflection point of the O_2 and N_2O profiles with depth; the layer between the first and second inflection points (i.e. immediately above the biofilm surface) represents the diffusive boundary layer (Jørgensen and Revsbech, 1985). In the oxic surface zone, from 0 to 0.3 mm depth, the curvature of the O_2 profile indicated a consumption (O_2 respiration), but here the N_2O profile was linear marking that only diffusion and neither production nor consumption of N_2O (denitrification)

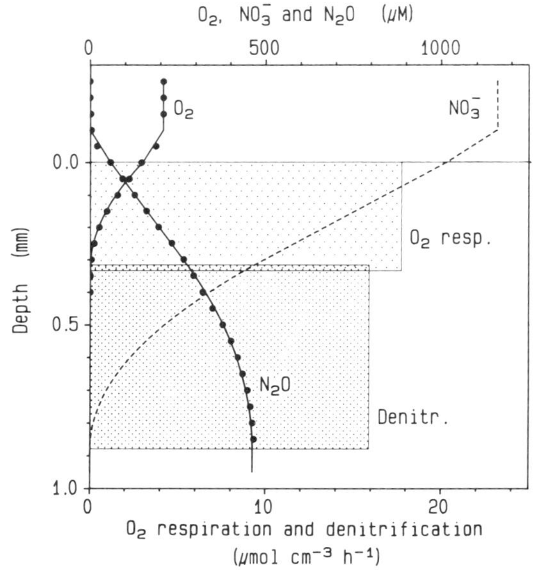

Fig. 1. Steady state profiles of O_2, NO_3^- and N_2O in acetylene-treated biofilm (Døde Å). Oxygen respiration rates were calculated from the O_2 profile while both the denitrification rate and the NO_3^- profile were determined from the N_2O profile. (Modified from Nielsen et al., in press (Microbial Ecol.)).

occurred. The steady state N_2O profile finally predicted a NO_3^- profile as shown in Fig. 1; In this profile, the estimated NO_3^- concentration at the bottom of the denitrification zone was almost zero, confirming that denitrification was the only important process of NO_3^- consumption in this biofilm.

It was evident from the microprofiles that O_2 is an important control factor of denitrification by limiting the latter to depths below the O_2-containing surface zone. We could safely conclude that, at the oxic-anoxic interface, denitrification did not occur at O_2 concentrations above 10 μM. The specific activity was constant, about 16 μmol cm^{-3} h^{-1}, and thus followed zero-order kinetics within most of the denitrification zone. Towards the lower edge of the denitrification zone, however, the process must be controlled by the NO_3^- availability. Thus, if Michaelis-Menten kinetics is assumed for the denitrification activity, the K_m value (half-saturation constant) was found to be less than 25 μM NO_3^- (Nielsen et al., in press) and a NO_3^- control of denitrification occurred in less than 15% of the denitrification zone. Figure 1 further shows a specific activity of O_2 respiration of 18 μmol cm^{-3} h^{-1}, which was only slightly higher than that of denitrification. Based on the stoichiometries of carbon oxidized per oxidant consumed in O_2 respiration (1:1) compared to denitrification when acetylene is present and N_2O is the end product (1:1) (Sørensen et al., 1979), the oxidant consumptions indicated a relatively uniform distribution of available organic substrate in the two activity zones.

Effect of NO_3^- on denitrification activity

The low K_m value for NO_3^- consumption by denitrification indicates that changes of NO_3^-concentration in the water phase will affect the thickness of the

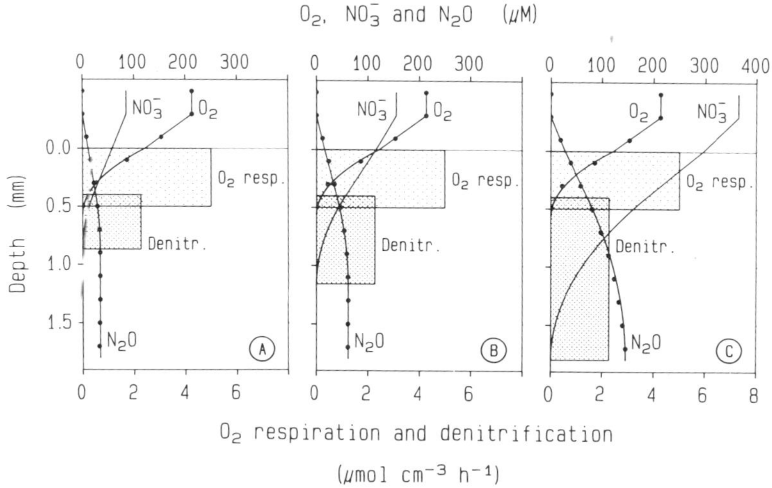

Fig. 2. Effect of NO_3^- concentration on O_2 respiration and denitrification activity in biofilm (Giber). A: 114 μM, B: 219 μM and C: 570 μM NO_3^- added to water phase. Steady state profiles as in Fig. 2. (Modified from Nielsen et al., in press (Microbial Ecol.)).

denitrification zone rather than the specific activity in the denitrification zone. Figure 2 shows an example of data obtained in a biofilm at constant O_2 concentration (about 0.2 mM) but different NO_3^- concentrations in the water phase of about 0.1 mM, 0.2 mM, and 0.6 mM, respectively. As judged from the steady state N_2O profiles, the NO_3^- amendment clearly stimulated denitrification by extending the denitrification zone with depth, and not by increasing the specific activity; the lower edge of the denitrification zone was observed at 0.9 mm, 1.2 mm, and 1.8 mm depth, respectively, at the three NO_3^- treatments. Interestingly, the oxic - anoxic interface was unaffected and remained at about 0.4 mm depth; NO_3^- had no effect on the specific activity of O_2 respiration.

Effect of carbon availability on denitrification activity

The effect of an uneven distribution of organic substrate on biofilm denitrification is shown in Fig. 3. In the control experiment without carbon addition to the water phase (Fig. 3A) is shown an acitivity pattern similar to that already described in Fig. 2: addition of 2.2 mM NO_3^-extended the denitrification zone to 1.8 mm depth, but had no effect on specific activity within this zone. The specific activity was higher for O_2 respiration (about 4 μmol cm^{-3} h^{-1}) than for denitrification (about 2 μmol cm^{-3} h^{-1}), but both were similar to the values recorded in the same biofilm at the 0.6 mM NO_3^- amendment (Fig. 2C). Figure 3B shows that addition of carbon substrate (0.2% yeast extract) in the water resulted in a narrower oxic surface zone and a doubling of the specific activity of O_2 respiration to a value of about 9 μmol cm^{-3} h^{-1}. Denitrification was not affected in the deepest part of the activity zone, but elevated activity, about 7 μmol cm^{-3} h^{-1}, was now observed in a 0.2-mm thick layer, immediately below the new oxic - anoxic interface. This narrow layer previously comprised the lower part of the oxic zone, but was now at the upper part of the denitrification zone. The elevated specific activity in the uppermost part of the denitrification zone was only slightly lower than that of O_2 respiration in the new oxic surface zone,

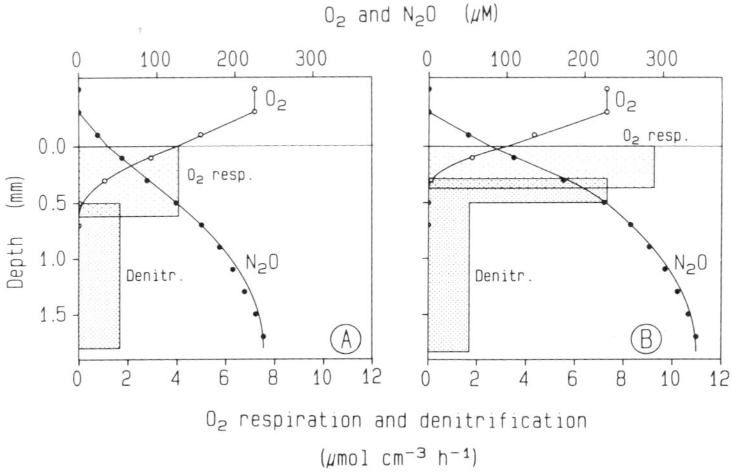

Fig. 3. Effect of organic substrate concentration on O_2 respiration and denitrification activity in biofilm (Giber). A: control and B: 0.2 % yeast extract added to water phase. Steady state profiles as in Fig. 2. (Modified from Nielsen et al., in press (Microbial Ecol.)).

indicating that availability of organic substrate was non-limiting until 0.5 mm or so. Further below, the denitrification remained low as in the control without carbon addition. The results demonstrated that an artificial gradient of organic substrate availability imposed from the surface of the biofilm significantly altered the specific activities of both O_2 respiration and denitrification, which were otherwise relatively similar and constant with depth.

The specific activity in the lower part of the denitrification zone may sometimes be supported by upwards diffusion of reduced substrates. The layers below the denitrification zone may thus be characterized by bacterial fermentation and respiration with SO_4^{2-}, which allow for less efficient substrate utilization and an efflux of dissolved organic carbon (and inorganic sulfide) from their activity zones. Results, which indicate that such a mechanism may be important for the microzonation of denitrification, were found in a study of a trickling filter biofilm (data not shown) (Revsbech et al., 1989). A narrow denitrification zone of relatively high specific activity (7 μmol cm^{-3} h^{-1}) was here observed at the in situ concentration of NO_3^-in the water phase. After more NO_3^- was added, denitrification activity could be observed at progressively greater depth as in the other biofilms studied, but surprisingly the specific activity dropped to a constant level of about 2 μmol cm^{-3} h^{-1} in the extended activity zone. The higher specific activity occurring initially in the narrow reaction zone was probably due to significant diffusion of carbon substrate (and/or inorganic sulfide) from below. When NO_3^- and denitrification activity penetrated deeper into the biofilm, this diffusion flux from the reduced layers must decrease, resulting in a more uniform substrate availability and constant specific activity throughout the denitrification zone.

Effect of light exposure on denitrification activity

As mentioned above, O_2 is a strong inhibitor of denitrification in biofilms and sediments, and we never found denitrification activity at O_2 concentrations above

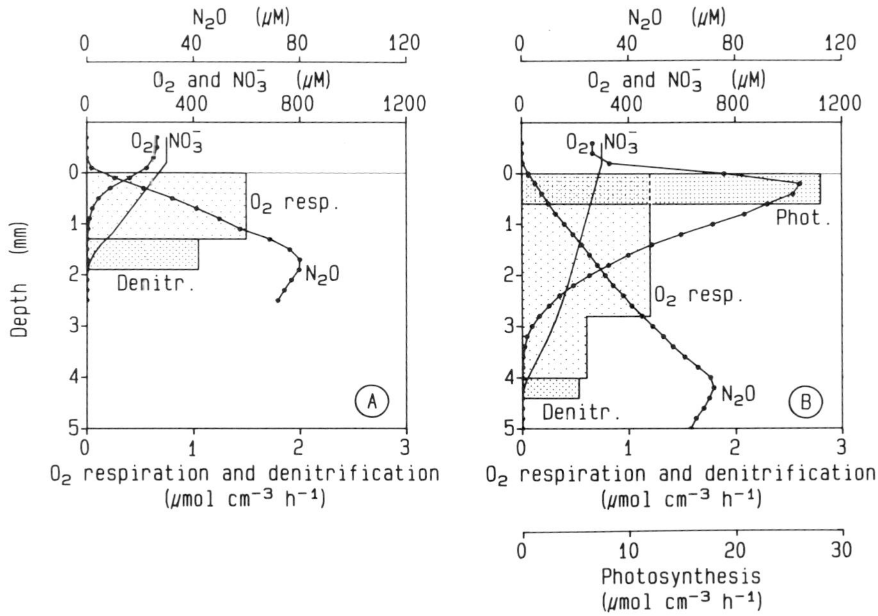

Fig. 4. Effect of light exposure on photosynthesis, O_2 respiration and denitrification activity in sediment (Gelbæk). A: dark incubation and B: light incubation (220 μEinstein m^{-2} s^{-1}). Steady state profiles as in Fig. 2. (Modified from Nielsen et al., in press (Limnol. Oceanogr.)).

10 μM. A deeper extension of the oxic surface zone during light exposure was therefore a plausible and simple explanation for the numerous observations of lower denitrification rates in the day than at night (Andersen et al., 1984; Christensen and Sørensen, 1986; Christensen et al., 1989; Christensen et al., in press; Jørgensen and Sørensen, 1988; Nielsen et al., in press; Sørensen et al., 1988).

In both biofilms and stream sediments, the microsensor studies have revealed the expected changes in microzonation and overall rate of denitrification after light exposure. One example is shown in Fig. 4, which compares a dark incubation (Fig. 4A) with an incubation in light at about 200 μEinstein m^{-2} s^{-1} (Fig. 4B). Denitrification was displaced to a deeper position in the sediment by incubation in light, but the profiles showed that lower overall denitrification rate in the light could be due to both a lower specific activity and a narrower denitrification zone at depth. Thus, as judged from the specific activity of O_2 respiration, availability of organic carbon substrate (or size of O_2-respiring population of bacteria) decreased below 3 mm, where the organic content was lower than at the surface (Nielsen et al., in press). Finally, also the rate of NO_3^- supply by diffusion will be lower in the light than in the dark: The water phase contained about 0.3 mM NO_3^- in both incubations, and with the upper edge of the denitrification zones starting at 1.3 mm and 4.0 mm in dark and light, respectively, a 3-fold lower activity in the light than in the dark should be expected because of the larger diffusion distance; as shown in Fig. 4, this is indeed very close to degree of light inhibition observed.

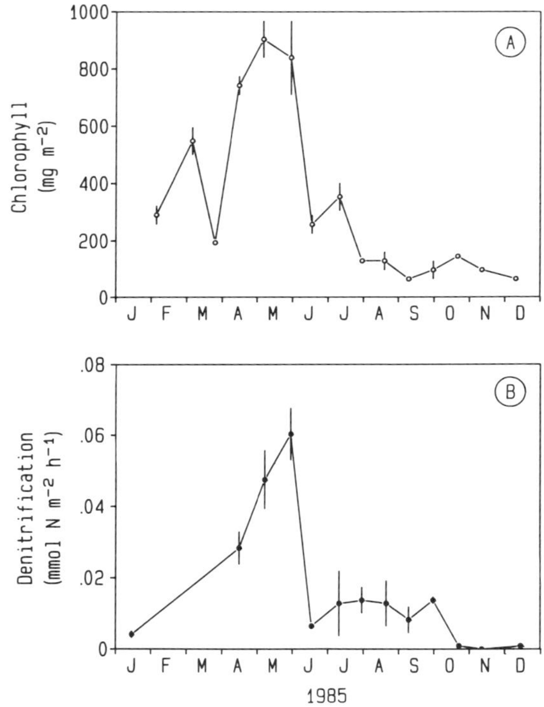

Fig. 5. Seasonal variation of denitrification rate in biofilm (Gelbæk, 1985). A: NO_3^- concentration in water phase, B: chlorophyll content in biofilm, and C: denitrification (dark incubation). (Modified from Sørensen et al., 1988).

SEASONAL AND DIEL VARIATIONS OF DENITRIFICATION ACTIVITY

The epilithon community

Occurrence of significant denitrification is strongly dependent on the thickness of the algal - bacterial community in epilithic biofilms. This is most clearly seen in a seasonal study of biofilm denitrification in the Danish stream Gelbæk (Sørensen et al., 1988). As shown in Fig. 5, there was a rapid development of the algal community (chlorophyll content) in the early spring (March -April) (Fig. 5A) and the maximum of denitrification (Fig. 5B) appearing in May was clearly associated with a maximum density of algal biomass. As the epilithon grows in thickness, the increasing rate of O_2 respiration leads to anoxia and onset of denitrification activity in the innermost layer of the biofilm. The subsequent dramatic decrease of denitrification rates in early summer (June) was due to a sudden loosening of the biofilm after a maximum thickness had been reached. This event has been explained by progressive degradation of the innermost layers of the biofilm, weakening the attachment to the stone surface (Lock, 1981). Increasing grazing rates by the benthic fauna seemed to prevent any recolonization of the epilithic community at this time, and denitrification remained low throughout the summer.

Despite of the complex regulation of denitrification described in the preceeding paragraphs we may suggest the following factors to be important for the increase of denitrification activity in the spring. In addition to the general

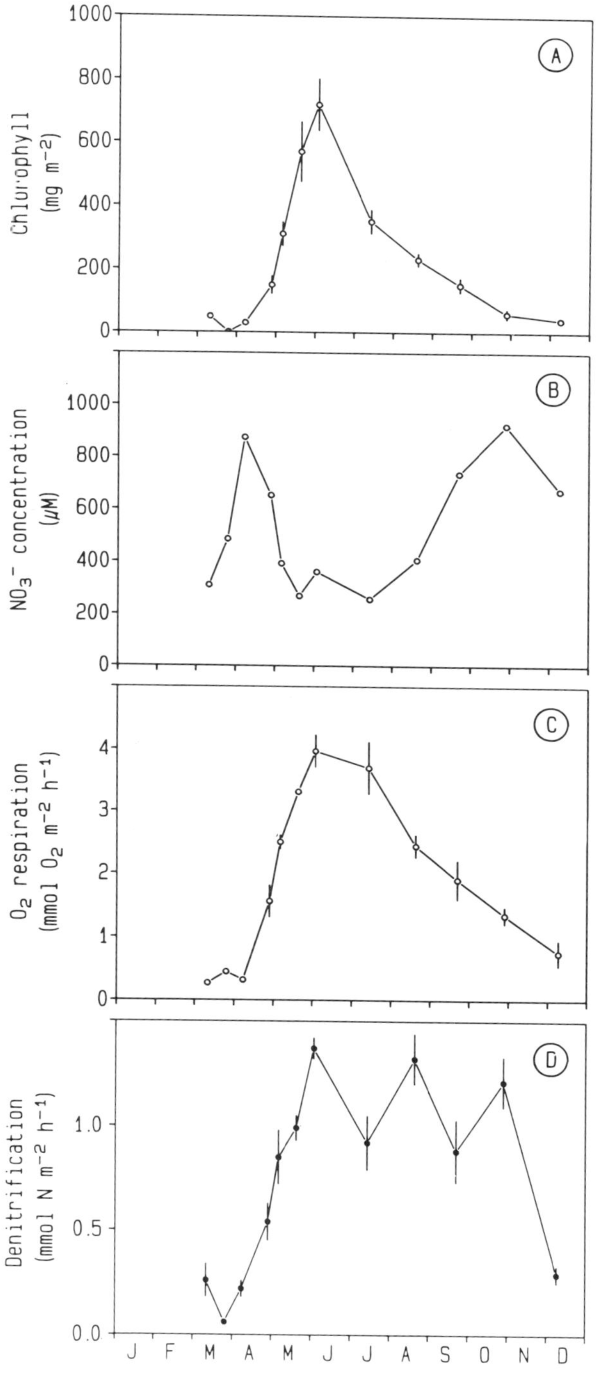

Fig. 6. Seasonal variation of denitrification rate in sediment (Gelbæk, 1987). A: chlorophyll content in sediment (upper 5 mm), B: NO_3^- concentration in water phase, C: O_2 respiration (incubation in dark), and D: denitrification (incubation in dark). (Modified from Christensen et al., in press).

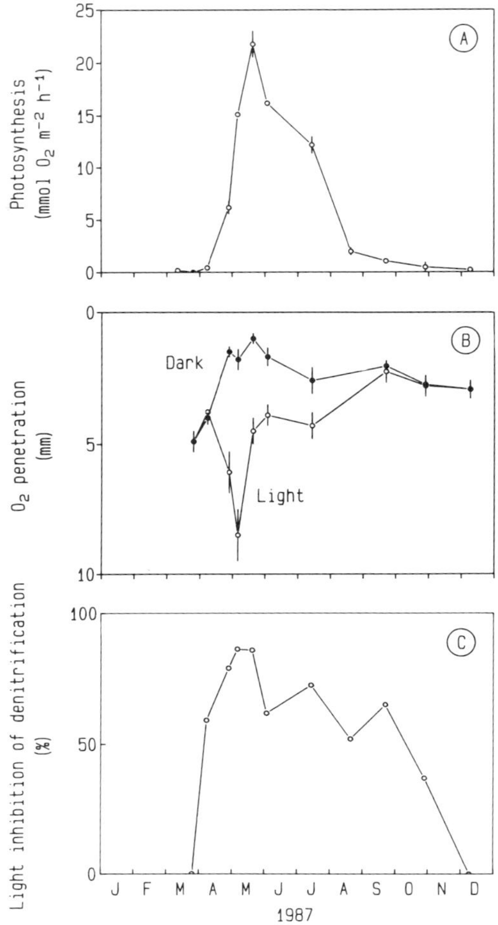

Fig. 7. Seasonal variation of photosynthesis rate in light, O_2 penetration in dark and light, and inhibition of denitrification rate in light-incubated sediment (Gelbæk, 1987). Light exposure was 220 μEinstein m^{-2} s^{-1}. A: photosynthesis in light, B: O_2 penetration in dark and light, and C: per cent inhibition of denitrification in light relative to dark incubation. (Modified from Christensen et al., in press (Limnol. Oceanogr.)).

increase of substrate availability from a growing population of microalgae, improved conditions for activity and growth of denitrifying bacteria may also result from the increasing O_2 respiration which gradually reduced the depth of O_2 penetration into the dark-incubated biofilm. This in turn provided a shorter diffusion distance for NO_3^- into the denitrification zone. When the denitrification occurs closer to the biofilm surface, relatively higher availability of both NO_3^- and organic substrate must therefore stimulate the activity and growth of the denitrifying bacteria. It should also be remembered, however, that development of the denitrifying bacterial population (i.e. bacterial numbers) may also closely follow that of the O_2-respiring heterotrophs, since the denitrifiers may be "born" aerobes utilizing their capacity for denitrification by chance (Koike and Sørensen, 1988).

The sediment community

In the following, we describe results from a seasonal study of denitrification in a stream sediment in which several factors were followed to determine their importance in controlling the full seasonal cycle of denitrification.

Figure 6(A-D) shows the seasonal variation of chlorophyll content (upper 5 mm of sediment), NO_3^- concentration in the water phase, and rates of O_2 respiration (incubation in dark) and denitrification (incubation in dark or in light at 200 μEinstein m^{-2} s^{-1}) in the sediment. In April -May, which is of particular interest due to the significant increase of denitrification activity, the chlorophyll content increased rapidly and the NO_3^- concentrations decreased from 0.9 mM to 0.3 mM. After the maximum development of the algal community, the chlorophyll content soon decreased, and the NO_3^- levels remained low until fall (August-October) when a new increase of the NO_3^- concentration in the stream was observed.

In the dark-incubated sediment cores (Fig. 6D), denitrification in the spring clearly followed the rapid increase of both the density of algal biomass (chlorophyll content) and the heterotrophic metabolism (O_2 respiration) associated with accumulated organic matter. As for the biofilm community described above (Fig. 5), the denitrification maximum thus related well to the development of the algal community in the spring. A poorer correlation was observed later in the summer and fall; the relatively high denitrification rates at this time were possibly due to the increasing NO_3^- concentrations in the stream. A similar mechanism seemed to explain the high denitrification rates recorded in the fall in an estuarine sediment at low water depth (Jørgensen and Sørensen, 1988).

The inhibition of denitrification activity by light must depend on the difference between O_2 penetration into the sediment in light and dark. The seasonal patterns of benthic photosynthetic capacity and O_2 penetration in both dark and light incubations of the sediment are shown in Fig. 7(A-B). As expected, photosynthesis increased rapidly from the onset of algal growth in April and reached a maximum in May. In the same period, the difference between O_2 penetrations into the sediment in dark and light also reached a maximum; in early May, the oxic - anoxic interface was thus observed at depths of about 2 mm and 9 mm for dark and light incubations, respectively. The difference in penetration depth was negligable before the growth season and became also smaller later in the summer when the photosynthesis rate decreased. Figure 7C finally shows the inhibition of denitrification caused by light exposure (in per cent of activity of dark control) during the seasonal cycle. There is generally a trend of increasing light inhibition during the spring, when the difference in O_2 penetration is largest, followed by decreasing inhibition during the rest of the summer and fall.

SUMMARY

The development of a new N_2O-sensitive microelectrode has greatly improved the understanding of spatial distribution and regulation of denitrification activity in biofilms and sediments. In combination with the acetylene inhibition technique, the sensor has made it possible to determine microprofiles of denitrification in the sub-mm scale. Controlled treatments with NO_3^-, organic substrate, and light in laboratory experiments have revealed: 1. High NO_3^- concentrations in the water phase of epilithic biofilms generally controls the overall depth of the denitrification zone rather than affecting the specific activity within this zone; the apparent K_m value for NO_3^- consumption by denitrification was less than 25 μM. 2. O_2 controlled denitrification by complete inhibition at O_2 concentrations higher than 10 μM. Denitrification was limited to deeper strata when the biofilms and sediments were exposed to light and the benthic photosynthesis resulted in relatively deep penetration of O_2. In such cases, the specific activity of denitrification could be low at depth due to the longer diffusion distance for NO_3^- through the oxic surface layer. 3. Organic substrate greatly stimulated the specific activity and overall rate of denitrification when applied to the surface water. In particular systems, however, elevated denitrification activity was also observed at the lower edge of the denitrification zone, probably because of an additional supply of carbon or inorganic substrate such as sulfide from the underlying, reduced layers.

Both diel and seasonal studies of denitrification should improve from applications of the new microsensor technique. One example was the correlation observed in a stream sediment during a seasonal cycle between increased benthic photosynthesis and reduced denitrification resulting from the extended O_2 penetration.

ACKNOWLEDGMENTS

We wish to thank our graduate students Peter Bondo Christensen and Lars Peter Nielsen who enthusiastically performed most of the work presented. Thanks are also due to Søren Brandt and Torben Jørgensen, who contributed to the biofilm work at an early stage, and to Karen Schacht Christiansen and Dorthe Olsson who skillfully performed a large number of the chemical analyses. The work was funded by the National Agency of Environment Protection under the Danish Ministry of the Environment and by the Center for Environmental Biotechnology.

REFERENCES

Andersen, T. K., Jensen, M.H., and Sørensen, J., 1984, Diurnal variation of nitrogen cycling in coastal marine sediments. I. Denitrification, Mar. Biol., 83: 171-176.

Christensen, P. B., and Sørensen, J., 1986, Temporal variation of denitrification activity in plant covered littoral sediment (lake Hampen, Denmark), Appl. Environ. Microbiol., 51: 1174-1178.

Christensen, P. B., and Sørensen, J., 1988, Denitrification in sediment of lowland streams: Regional and seasonal variation in Gelbæk and Rabis Bæk, Denmark, FEMS Microbiol. Ecol., 53: 335-344.

Christensen, P. B., Nielsen, L.P., Revsbech, N.P., and Sørensen, J., 1989, Microzonation of denitrification activity in stream sediments as studied with a combined oxygen and nitrous oxide microsensor, Appl. Environ. Microbiol., 55: 1234-1241.

Christensen, P. B., Nielsen, L.P., Revsbech, N.P., and Sørensen, J., in press, Denitrification in nitrate-rich streams: Diel and seasonal variation related to benthic oxygen metabolism, Limnol. Oceanogr.

Duff, J. H., Triska, F.J., and Oremland, R.S., 1984, Denitrification associated with stream periphyton: Chamber estimates from undisrupted communities, J. Environ. Qual., 13: 514-518.

Jørgensen, B. B., and Revsbech, N.P., 1985, Diffusive boundary layers and oxygen uptake of sediments and detritus, Limnol. Oceanogr., 30: 111-122.

Jørgensen, K. S., and Sørensen, J., 1988, Two annual maxima of denitrification in estuarine sediment (Norsminde Fjord, Denmark), Mar. Ecol. Prog. Ser., 48: 147-154.

Haack, T. K., McFeters, G.A., 1982, Nutritional relationships between microorganisms in epilithic biofilm community, Microbial Ecol., 8: 115-126.

Koike, I., and Sørensen, J., 1988, Nitrate reduction and denitrification in marine sediments, pp. 251-273, in: "Nitrogen cycling in coastal marine environments," T. H. Blackburn and J. Sørensen, eds., John Wiley and Sons, Chichester.

Lock, M. A, 1981, River epilithon - a light and organic energy transducer, p. 3-40, in: Perspectives of running water ecology, M. A. Lock and D. D. Williams, eds., Plenum Press, New York.

Lomstein, E., Jensen, M.H., and Sørensen, J., in press, Particulate NH_4^+ and NO_3^- pools associated with phytoplankton sedimentation in a marine sediment (Aarhus Bight, Denmark), Mar. Ecol. Prog. Ser.

Nielsen, L. P., Christensen, P.B., Revsbech, N.P., and Sørensen, J., in press, Denitrification and oxygen respiration in biofilms studied with a combined microsensor for nitrous oxide and oxygen, Microbial Ecol.

Nielsen, L. P., Christensen, P.B., Revsbech, N.P., and Sørensen, J., in press, Denitrification and photosynthesis in stream sediment studied with microsensor and whole core techniques, Limnol. Oceanogr.

Revsbech, N. P., Madsen, B., and Jørgensen, B.B., 1986, Oxygen production and consumption in sediments determined at high spatial resolution by computer simulation of oxygen microelectrode data, Limnol. Oceanogr., 31: 293-304.

Revsbech, N. P., Nielsen, L.P., Christensen, P.B., and Sørensen, J., 1988, Combined oxygen and nitrous oxide microsensor for denitrification studies, Appl. Environ. Microbiol., 54: 2245-2249.

Revsbech, N. P., Christensen, P.B., Nielsen, L.P., and Sørensen, J., 1989, Denitrification in a trickling filter biofilm studied by a microsensor for oxygen and nitrous oxide, Water Res., 23: 867-871.

Sand-Jensen, K., Møller, J., and Olesen, B.H., 1988, Biomass regulation of microbenthic algae in Danish lowland streams, Oikos 53: 332-340.

Sørensen, J., Jørgensen, B.B., and Revsbech, N.P., 1979, A comparison of oxygen, nitrate, and sulfate respiration in coastal marine sediments, Microbial Ecol. 5: 105-115.

Sørensen, J., Jørgensen, T., and Brandt, S., 1988, Denitrification in stream epilithon: Seasonal variation in Gelbæk and Rabis Bæk, Denmark, FEMS Microbiol. Ecol., 53: 345-354.

Walter, H. M., Keeney, D.R., and Fillery, I.R., 1979, Inhibition of nitrification by acetylene, Soil Sci. Soc. Amer. J., 43: 195-196.

MEASUREMENT OF SEDIMENT DENITRIFICATION USING 15-N TRACER METHOD

Isao Koike

Ocean Research Institute
University of Tokyo
Nakano, Tokyo 164
Japan

INTRODUCTION

Because of the rather expensive equipment and complicated analytical procedures, application of 15-N tracer for denitrification in sediment has been rather restricted (Hattori, 1983; Koike and Sørensen, 1988; Seitzinger, 1988). Use of a 15-N tracer for denitrification studies in sediments, however, has several advantages over other techniques, including the acetylene blockage method. Primarily, enrichment of 15-N in the N_2 pool after introduction of 15-N NO_3^- into the sediments is a direct proof of denitrification.

In Table 1, several methodologies currently applied for denitrification studies are summarized together with the detection limit of each method. A mass spectrometer equipped with a double collector has a precision better than 0.0005 atom % difference in 15-N, which allows detection of ca. 10 nmole of N_2 produced by denitrification per liter of sea water (Koike et al., 1978). Since ambient concentrations of N_2O in sediment are usually quite low (ca. 5-20 nmol/l), the acetylene blockage technique using N_2O detection with a chromatograph equipped with an ECD detector is actually the most sensitive method together with 13-N radioactive tracer technology (Cohen, 1978; Sørensen, 1978b).

In this paper I will concentrate on two aspects of 15-N tracer methodology for denitrification studies in sediments, i.e., (1) methodology relating to _in situ_ estimation of denitrification rates, (2) simultaneous determination of various steps of nitrogen cycling including denitrification. Most of the data referred to in this chapter was obtained in our laboratory during the last 10 years. Other reviews of denitrification in sediments are those of Hattori (1983) and Seitzinger (1988).

POTENTIAL AND _IN SITU_ RATES OF DENITRIFICATION

The first evaluation of denitrification in sediment using 15-N was the work in Smith Lake in Alaska some 20 years ago (Goering and Dugdale, 1966). Here 15-N NO_3^- was added to an anoxic slurry of mud and water, and the 15-N enrichment in N_2 was measured after incubation for several

Denitrification in Soil and Sediment, Edited by N. P. Revsbech and J. Sørensen, Plenum Press, New York, 1990

Table 1. Experimental approaches for estimates of denitrification in sediments.

Technique	Detection limit (ng at. N/1)
1) N_2 increase after degassing	ca. 1000
2) Colorimetric measurement of NO_3^- decrease	100
3) 15-N tracer methods	
a) optical emission spectrometer	ca. 400
b) mass spectrometer with double collector	ca. 10
4) C_2H_2 inhibition method	
a) head space method with direct injection	1-2
b) cold trap concentration	ca. 0.5
5) 13-N radio-isotope measurement	?

days. This procedure has basically been followed by several later investigators (e.g., Koike and Hattori, 1978a; Madsen, 1979). The bottle incubation method has also applied to estimate denitrification in the water column, where physical mixing of the samples has little effect on the microbial activity (Chan and Campbell, 1980).

An inevitable increase of the NO_3^- concentration in the sample can be corrected for by measuring a correlation between denitrification rates and the concentration of NO_3^- in the sediment (Koike and Hattori, 1979, Oren and Blackburn, 1979). However, even after this correction there are several other uncertainties which prevent a detection of realistic denitrification rates in sediment from the slurry type experiment (Koike and Sørensen, 1988; Seitzinger, 1988). Heterogeneity including microgradients of both the oxygen tension and substrate concentrations may thus be destroyed by the manipulation of the intact sediment core and the rate obtained in the slurries is usually higher than under *in situ* conditions.

From a geochemical point of view, denitrification activity measured per unit of surface area of sediment is preferable, since other biogeochemical processes, such as oxygen and nutrient fluxes across the sediment-water interface are often estimated from bell jar experiments (Smith, 1978). One of the approaches in this direction was the late study of Chan and Campbell (1980), in which a closed chamber of Plexiglas with circulating bottom water received a spike of 15-N NO_3^- and was incubated for 1-2 days. The authors found considerable decrease in NO_3^--N concentrations after 2 days and ca. 20% of the consumption was recovered as 15-N N_2.

A continuous flow incubation with 15-N tracer added to an intact marine sediment-water system was later introduced (Nishio, 1982; Nishio et al. 1982; Nishio et al. 1983). Filtered bottom water spiked with a small amount of 15-N NO_3^- flowed continuously over the sediment (Fig. 1) the 15-N content of N_2 in the effluent water was monitored. Steady state evolution of 15-N N_2, usually obtained 10-20 hours after the addition of 15-N NO_3^-, was used to estimate the *in situ* rate of denitrification. The assumption here was that NO_3^- in the overlying water was a primary source of denitrified N_2, which seems true for an organic-rich sediment with a high content of NO_3^- (more than 30-50 μg at. N/1) in the overlying water. By replacing 15-N NO_3^- with 15-N NH_4^+ as the tracer, a

Table 2. Estimated rates of nitrification, and denitrification fed by NO_3^- from the overlying water (A), or from nitrification in the sediment (B) in Tama Estuary and Odawa Bay sediments (Nishio, 1982).

Location	Sampling Date	Nitrification	Denitrification	
			A	B
		(ng at. N cm^{-2} h^{-1})		
Tama Estuary	May 1981	7.3	55.4	3.9
Odawa Bay	May 1981	11.3	1.02	1.34

significant and steady production of 15-N N_2 was also demonstrated in sediments having relatively low organic content and low NO_3^- in the overlying water (Nishio et al. 1983). In those sediments the nitrification activity within the surface sediment was a significant source of denitrified nitrogen (Table 2). Since the addition of acetylene strongly inhibits the process of nitrification, the 15-N tracer methcd has obvious advantages in such sediments (Nishio et al. 1983).

So far there has been no ideal technique to evaluate <u>in situ</u> rates of denitrification in sediment and all the proposed techniques need certain assumptions. It is therefore preferable to apply two independent techniques at the same time to check each result obtained. A geochemical method, using a diffusion-advection model based on the vertical profiles of NO_3^- or N_2 /Ar ratios in sediments, is another technique to evaluate <u>in situ</u> rates of denitrification and has been applied to various types of sediments, (Barnes et al. 1975; Bender et al. 1977; Vanderborght and Billen, 1977; Wilson, 1978).

Using Bering Sea shelf sediments, denitrification rates measured by slurry incubation with 15-N NO_3^- and correction to fit the correct concentration of NO_3^- in the pore water were compared with the rates state (Fig. 2).

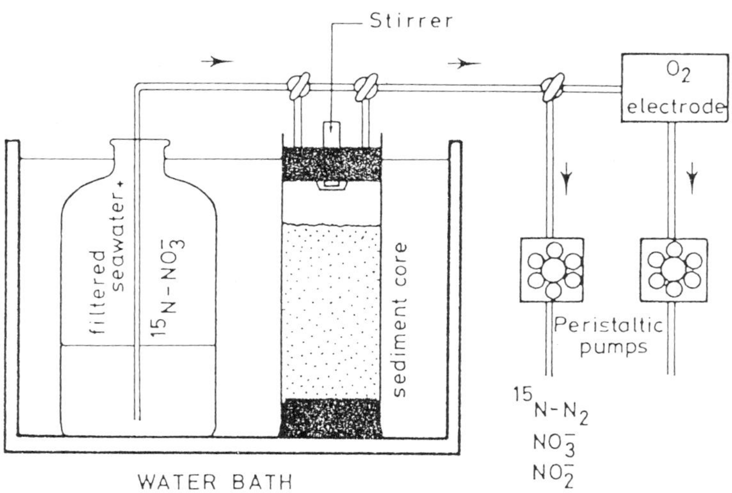

Figure 1. Continuous-flow system using 15-N tracer for determination of inorganic nitrogen cycling in sediments.

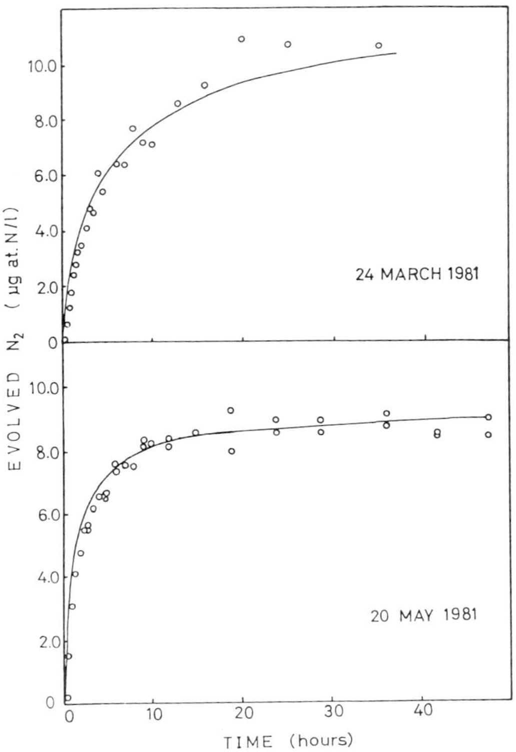

Figure 2. Change with time in concentrations of evolved N_2 in the overlying water of the continuous-flow sediment-water system. The sediments were obtained from Tama Estuary on March 1981 (upper) and on May 1981 (lower).

estimated from a one-dimensional diffusion model of the NO_3^- profiles (Koike and Hattori, 1979). The two independent estimates of denitrification agreed within a factor of two when an apparent diffusion coefficient of $2x10^{-5}$ cm^2 s^{-1} was assumed for the top layer. A diffusion constant in the order of 10^{-5} is often reported in coastal sediments (Billen, 1982). However, different results were obtained in Tama Estuary sediments, where the same 15-N tracer technique was compared with a one-dimensional diffusion model using N_2/Ar ratio profiles (Nishio et al. 1981). To fit the two estimated rates, an apparent diffusion coefficient of 10^{-3} cm^2 s^{-1} was required. Apparent diffusion coefficients of dissolved gases and nutrients in the interstitial waters of coastal and estuarine sediments may be considerably higher than in shelf sediments, primarily because of physical turbulence, such as tidal movements, and biological processes, such as bioturbation or reworking (Aller, 1978; Billen, 1982).

As described above, the measurement of denitrification using a continuous flow sediment-water system could be carried out after the establishment of steady state 15-N N_2 production (Nishio et al. 1982). In these experiments, 15-N NO_3^- added to the overlying water gradually diffused downward into the sediment. At the sites of denitrification, 15-N NO_3^- was reduced to 15-N N_2 and emission of 15-N N_2 increased during the initial period of the incubation and then approached steady

Applying a first order kinetic model, the observed increase of 15-N N_2 in the overlying water could be described (Nishio, 1982). The following four basic equations concerning the change of 15-N NO_3^- (C) and N_2 (B) with time and depth in the sediment (Z) were used.

$$\delta C/\delta t = D \cdot \delta^2 C/\delta Z^2 - K_1 C_1 \quad (1)$$

$$\delta B/\delta t = D \cdot \delta^2 B/\delta Z^2 + K_2 C_2 \quad (2)$$

$$Fc = -D(dC/dZ)_{z=0} \quad (3)$$

$$Fb = -D(dB/dZ)_{z=0} \quad (4)$$

where D is the apparent diffusion coefficient of NO_3^- and N_2, K_1 and K_2 are rate constant for bacterial NO_3^- reduction and denitrification, respectively. Fc and Fb are the flux of 15-N labelled NO_3^- and N_2 across the sediment-water interface, following Fick's first law. The apparent diffusion constant (D) can be estimated from the computer selected best fit curves derived from eqs. (1), (2), (3) and (4) to the experimental data of 15-N N_2 production. The rate constants, K_1 and K_2, were also calculated using the estimated D and the flux of 15-N NO_3^- and N_2 under the steady state condition. Estimated constants from the data (May 1981) in Fig. 2 were 2.5×10^{-4} s^{-1} for K_1, 1.2×10^{-4} s^{-1} for K_2 and 5.4×10^{-4} cm^2 s^{-1} for D (Nishio, 1982).

One advantage of this kinetic model using the initial phase of a continuous flow system is that it allows us to estimate the rate constant of denitrification and the diffusion constant of N_2 simultaneously. Previous one-dimensional diffusion models for denitrification could estimate K/D, but not K and D separately (Berner, 1974). Calculated profiles of NO_3^- in the interstitial water using K_1 and D suggest that the active site of denitrification, estimated from the depth of NO_3^- disappearance, in organic-rich sediments was in the top millimeters (Nishio, 1982). The above finding was supported by the fine vertical profiles of NO_3^- in sediments, obtained by intact core squeezing techniques or microelectrode methods (Bender et al. 1987; Sweerts and Beer, 1989).

ESTIMATES OF SEDIMENT NITROGEN CYCLING INCLUDING DENITRIFICATION WITH A SINGLE ADDITION OF 15-N TRACER

Denitrification is one of the key processes of nitrogen cycling in coastal sediments and evaluation of its relative importance in the overall nitrogen cycle is of primary importance. By the use of 15-N tracer and a suitable mathematical model, multiple processes of nitrogen cycling including denitrification can be measured simultaneously. In soil systems, several mathematical models to simulate experimental data of nitrogen cycling using 15-N were developed (Myrold and Tiedje, 1986). In sediments, however, relatively few 15-N studies have been addressed to differentiate several processes of the nitrogen cycle.

Using 15-N NO_3^-, the competitive anaerobic reduction of NO_3^- to either N_2 (denitrification) or NH_4^+ (NO_3^- ammonification) may be distinguished by measuring production of 15-N N_2 and 15-N NH_4^+ from 15-N NO_3^- in the sample (Koike and Hattori, 1978a; Sørensen, 1978a; Goeyens et al. 1987). By addition of 15-N NH_4^+ to the surface layer of sediment, simultaneous estimates of nitrification and denitrification were possible by measuring the increase of 15-N NO_3^- and 15-N N_2 with time

(Nishio et al. 1983; Jenkins and Kemp, 1984). Another application of 15-N NO_3^- was the simultaneous evaluation of nitrification and NO_3^- reduction (denitrification) in a surface sediment, using 15-N NO_3^- isotope dilution. The rate of nitrification is here estimated by the production of relatively 14-N rich NO_3^- from the unlabelled NH_4^+ pool resulting in a change in the 15-N/14-N ratio of NO_3^- with time (Koike and Hattori, 1978b). Similar isotope dilution technique with 15-N NH_4^+ was also used for simultaneous determination of ammonification and NH_4^+ assimilation rates in anaerobic sediments (Blackburn, 1979; Iizumi et al. 1980; Blackburn and Henriksen, 1983; Bowden, 1984).

One important drawback of these 15-N studies is that it is difficult to evaluate the extent of 15-N cycling during the incubation. In the case of anaerobic 15-N NO_3^- reduction, the 15-N N_2 and 15-N NH_4^+ released are substrates of other processes, i.e., N_2 fixation and NH_4^+ assimilation (Blackburn, 1979; Seitzinger, 1988). If significant transfer of 15-N into other than the expected pools occurs during the incubation, simple measurement of 15-N in N_2 and NH_4^+ gives an underestimated value. This is also true for the simultaneous measurement of nitrification and NO_3^- reduction by 15-N NO_3^-isotope dilution, where the 15-N content of NH_4^+, (substrate pool of nitrification) is assumed to be of natural 15-N abundance (0.37%). If NO_3^- reduction of 15-N NO_3^- to NH_4^+ is significant to affect the 15-N content of ambient NH_4^+ in the interstitial water, the assumption made above must be corrected by measuring 15-N NH_4^+ during the incubation (Koike and Hattori, 1978a; Sørensen, 1978a).

Since all the known processes of nitrogen cycling occurs in the top layer of sediment which contains both aerobic and anaerobic microenvironments, 15-N in a single nitrogen pool added into the sediment sooner or later must distribute itself between other nitrogen pools in the sediment. Therefore, an improvement of the previous method developed for single processes of the nitrogen cycle may lead to the

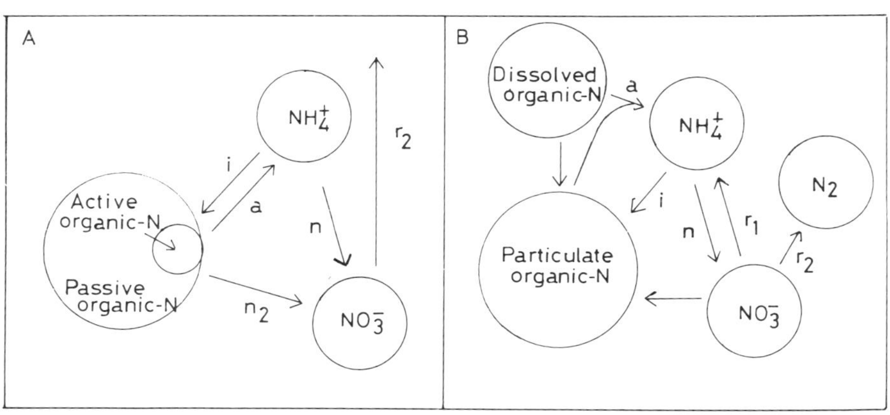

Figure 3. Compartmental models of nitrogen cycling in soils (A) and sediments (B). Concerned processes in the models are mineralization or ammonification (a), immobilization or NH_4^+ assimilation (i), nitrification (n), heterotrophic nitrification (n_2), denitrification (r_2), and NO_3^- reduction to NH_4^+ (r_1).

simultaneous evaluation of all the major steps of the nitrogen cycle. In soil systems, four major nitrogen metabolisms, i.e., mineralization, immobilization, nitrification, and denitrification, were thus successfully estimated by measuring nitrogen pools and their 15-N contents with a compartmental model of the nitrogen cycle (Myrold and Tiedje, 1986). A model used for uncultivated soil contained three basic compartments with five processes of the nitrogen cycle (Fig. 3A).

Recently we developed a model for evaluating several major steps of nitrogen cycling in the surface layer of sediment, in which a close coupling between nitrification and NO_3^- reduction has been demonstrated (Sumi and Koike, in press). The compartments and major processes concerned here are shown in Fig. 3 B. A primary difference between the compartmental model of the sediment and that of soil was that the former included NO_3^- reduction to NH_4^+, but ignores heterotrophic nitrification (Fig. 3 A and B). If NO_3^- reduction to NH_4^+ is significant, as was demonstrated in marine sediments, this term must be included in the equation of NH_4^+ mass balance during the incubation.

The following four equations derived from the mass balance of nitrogen and its 15-N content in each compartment, together with the experimental data obtained after a small addition of 15-N NH_4^+, were used to evaluated the four steps in Fig. 3 B:

$$dCa/dt = (a+r_1)-(i+n) \tag{3}$$

$$d(Ca\cdot Ra)/dt = a\cdot\alpha + r_1\cdot Rn-(i+n)Ra \tag{4}$$

$$dCn/dt = n-(r_1+r_2) \tag{5}$$

$$d(Cn\cdot Rn)/dt = n\cdot Ra-(r_1+r_2)\cdot Rn \tag{6}$$

where Ca and Ra are the pool size of KCl extractable NH_4^+ and its 15-N abundance, respectively, and Cn and Rn are the pool sizes of NO_3^- plus NO_2^- and its 15-N abundance, respectively. α is the natural abundance of 15-N (0.37%). The difference between the initial enrichment as 15-N in NH_4^+ and the sum of that in NH_4^+, NO_3^- plus NO_2^- and particulate organic nitrogen (PON) after the incubation was assumed to be denitrified nitrogen (r_2). Finally, using the modelling approach, NH_4^+ assimilation (i) by microorganisms was also directly estimated by measuring the 15-N increase in PON with time based on the method of Laws (1984), which corrects for isotope dilution of the 15-N NH_4^+ pool during the incubation.

The comparison of two independent estimates of NH_4^+ assimilation and the four other processes of nitrogen cycling, including denitrification, in the top layer of a number of coastal and estuary sediments is shown in Table 3. For these experiments, a slurry from the top 1 cm layer of sediment was used and the estimated rates may therefore be different from the <u>in situ</u> rates because of the destruction of microgradients. Several interesting features prevailed from these experiments, however. Except for one experiment (Tama-M: Oct.1985), two independent estimates of NH_4^+ assimilation (i_a and i_b) agreed well, strongly suggesting the suitability of the proposed model. Also, simultaneous occurrence of nitrification and NO_3^- reduction was observed in more than half of the sediments examined and both rates were roughly balanced as indicated in previous experimental studies (Nishio et al. 1983; Jenkins and Kemp, 1984) . The data given in Table 3 suggests that NO_3^- reduction and

Table 3. Determinations of nitrification (n), total NO_3^- reduction (r), denitrification (r_2), NH_4^+ assimilation (i_a), and ammonification (a) by modelling, and comparison with NH_4^+ assimilation (i_b) obtained from 15-N NH_4^+ incorporation into particulate organic materials in the surface sediments of Japanese coastal areas (Sumi and Koike, in press).

Location & Data		Direct assay (i_b)	Mathematical modelling (n)	(r)	(r_2)	(i_a)	(a)
			(ng at. N gr^{-1} h^{-1})				
Otsuchi-A,	May 1985	9.1	4.6	-0.9	0.4	12	15*
Tama-M,	Oct. 1985	10.8	40	29	43	30	28*
Misaki-A,	Apr. 1986	57.9	8.2	-9.2	65	64	35*
Misaki-B		60.2	78	54	12	69	72
Otsuchi-A,	May 1986	19.9	2.3	11	54	26	23*
Orsuchi-B		15.8	153	142	34	16	64
Tama-T,	Jul. 1986	24.2	244	180	130	26	122
Tokyo-A,	Aug. 1986	114	1444	1310	164	131	217
Tokyo-B		111	509	349	166	116	220

*: calculated assuming r_1 (NO_3^--ammonification) = 0.

denitrification are major processes in the nitrogen cycle, even in the top layer of the sediment.

In order to obtain the rate of denitrification (r_2) we assumed that the accumulation of 15-N in the dissolved organic nitrogen fraction was negligible within the short time of incubation (max. 60 h) in order to obtain the rate of denitrification (r_2). However, it is preferable to measure 15-N N_2 as well as the 15-N dissolved organic nitrogen fraction together with other components to check the complete recovery of added 15-N. The experimental system described in the first section (Fig. 1) was primarily designed to evaluate the oxidation and reduction steps of the nitrogen cycle, i.e., nitrification and NO_3^- reduction (denitrification). Development of similar techniques to measure all the major steps of the nitrogen cycle in the sediment at _in situ_ conditions which is possible only by using a 15-N tracer technique should be addressed in future studies of sediment denitrification.

SUMMARY

Two aspects of 15-N tracer methodology for denitrification studies in sediments were summarized. Concerning in situ estimation of denitrification, some results obtained from a continuous flow incubation with 15-N tracer added to an intact marine sediment-water system were presented. Tight coupling between nitrification and denitrification was demonstrated for surface sediments of relatively low organic content. The 15-N tracer methodology combined with modelling was also applied to evaluate multiple processes of nitrogen cycling in surface sediments.

Five major processes of the nitrogen cycle, i.e., mineralization, ammonium assimilation, nitrification, denitrification, and nitrate reduction to ammonium were estimated by the model.

REFERENCES

Aller, R. C., 1978, Experimental studies of changes produced by deposit feeders on pore water, sediment and overlying water chemistry, Amer. J. Sci., 278: 1185.

Barnes, R. O., Bertine, K. K., and Goldberg, E. D., 1975, N_2:Ar, nitrification and denitrification southern California borderland basin sediments, Limnol. Oceanogr., 20: 962.

Bender, M. L., Ganning, K. A., Froelich, P. N., Heath, G. R., and Maynard, V., 1977, Interstitial nitrate profiles and oxidation of sedimentary organic matter in the eastern equatorial Atlantic, Science, 198: 605.

Bender, M., Martin, W., and Hess, J., 1987, A whole-core squeezer for interfacial pore-water sampling, Limnol. Oceanogr., 32: 1214.

Berner, R. A., 1974, Kinetic models for the early diagenesis of nitrogen, sulphur, phosphorus and silicon in anoxic marine sediment, in: "The Sea Vol. V," E. D. Goldberg, ed., Wiley, New York.

Billen, G., 1982, Modeling the processes of organic matter degradation and nutrients recycling in sedimentary systems, in: "Sediment Microbiology," D. B. Nedwell and C. M. Brown, eds., Academic Press, London.

Blackburn, T. H., 1979, A method for measuring rates of NH_4^+ turnover in anoxic marine sediments using a 15-N-NH_4^+ dilution technique, Appl. Environ. Microbiol., 37: 760.

Blackburn, T. H., and Henriksen, K., 1983, Nitrogen cycling in different types of sediments from Danish Waters, Limnol. Oceanogr., 28: 477.

Bowden, W. B., 1984, A nitrogen-15 isotope dilution study of ammonium production and consumption in a marsh sediment, Limnol. Oceanogr., 29: 1004.

Chan, Y. K., and Campbell, N. E. R., 1980, Denitrification in Lake 227 during summer stratification, Can. J. Fish Aquat. Sci., 37: 506.

Cohen, Y., 1978, Shipboard measurement of dissolved nitrous oxide in seawater by electron capture gas chromatography, Anal. Chem., 49: 1238.

Goering, J. J., and Dugdale, V. A., 1966, Estimates of the rates of denitrification in a subarctic lake, Limnol. Oceanogr., 11: 113.

Goeyens, L., de Vries, R. T. P., Bakker, J. F., and Helder, W., 1987, An experiment on the relative importance of denitrification, nitrate reduction and ammonification in coastal marine sediment, Nethr. J. Sea Res., 21: 171.

Hattori, A., 1983, Denitrification and dissimilatory nitrate reduction, in: "Nitrogen in Marine Environment," E. J. Capenter and D. C. Capone, eds., Academic Press, New York.

Iizumi, H., Hattori, A., and McRoy, C. P., 1980, Nitrate and nitrite in interstitial water of eelgrass beds in relation to the rhizosphere. J. Exp. Mar. Biol. Ecol., 47: 191.

Jenkins, M. C., and Kemp, W. M., 1984, The coupling of nitrification and denitrification in two estuarine sediments, Limnol. Oceanogr., 29: 609.

Koike, I., and Hattori, A., 1978a, Denitrification and ammonium formation in anaerobic coastal sediments, Appl. Environ. Microbiol., 35: 278.

Koike, I., and Hattori, A., 1978b, Simultaneous determination of nitrification and nitrate reduction in coastal sediments by a ^{15}N dilution technique, Appl. Environ. Microbiol., 35: 853.

Koike, I., Hattori, A., and Goering, J. J., 1978, Controlled ecosystem pollution experiment: effect of mercury on enclosed water column, VI, Denitrification by marine bacteria, Mar. Sci. Commun., 4: 1.

Koike, I., and Hattori, A., 1979, Estimates of denitrification in sediments of the Bering Sea shelf, Deep-Sea Res., 26: 409.

Koike, I., and Sørensen, J., 1988, Nitrate reduction and denitrification in marine sediments, in: "Nitrogen Cycling in Coastal Marine Environments," T. D. Blackburn and J. Sørensen, eds., John Wiley & Sons Ltd, Chichester.

Laws, E., 1984, Isotope dilution models and the mystery of the vanishing ^{15}N, Limnol. Oceanogr., 29: 379.

Madsen, P. P., 1979, Seasonal variation of denitrification rate in sediment determined by use of ^{15}N, Water Res., 13: 461.

Myrold, D. D., and Tiedje, J. M., 1986, Simultaneous estimation of several nitrogen cycle rates using ^{15}N: Theory and application, Soil Biol. Biochem., 18: 559.

Nishio, T., Koike, I., and Hattori, A., 1981, N_2/Ar and denitrification in Tama estuary sediments, Geomicrobiol. J., 2: 193.

Nishio, T., 1982, Nitrogen cycling in coastal and estuarine sediments with special reference to nitrate reduction, denitrification and nitrification, Doctoral dissertation, University of Tokyo, Tokyo., pp. 115.

Nishio, T., Koike, I., and Hattori, A., 1982, Denitrification, nitrate reduction, and oxygen consumption in coastal and estuarine sediments, Appl. Environ. Microbiol., 43: 648.

Nishio, T., Koike, I., and Hattori, A., 1983, Estimates of denitrification and nitrification in coastal and estuarine sediments, Appl. Environ. Microbiol., 45: 444.

Oren, A., and Blackburn, T. H., 1979, Estimation of sediment denitrification rates at in situ nitrate concentrations, Appl. Environ. Microbiol., 37: 174.

Seitzinger, S. P., 1988, Denitrification in freshwater and coastal marine sediments: Ecological and geochemical significance, Limnol. Oceanogr., 33: 702.

Smith, K. L., 1978, Benthic community respiration in the N. W. Atlantic Ocean: in situ measurements from 40 to 5200 m, Mar. Biol., 47: 337.

Sørensen, J., 1978a, Capacity for denitrification and reduction of nitrate to ammonia in a coastal marine sediment, Appl. Environ. Microbiol., 35: 301.

Sørensen, J., 1978b, Denitrification rates in a marine sediment as measured by the acetylene inhibition technique, Appl. Environ. Microbiol., 36: 139.

Sumi, T., and Koike, I., 1990, Estimation of ammonification and ammonium assimilation in surficial coastal and estuarine sediments, Limnol. Oceanogr., (in press).

Sweerts, J-P. R. A., and Beer, D., 1989, Microelectrode measurements of nitrate gradients in the littoral and profundal sediments of a mesoeutrophic lake (Lake Vechten, The Netherlands), Appl. Environ. Microbiol., 55: 754.

Vanderborght, J.-P., and Billen, G., 1977, Vertical distribution of nitrate concentration in interstitial water of marine sediments with nitrification and denitrification, Limnol. Oceanogr., 20: 953.

Wilson, T. R. S., 1978, Evidence for denitrification in aerobic pelagic sediments, Nature (Lond.), 274: 354.

DENITRIFICATION IN AQUATIC SEDIMENTS

Sybil P. Seitzinger

Academy of Natural Sciences
Division of Environmental Research
19th St and the Parkway
Philadelphia, PA 19103
U.S.A.

INTRODUCTION

Nutrients (N and P) are important in controlling primary production in aquatic ecosystems. The availability of N or P within an ecosystem is partly a function of the rates of external inputs to the system and permanent removal within the system by biological, chemical and/or physical processes. N is a limiting nutrient in many estuaries, coastal and continental shelf marine systems (Ryther and Dunstan, 1971), some lakes (Keeney, 1973; Gerhart and Likens, 1975), and some streams and rivers (Grimm and Fisher, 1986). Denitrification in the sediments or anoxic water removes N from aquatic systems as the gaseous nitrogen produced diffuses into the atmosphere. Denitrification in aquatic systems, therefore, decreases the amount of nitrogen available for primary production, as well as decreases the transport of N to downstream or offshore waters. In aquatic systems that receive N from pollution sources, the removal of a portion of that N via denitrification may help control the degree of eutrophication of the system.

Aquatic sediments present an ideal environment for denitrification. They are a concentration site for organic matter from the water column, which upon decomposition releases NH_4^+. The NH_4^+ is then available for nitrification and subsequent denitrification (Blackburn and Henriksen, 1983). In addition, NO_3^- from the overlying water can diffuse into the sediments (Andersen et al., 1984), especially in relatively eutrophic systems where NO_3^- concentrations in the water are high. These characteristics, in combination with the close juxtaposition of aerobic and anaerobic sites within a few mm of the sediment-water interface, lead to high denitrification activity in aquatic sediments.

The present chapter examines ambient rates of denitrification in freshwater and marine sediments, factors controlling denitrification rates, differences in denitrification in freshwater and marine sediments, and ecosystem level considerations of denitrification. This chapter is restricted to submerged sediments in streams, rivers, lakes, estuaries, coastal and continental shelf systems, and the deep sea. Other recent reviews of denitrification in aquatic systems

Denitrification in Soil and Sediment, Edited by N. P. Revsbech and
J. Sørensen, Plenum Press, New York, 1990

include Knowles (1982), Hattori (1983), Kuenen and Robertson (1987), Koike and Sørensen (1988) and Seitzinger (1988), among others.

RATES OF DENITRIFICATION IN FRESHWATER AND MARINE SEDIMENTS

A variety of techniques has been used to measure denitrification in aquatic sediments (Table 1), including the acetylene inhibition technique (Sørensen, 1978), ^{15}N tracer techniques (Nishio et al., 1983; Jenkins and Kemp, 1984), N_2 production measurements (Seitzinger et al., 1984; Gardner et al., 1987), NO_3^- consumption measurements (Andersen, 1977), diagenetic models of pore water profiles (Vanderborght and Billen, 1975; Madsen, 1979; Bender et al., 1977), and the N mass balance approach (Andersen, 1971; Owens et al., 1972, Hill, 1983; Likens and Loucks, 1978), among others. Reviews of the various techniques can be found in Knowles (1982), Taylor (1983), Koike and Sørensen (1988), and Kuenen and Robertson (1987). In this volume the acetylene inhibition technique (Chapters 9 and 16) and the ^{15}N isotope technique (Chapter 18) are reviewed.

Denitrification is a relatively ubiquitous process in aquatic sediments (Table 1). Lowest denitrification rates generally occur in deep sea sediments (0.03 to 2.4 μg-at m^{-2} h^{-1}), with rates in continental shelf sediments being about an order of magnitude higher (up to 20 μg-at m^{-2} h^{-1}) (Table 1). In oligotrophic to moderately eutrophic lakes, denitrification rates generally range from 2 to 60 μg-at m^{-2} h^{-1}, and generally higher rates exist in eutrophic lakes (20 to 292 μg-at m^{-2} h^{-1}). Some of the highest denitrification rates occur in highly polluted estuarine sediments (>500 μg-at m^{-2} h^{-1}); however, rates in most estuaries range from 5 to 250 μg-at m^{-2} h^{-1}. Denitrification is also active in streams and rivers. Most studies have been conducted in relatively eutrophic streams or rivers with denitrification rates generally ranging from 40 up to 2121 μg-at m^{-2} h^{-1}. In fact, most denitrification measurements in shallow water systems are from temperate systems that are currently receiving considerable inputs of anthropogenic nutrients; little information is available at present on denitrification in tropical or unpolluted aquatic environments.

FACTORS CONTROLLING DENITRIFICATION RATES

A number of factors influence denitrification rates in aquatic sediments including oxygen, NO_3^- and organic matter concentrations, pH, temperature, and toxic compounds from anthropogenic sources. Many of these factors are interrelated, resulting in complex spatial and temporal patterns of denitrification.

Nitrate Concentrations and Denitrification Rates

The availability of NO_3^- is an important factor controlling denitrification rates in many aquatic sediments. Denitrification rates, as measured in homogenized sediment slurries, increase as a function of NO_3^- concentration and generally can be described by Michaelis-Menten-type kinetics (Billen, 1978; Koike et al., 1978; Messer and Brezonik, 1984). Half-saturation values (K_m) for denitrification measured in sediment slurries are usually between 25 and 50 μM NO_3^- (Billen, 1978; Koike et al., 1978; Iizumi et al., 1980; Oremland et al., 1984), although K_m values as high as 219 μM (Messer and Brezonik, 1984) and 344 μM (Oren and Blackburn, 1979) occur in some systems. Recent measurements of

Table 1. Denitrification rates measured at near ambient conditions in sediments from rivers, streams, lakes, coastal marine systems, continental shelf and deep sea sediments. (Modified from Seitzinger 1988).

Location	Rate (μmol N $m^{-2}h^{-1}$)	Method	References
RIVERS/STREAMS			
Gelbæk	42-375	acetylene blockage	Christensen and Sørensen 1988
Rabis Bæk, unvegetated	42-167	acetylene blockage	Christensen and Sørensen 1988
Rabis Bæk, vegetated	42-458	acetylene blockage	Christensen and Sørensen 1988
San Francisquito Creek, CA	54	acetylene blockage	Duff et al. 1984
Little Lost Man Creek	0	acetylene blockage	Duff et al. 1984
Swift Brook, Ontario	121-302*	NO_3 uptake	Robinson et al. 1979
Potomac River	210-235	N_2 flux	Seitzinger 1987a
Delaware River	166-345	N_2 flux	Seitzinger 1987b
Århus Å	400	acetylene blockage	Christensen et al. 1989
Gudenå	900	acetylene blockage	Christensen et al. 1989
Purukohukohu Stream,			
forested basin	14-364	acetylene blockage	Cooper and Cooke 1984
agricultural basin	364-2121	acetylene blockage	Cooper and Cooke 1984
LAKES			
Oligotrophic/Mesotrophic			
Michigan	12-51	N_2 flux	Gardner et al. 1987
Hampen, Denmark	5-50	acetylene blockage	Christensen and Sørensen 1986
Lacawac, Pennsylvania	50	N_2 flux	Seitzinger unpubl. data
Ernest, Pennsylvania	56	N_2 flux	Seitzinger unpubl. data
Lake Titicaca	0.3	acetylene blockage	Vincent et al. 1985
Moderately Eutrophic			
Okeechobee	2-25	acetylene blockage	Messer and Brezonik 1983
Mendota, Wisconsin	10*	^{15}N	Keeney et al. 1971
Arresø	11	^{15}N	Madsen 1979
Vechten	11-55*	pore water NO_3 profile	Sweerts and deBeer 1989
Eutrophic			
ELA Lake 227	42-58*	^{15}N	Chan and Campbell 1980
Bryrup Langsø	171*	NO_3 uptake	Andersen 1977
Kvind Sø	102*	NO_3 uptake	Andersen 1977
Lake Erie	20-292	N_2 pore water profile	Fendinger and Adams 1987
COASTAL MARINE			
Patuxent R. Estuary	77-89	^{15}N	Jenkins and Kemp 1984
Izembok Lagoon,Alaska	3.5-34.5	^{15}N	Iizumi et al. 1980
Delaware Inlet,N.Z.	6-12	acetylene blockage	Kaspar 1983
Kenepuru Sound,N.Z.	6-35	acetylene blockage	Kaspar et al. 1985b
Princess Louisa Inlet	216	model water column prof.	Smethie 1987
Århus Bight, Denmark	<2-42	acetylene blockage	Jensen et al. 1988
Tama Estuary, Japan	145-594	^{15}N	Nishio et al. 1983
Odawa Bay, Japan	37	^{15}N	Nishio et al. 1983
Tokyo Bay, Japan	16-33 *	^{15}N	Nishio et al. 1982
Kysing Fjord, Denmark	7	^{15}N	Oren and Blackburn 1979
Tejo Estuary, Portugal	107-1067 (232)	N_2 flux	Seitzinger unpubl. data
Delaware Bay	0-260 (98)	N_2 flux	Seitzinger unpubl. data
Ochlockonee Bay, FL	0-210 (73)	N_2 flux	Seitzinger 1987c
Narragansett Bay, R.I.	39-109 (59)	N_2 flux	Seitzinger et al. 1984
Great Bay, Long Island	3-16	acetylene blockage	Slater and Capone 1987
Randers Fjord	6-42	acetylene blockage	Sørensen 1978, et al. 1979
Kysing Fjord	1-330	acetylene blockage	Sørensen 1978, et al. 1979
Coastal North Sea	0-84 (50)	model pore water profile	Billen 1978
Lendrup Vig	12-213	acetylene blockage	Andersen et al. 1984
Four League Bay	2-74	acetylene blockage	Smith et al. 1985
MERL mesocosms	305-888	N_2 flux	Seitzinger and Nixon 1985
W. Coast, N.Z.	0.4-141	acetylene blockage	Kaspar et al. 1985a
San Francisco Bay	0.8-1.2	acetylene blockage	Oremland et al. 1984
Norsminde Fjord	83-417	acetylene blockage	Jørgensen and Sørensen 1988
CONTINENTAL SHELF			
Gulf of Maine	3.5	model pore water profile	Christensen et al. 1987a
Mexican Shelf	5.4	model pore water profile	Christensen et al. 1987a
Washington Shelf	20.5	model pore water profile	Christensen et al. 1987b
East China Sea	0-54	NH_4 production-DIN efflux	Aller et al. 1985
Bering Sea, Shelikaf Strait	2.1	acetylene blockage	Haines et al. 1981
Bering Sea	9.0	acetylene blockage	Koike and Hattori 1979
DEEP SEA SEDIMENTS			
NE Atlantic	0.42	N_2 pore water profile	Wilson 1978
NW Atlantic	0.33-0.37	model pore water profile	Christensen and Rowe 1984
E. Equatorial Atlantic	0.15-0.33	model pore water profile	Bender et al. 1977
E. Equatorial Atlantic	0.03	acetylene blockage	Goloway and Bender 1982
E. Equatorial Atlantic	0.11-0.39	model pore water profile	Sørensen et al. 1984
E Equatorial Pacific	0.06-2.4	model pore water profile	Goloway and Bender 1982

*May be underestimate of ambient rate as technique used did not include denitrification fed by sediment nitrification

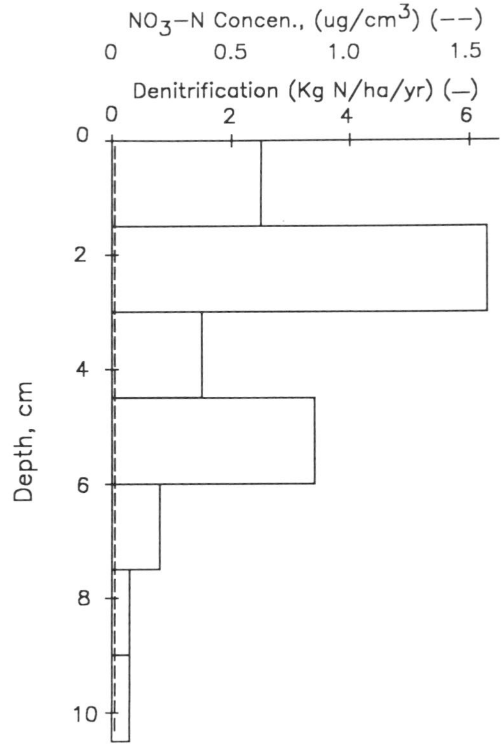

Figure 1. Vertical profile of in situ denitrification rates and interstitial water nitrate concentrations measured at 1.5-cm depth intervals in Delaware Inlet, New Zealand mud-flat, March 1981 (Redrawn from Kaspar, 1982; 1983).

denitrification rates within NO_3^- concentration profiles in intact sediment cores indicated K_m values less than 10 μM (Christensen et al., 1989). NO_3^- concentrations in aquatic sediments are generally considerably less than 20 μM and thus within the range in which denitrification would be expected to be first order with respect to NO_3^- concentrations. Highly polluted systems with high NO_3^- concentrations in the water column are an exception; in such systems NO_3^- concentrations of tens to hundreds of μM occur in surface sediments.

The in situ rate of denitrification at a given NO_3^- concentration is spatially variable. This spatial (vertical and horizontal) variability makes extrapolation of denitrification rates from measurements of NO_3^- concentrations problematic. For example, denitrification rates within the top 7.5 cm of sediments from the South Island West Coast, New Zealand, varied over seven fold at saturated NO_3^- concentrations (Kaspar et al., 1985a). In Kenepuru Sound, New Zealand, NO_3^- concentrations were below measurable concentrations throughout the top 10.5 cm of sediment, while denitrification rates measured at 1.5 cm depth intervals over the top 10.5 cm varied by a factor of greater than 30 (Kaspar, 1983) (Fig. 1). The relationship between denitrification rate and NO_3^- concentration also varies among sites. For example, while NO_3^- concentrations in the sediments at two different sites in Kenepuru Sound were similar, denitrification rates differed by a factor of seven (Kaspar et al., 1985b). Similarly, at nine locations off the coast of New Zealand, NO_3^- concentrations varied by a factor of two among sites, while denitrification rates varied by a factor of greater than 10 (Kaspar et al., 1985a). This variability suggests that factors in addition to NO_3^- concentration (e.g., organic content, population of denitrifiers, etc.) are important in controlling denitrification rates. In addition, some of the apparent variability in the relationship between NO_3^- concentrations and denitrification rates may be a result of the spatial interval over which the measurements are made. The NO_3^- concentration in microenvironments may be controlling denitrification rates; not the bulk NO_3^- concentration as measured at 1-cm depth intervals.

Distribution of Denitrification within Sediments

An important technological advance which led to new insights into nitrogen cycling in aquatic sediments, in particular the spatial relationships of the various nitrogen processes, is the development of microelectrodes, starting with the oxygen microelectrode by Revsbech et al. (1980). Microelectrodes are also now available for nitrous oxide (Revsbech et al., 1988) and NO_3^- (Sweerts and de Beer, 1989). These electrodes allow measurement of chemical gradients at intervals of tens of microns in contrast to previous measurements of chemical gradients from extracted pore waters with a vertical resolution of several orders of magnitude less (0.5- to 1-cm intervals).

Development of the oxygen microelectrode has been particularly useful because the distribution of oxygen controls the distribution and, indirectly, the rate of denitrification in aquatic sediments. While anoxic or low oxygen (<0.2 mg O_2/L) conditions are generally considered necessary for denitrification, a major source of NO_3^- for denitrification in many sediments is from nitrification, which requires oxygen. (Aerobic denitrification can also occur and is covered by Robertson and Kuenen in Chapter 5 of this volume.) To explain the co-occurrence of these two processes in sediments, investigators often considered that they are separated vertically within the sediment (Vanderborght et al., 1977; Billen, 1978). Alternatively, denitrification may occur within reduced microzones in the aerobic surface layer of sediments (Sørensen et al., 1979; Jenkins and Kemp, 1984). Measurement of dissolved chemical species by use of few μm thick microelectrodes have clarified the zones of nitrification and denitrification in aquatic sediments.

Oxygen microelectrodes, for the first time, provided reliable measurements of the vertical profiles of oxygen in aquatic sediments. In shallow aquatic systems, free oxygen concentrations rapidly decrease to zero within the top 2 to 8 mm of most surface sediments or the lining of infauna burrows (Revsbech et al., 1980; Jørgensen and Revsbech, 1985). Thus, nitrification, which requires free O_2, would be expected to be restricted to this thin aerobic surface layer of sediments.

The recent development of a combined oxygen and nitrous oxide microelectrode now makes it possible to simultaneously measure fine scale vertical profiles of O_2 and of denitrification using the acetylene blockage technique (Revsbech et al., 1988). Use of this electrode in sediments of Gudena, a stream in Denmark with 550 μM NO_3^- in the stream water (Christensen et al. 1989), demonstrated that denitrification in the surface sediments occurred in a zone about 3.6-mm thick, starting about 0.2 mm below the sediment surface (Fig. 2). In Århus Å, with 115 μM NO_3^- in the water, denitrification was restricted to an even narrower zone in the surface sediments, 0.7 mm thick, and beginning approximately 0.7 mm below the interface. In both streams, the denitrification

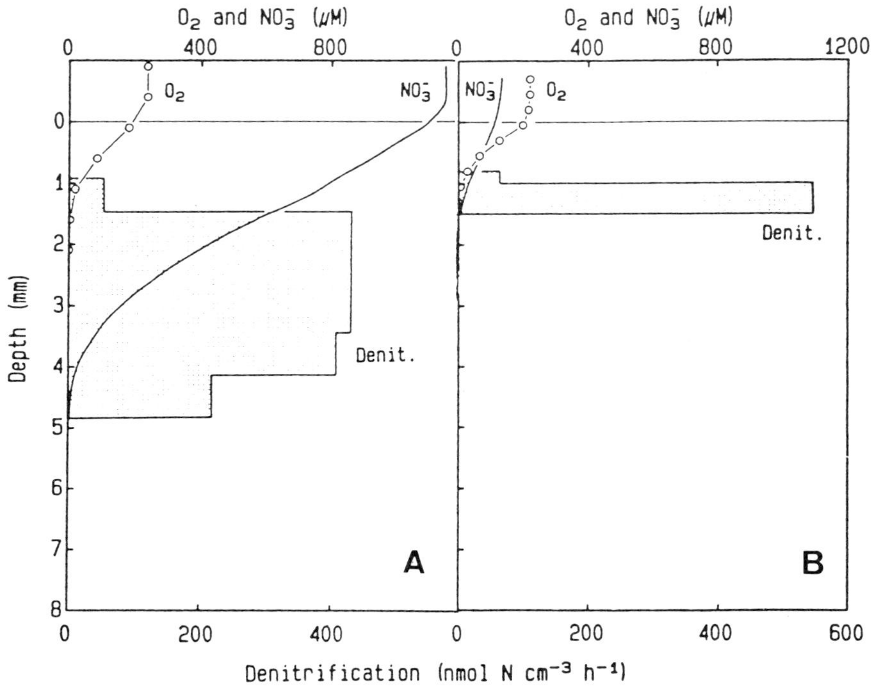

Figure 2. Vertical profiles of oxygen (measured), denitrification (measured) and nitrate (simulated) in sediment cores from two streams in Denmark (a) Gudenå and (b) Århus Å (from Christensen et al., 1989).

zone began just above the oxic/anoxic interface, with denitrification primarily occurring in the anoxic zone in which NO_3^- was present from diffusion from the overlying water (Christensen et al., 1989). Measurements of NO_3^- profiles in lake sediments using a newly developed NO_3^- microelectrode have been reported (Sweerts and de Beer, 1989). In Lake Vechten, NO_3^- concentrations decreased to zero within the top 1.3 mm of silty sediments, and within the top 5 mm of sandy sediments, with NO_3^- concentrations of 30 μM in the overlying water. Coupled with oxygen microelectrode profiles, these data have shown that denitrification occurred in a thin zone, approximately 500 μm thick, immediately adjacent to the aerobic zone (Sweerts and de Beer, 1989). In both of the above studies, NO_3^- from the water was diffusing into the sediments to supply the NO_3^- for denitrification. In many sediments, nitrification in the sediments is a major source of NO_3^- for denitrification. Future application of microelectrode techniques in such sediments should clarify the spatial relationship between nitrification and denitrification.

In continental shelf sediments denitrification occurs within the top few cm of sediment, as indicated by pore water profiles of NO_3^- concentration measured from extracted pore waters (Christensen et al., 1987a). Microelectrode techniques would be useful to investigate the fine scale vertical resolution of the zone of denitrification within this region. Sediment nitrification is equal to or exceeds the NO_3^- influx from the bottom waters (20-30 μM NO_3^-) as a source of NO_3^- for denitrification in Gulf of Maine sediments (Christensen et al., 1987a).

In deep sea sediments, the zone of denitrification is thicker and appears to start deeper than in shallow water or continental shelf sediments. Diagenetic modelling of pore water profiles in Eastern Equatorial Atlantic sediments demonstrated that the zone of denitrification was between approximately 35 to 40 cm, with aerobic oxidation of organic matter occurring in the top 5 cm (Bender et al., 1977). This is similar to the depth of oxygen penetration reported in other deep sea sediments using microelectrode techniques (Reimers et al., 1986). Nitrification of NH_4^+ produced within the sediments was the NO_3^- source for denitrification in Eastern Equatorial Atlantic sediments (Bender et al., 1977). However, in contrast with coastal and shelf sediments, the coupling of nitrification to denitrification was not very efficient as nitrification rates exceeded denitrification rates by an order of magnitude.

Effect of Infauna and Macrophytes

As discussed above, oxygen microelectrode studies repeatedly demonstrated that shallow water sediment oxygen concentrations go to zero within the top few mm (Revsbech et al., 1980; Jørgensen and Revsbech, 1985; Sweerts and de Beer, 1989). However, measurable NO_3^- concentrations occur at considerable depths (top 10 cm) in aquatic sediments in which overlying water NO_3^- concentrations are relatively low. Vertical profiles of denitrification using the intact core acetylene blockage technique generally show denitrification occurring to considerable depths within the sediments (Sørensen, 1978; Kaspar et al., 1985a,b). What is the source of this NO_3^- measured in the sediments that supplies the measured denitrification?

Animals can have a significant effect on the NO_3^- supply for denitrification. Bioturbation mixes NO_3^- and oxygen for nitrification from the overlying water or from the surface sediments to deeper layers within the sediments (Grundmanis and Murray, 1977; Chatarpaul et al., 1980). Irrigation of burrows by certain infauna also increases the transport of NO_3^- from the overlying water to increased depths in the sediment. For example, in Washington continental shelf sediments where bottom water NO_3^- concentrations are typically ~30 μM, the transport of NO_3^- into the sediments by macrofauna mixing and irrigation is estimated to be two times the vertical molecular diffusive flux of NO_3^- across the sediment-water interface (Christensen et al., 1987b). The NO_3^- influx from the bottom water (0.82 pmol N cm^{-2} s^{-1}) on the Washington shelf is slightly less than the estimated supply of NO_3^- from sediment nitrification (~1 pmol N cm^{-2} s^{-1}) (Christensen et al., 1987b). Irrigation also extends the area of oxygenated sediments to the lining of burrows (Jørgensen and Revsbech, 1985), which also often have higher NH_4^+ concentrations than surface sediments due to animal excretion; this increases the zone within which nitrification occurs and thus increases overall rates of coupled nitrification/denitrification within the sediments (Henriksen et al., 1980; Henriksen et al., 1983; Aller et al., 1983). Direct measurements of denitrification rates in the lining of infauna burrows and in the bulk sediment would be useful to evaluate the overall importance of irrigation in enhancing denitrification. A further effect of macrofauna on coupled nitrification/denitrification occurs in the fecal pellets from surface deposit feeders and filter feeders in which potential nitrification rates in surface sediments are enhanced. For example, potential nitrification rates in surface sediments containing *Macoma balthica* increased by approximately a factor of two over control sediments; below the top cm, only relatively small differences in potential nitrification were found (Henriksen et al., 1983).

The presence of macrophytes within the euphotic zone of lakes, streams, rivers and coastal marine systems may affect denitrification rates. Oxygen transported through the leaves to the rhizomes and roots, and released in the otherwise anoxic zone of sediments, may stimulate nitrification, thus providing an additional source of NO_3^- for denitrification (Iizumi et al., 1980; Christensen and Sørensen, 1986; Christensen and Sørensen, 1988). For example, oxygen released from roots of a population of Potamogeton perfoliatus L. in Chesapeake Bay was estimated to be sufficient to support approximately five times ambient nitrification rates in unvegetated sediments (Kemp and Murray, 1986). Differences in morphological features and environmental factors affect oxygen release within a species (Kemp and Murray, 1986). However, different species of macrophytes differ in the percent of the total oxygen (leaf plus root release) released from their roots; relatively high percentages of oxygen were released from the roots of Isoetes (39%), Littorella (28%), and Lobelia (100%) compared to Zostera (1-4%) (Sand-Jensen et al., 1982). These differences in oxygen release likely result in differences in the effect of macrophytes on nitrogen cycling in sediments. Macrophytes may also stimulate denitrification by increasing the content of easily degradeable organic matter in the sediments through trapping of organic detritus from the water column (Christensen and Sørensen, 1988), or releasing dissolved organic carbon from the roots (Christensen and Sørensen 1986).

The complexity of the effect of macrophytes on denitrification rates in stream sediments was demonstrated in a comparison of denitrification rates in vegetated (Batrachium peltatum) and unvegetated sediments in Rabis Bæk, a lowland stream receiving inputs of high NO_3^- groundwater (Christensen and Sørensen, 1988). Denitrification rates during winter were similar; however, during April to July when plant growth was abundant, denitrification rates (max. July, 11 mmol N m^{-2} d^{-1}) in the organic-rich vegetated sediments exceeded those in the unvegetated sandy area. The vegetated area denitrification rates were probably higher because macrophytes increased the sediment organic content by trapping detritus and increased coupled nitrification/denitrification by transporting oxygen into the sediments.

Benthic algae may decrease denitrification rates in surface sediments by producing high concentrations of oxygen during photosynthesis (Andersen et al., 1984; Christensen and Sørensen, 1986; Jørgensen and Sørensen, 1988). Interpretation of the significance of light inhibition of denitrification requires consideration of diel and seasonal patterns as shown by Jørgensen and Sørensen (1988). While denitrification rates were 60% lower in light-incubated than dark-incubated sediments (maximum differences noted in spring), over an annual cycle denitrification rates were only ~13% lower due to light inhibition. In contrast to organic-rich sediments, benthic algae in sandy, low organic content sediments may increase denitrification by increasing the organic content of the sediments (Christensen and Sørensen, 1988).

Many lakes, streams, rivers and coastal marine systems with macrophytes and benthic algae receive increased rates of nutrient inputs from anthropogenic sources, which leads to eutrophic conditions. The effect of increased nutrient inputs on macrophytes and benthic algae is likely complex. Marked changes in submerged aquatic vegetation occur in many freshwater and marine areas including Chesapeake Bay (Orth and Moore, 1983), the Dutch Waddenzee (Den Hartog and Polderman, 1975), and numerous lakes. A number of factors may contribute to changes in the abundance of aquatic plants, including increased nutrient inputs. Increased nutrient inputs could adversely affect aquatic plant growth due to shading from either increased phytoplankton or epiphyte growth (Cattaneo and Kalff, 1980; Sand-Jensen and Søndergaard, 1981; Twilley et al.,

1985). Thus, it is likely that denitrification in macrophyte and benthic algae-covered areas of these systems is affected by the increased nutrient inputs. A number of scenarios is possible. For example, increased phytoplankton growth may decrease light levels reaching benthic microalgae, thus decreasing photosynthetic oxygen production; this may alleviate oxygen inhibition of denitrification in the surface sediments and result in increased denitrification activity. On the other hand, denitrification in rooted macrophyte areas may decrease due to eutrophication if the decreased light levels result in a decrease in the transport of oxygen by the plants to the deeper zones of the sediments. Further investigations are needed to clarify the relationship between eutrophication and denitrification in vegetated aquatic sediments.

Temporal Variation in Denitrification

In addition to temperature, the supply of NO_3^- and organic matter control seasonal variations in denitrification rates. The seasonal pattern in a system thus depends on which factors control denitrification. NO_3^- from the overlying water is a major source of NO_3^- for denitrification in highly polluted systems with high NO_3^- concentrations in the water in some coastal marine (Andersen et al. 1984) and stream ecosystems (Christensen and Sørensen, 1988). In these systems, NO_3^- concentrations can reach hundreds of μM in the sediment, depending on the overlying water NO_3^- concentration, with NO_3^- extending several cm into the sediment. Seasonal variations in water column NO_3^- concentration are an important determining factor of denitrification rates in such systems (Andersen et al., 1984; Christensen and Sørensen, 1988). In Norsminde Fjord, the peak fall denitrification activity was attributed to an increase in the NO_3^- concentrations in the water (~700 μM), while a spring activity peak was attributed to deposition and subsequent decomposition of organic matter from a bloom of microalgae while the NO_3^- concentrations were still at high winter values (Jørgensen and Sørensen, 1988). This contrasts with sediments in which nitrification is a major source of NO_3^- in the sediments; NO_3^- concentrations in those sediments are generally less than 20 μM in the top cm, and less than 10 μM below. The distribution of oxygen in the thin surface sediments, as affected by temperature, organic matter degradation rates, macrofauna and macrophytes (discussed above), appears to be important in controlling nitrification rates and, therefore, NO_3^- for denitrification in those sediments (Jenkins and Kemp, 1984). In Patuxent River estuary sediments, denitrification rates coupled to nitrification were high in the spring, but decreased by two orders of magnitude in the summer when redox potentials were low and nitrification rates were decreased due to reduced O_2 concentrations (Jenkins and Kemp, 1984). This pattern of highest denitrification rates in spring followed by lower rates in summer was also reported for Ochlockonee Bay, FL, a relatively unpolluted estuary (Seitzinger, 1987c), in which sediment nitrification was the major source of NO_3^- for denitrification. However, in Narragansett Bay, RI, coupled nitrification/ denitrification rates were highest in summer (Seitzinger et al., 1984).

Denitrification and Total Community Respiration

Denitrification rates, measured in a variety of river, lake and estuarine sediments, are relatively well correlated with total community respiration (sediment oxygen uptake) (Fig. 3). The coupling between these two processes is complex. For example, while oxygen is not used directly during denitrification, a considerable percentage (often 35%) of the benthic oxygen consumption is due to

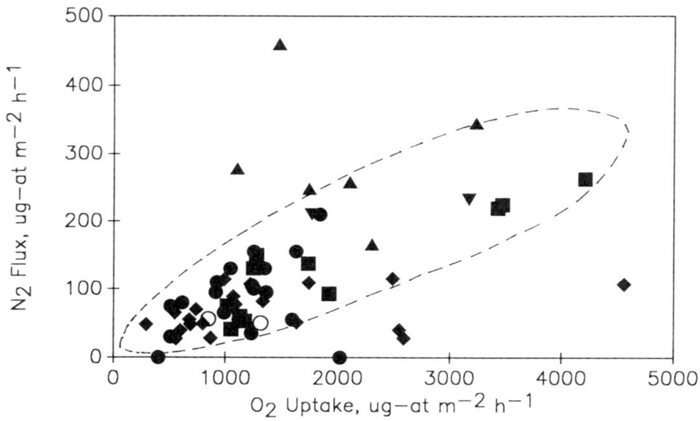

Figure 3. Denitrification rates vs oxygen consumption rates in sediments from Delaware Bay (■) (Seitzinger unpubl., data), Delaware River (▲) (Seitzinger, 1987b), Ochlockonee Bay, FL (●) (Seitzinger, 1987c), two Pennsylvania lakes (○) (Seitzinger unpubl., data), the Potomac River (▼) (Seitzinger, 1987a) and Narragansett Bay (◆) (Seitzinger et al., 1984).

nitrification (Christensen and Rowe, 1984; Henriksen and Kemp, 1988), which supplies the NO_3^- for denitrification in many sediments. The source of NH_4^+ for sediment nitrification is from decomposition of organic matter in the sediments. The rate of NH_4^+ production from both aerobic and anaerobic respiration is reflected in the rate of sediment oxygen uptake (reduced compounds produced by anaerobic respiration are oxidized in the aerobic zone of the sediments). Although the amount of organic matter oxidized in sediments via denitrification is generally quite small relative to other pathways (both aerobic and anaerobic) of organic matter decomposition (Jørgensen and Sørensen, 1985), all these pathways of decomposition depend on the availability of organic carbon and thus the correlation is expected, even in sediments in which NO_3^- is primarily supplied from the water column. The relationship between denitrification and total community respiration is explored in considerable detail elsewhere in this volume by Blackburn (Chapter 20) with a model of C, N, S and O cycling in sediments.

NITROUS OXIDE

In addition to N_2, N_2O and NO can also be produced as end products of denitrification (Knowles, 1982). The importance of understanding nitrous oxide production rates and factors controlling its production, including the effect of anthropogenic activities on production rates, is obvious in light of the role of N_2O in the destruction of stratospheric ozone (Crutzen, 1970; Hahn and Crutzen, 1982), and in the radiative heat budget of the atmosphere (Wang et al., 1976). A limited number of measurements of nitrous oxide fluxes from aquatic sediments has been made (Nishio et al., 1983; Jensen et al., 1984; Seitzinger, 1988). Net N_2O fluxes are generally less than 2 μmol N_2O-N m^{-2} h^{-1}, and less than 2% of N_2 production rates.

The net source of N_2O from aquatic sediments represents the combined production and consumption rates of nitrous oxide. Nitrous oxide is produced by denitrification (Knowles, 1982), nitrification (Goreau et al., 1980), and

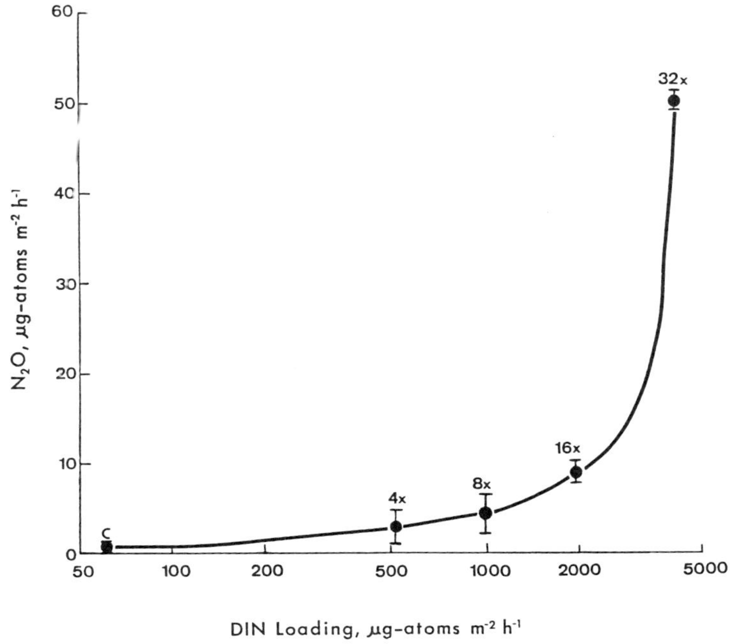

Figure 4. Nitrous oxide effluxes from coastal marine mesocosms (Marine Ecosystem Research Laboratory, URI) as a function of dissolved inorganic N loading rates (DIN) to the mesocosms (from Seitzinger and Nixon, 1985).

dissimilatory NO_3^- reduction to ammonia (Smith and Zimmerman, 1981), all of which are active processes in aquatic sediments. Denitrification can also be a sink for nitrous oxide (Knowles, 1982). In a series of well-designed experiments by Jørgensen et al. (1984), the relative magnitude of nitrous oxide production by denitrification and nitrification was measured in coastal marine sediments at varying oxygen concentrations. Their data suggest that denitrification is the major source of nitrous oxide at the low oxygen concentrations found in aquatic sediments.

The potential effect of anthropogenic activities on nitrous oxide production from aquatic sediments was demonstrated in a eutrophication experiment utilizing coastal marine mesocosms with coupled benthic-pelagic systems. The mesocosms received a range of inorganic nutrient (N, P and Si) loading rates. Nitrous oxide fluxes increased markedly as the rate of nutrient loading to the system increased (Fig. 4) (Seitzinger and Nixon, 1985). Total system production, benthic respiration, and denitrification increased (Nixon et al., 1984) as nutrient loading increased, although less dramatically than benthic nitrous oxide effluxes. The ratio of N_2O:N_2 released from the sediments increased from 0.2% to 5.8% as nutrient loading increased. The mechanism responsible for the increased ratio of N_2O:N_2 was not investigated; however, it may be a result of H_2S inhibition of nitrous oxide reduction to N_2 during denitrification (Sørensen et al., 1980). Higher rates of N_2O production and increased ratios of N_2O:N_2 also occur in more eutrophic areas of Narragansett Bay relative to areas receiving lower rates of nutrient loading (Seitzinger et al., 1984). The magnitude of the increase in nitrous oxide production on a global scale as a result of continued eutrophication (as well as other anthropogenic effects) of aquatic systems is uncertain, and warrants quantification.

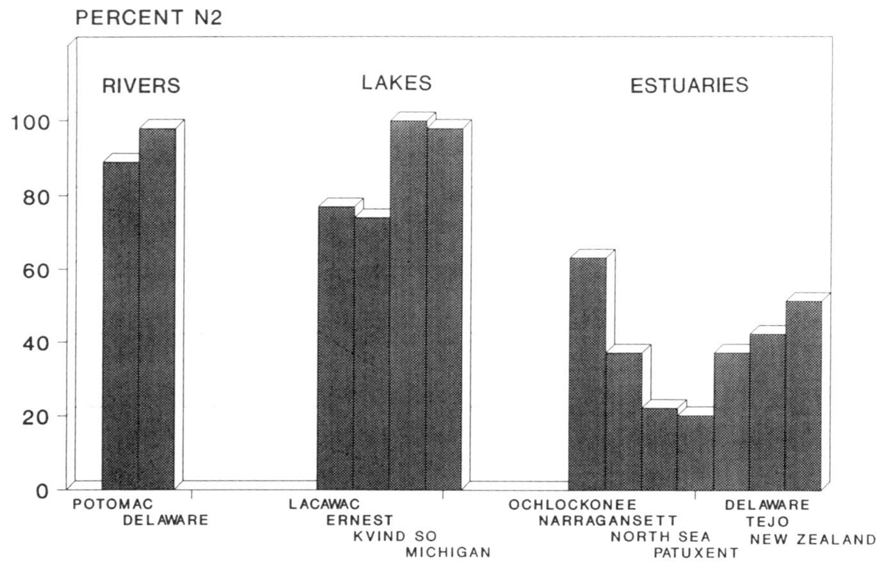

Figure 5. Percent of sediment-water nitrogen efflux that is N_2 (denitrification) in various freshwater and marine systems. Data from Seitzinger, 1988 (Potomac R., Delaware R., L. Ernest, L. Lacawac, Delaware Bay, Tejo Estuary); Andersen, 1971; 1977 (Kvind Sø); Gardner et al., 1987 (L. Michigan); Seitzinger, 1987c (Ochlockonee Bay); Seitzinger et al., 1984 (Narragansett Bay); Billen, 1978 (North Sea); Henriksen and Kemp, 1988 (Patuxent); Kaspar et al., 1985a (New Zealand coastal sediments) (from Seitzinger et al., submitted).

DENITRIFICATION AND BENTHIC NITROGEN RECYCLING

The sediments in many aquatic systems are an important source of recycled nitrogen (NH_4^+ and NO_3^-) for primary production (Nixon, 1981; Blackburn and Henriksen, 1983). For example, in estuaries and other nearshore marine systems recycling of N from the sediments often accounts for 25% to 60% of the phytoplankton N requirements (Nixon, 1981). However, a large portion of the organic nitrogen cycled through aquatic sediments is not returned to the water column as NH_4^+ or NO_3^-, but is removed from the recycling pathway by denitrification. Algal production in many estuaries and coastal and continental shelf marine systems, some lakes (Keeney, 1973; Gerhart and Likens, 1975), and some streams (Grimm and Fisher, 1986) is limited by nitrogen. The removal of nitrogen by denitrification in sediments of these systems may, therefore, be important in regulating algal and/or macrophyte production. Many aquatic systems receive large amounts of nitrogen from anthropogenic sources; the removal of nitrogen via denitrification in the sediments may help control the degree of eutrophication in those systems.

Examination of the data currently available for sediment-water N_2, NO_3^-, and NH_4^+ fluxes from freshwater and marine systems suggests that a larger percentage of the NH_4^+ produced during the mineralization of organic N in freshwater sediments is lost from the ecosystem as N_2 compared to marine sediments (Gardner et al., 1987; Seitzinger, 1988). In many freshwater systems (with aerobic bottom waters), essentially all (80%-100%) of the net NH_4^+

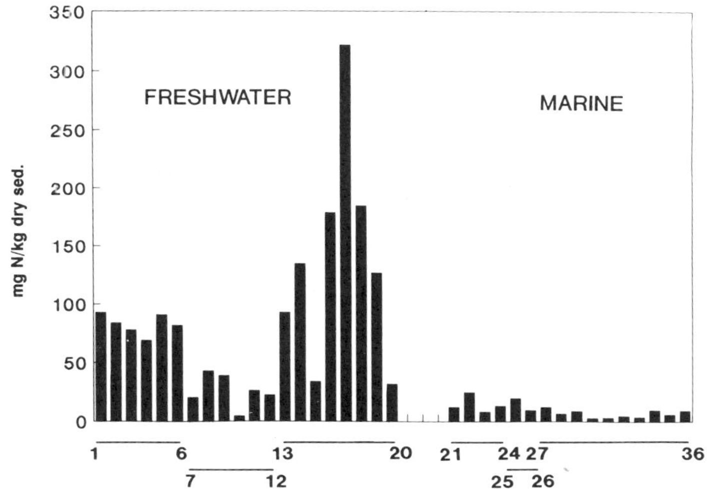

Figure 6. Ratio of exchangeable ammonium (mg N/kg dry sediment) to dissolved ammonium (mg N/L pore water) in a variety of freshwater and marine sediments. Data sources: (1-6) Bay Quinte, summer data 0-2 and 2-4 cm depth, Liao and Lean, 1978; (7-12) L. Michigan 6 sites, Chen et al., 1983; (13-20) lakes Tomahawk, Devils, Crystal, Trout, Little John, Mendota, Monona, Wingra, Keeney et al., 1970; (21-22) Long Island Sound and (23-24) Florida Bay, Rosenfeld, 1979; (25) Narragansett Bay, Garber, 1982; (26-36) 11 sites Danish coast, assumed dry sed. density 2.5 g/cm^3, Blackburn and Henriksen, 1983 (from Seitzinger et al., submitted).

produced in the sediments is nitrified and subsequently denitrified (Fig. 5). This contrasts with coastal marine sediments (with aerobic bottom waters) where only a portion (40% to 60%) of the NH_4^+ produced is nitrified and denitrified (Fig. 5). The mechanism(s) responsible for this apparently higher efficiency of nitrification in freshwater sediments is not known. However, one current hypothesis is that NH_4^+ concentrations (dissolved plus exchangeable) are higher in freshwater sediments relative to marine sediments, for the same rate of NH_4^+ production. Higher total NH_4^+ concentrations are predicted in freshwater sediments because of decreased competition for cation exchange sites and decreased ion pairing (compared to sediments in saltwater) resulting in increased exchangeable NH_4^+ concentrations in freshwater sediments. If both dissolved and exchangeable NH_4^+ are available to nitrifying bacteria, then nitrification rates would be greater in freshwater sediments (Gardner et al., submitted; Seitzinger et al., submitted). Thus, relatively less of the NH_4^+ produced in freshwater sediments should be recycled to the water, and more nitrified and available for denitrification.

A comparison of data from a range of freshwater and marine sediments suggests that exchangeable NH_4^+ concentrations in freshwater sediments are generally considerably greater than in marine sediments (Seitzinger et al., submitted, Fig. 6); the ratio of exchangeable to dissolved NH_4^+ in freshwater sediments is also generally higher than in marine sediments (Seitzinger et al., submitted). These data support the hypothesis that total NH_4^+ concentrations are higher in freshwater sediments. Direct experiments examining the availability of exchangeable NH_4^+ to nitrifiers in aquatic sediments are needed. However, it is likely that exchangeable NH_4^+ is available to benthic nitrifiers since: (1)

nitrifying bacteria are generally associated with particles, (2) exchangeable NH_4^+ is only loosely sorbed to the particles, and (3) exchangeable NH_4^+ appears to be efficiently nitrified in other systems (e.g., in dry forest soils; Matson and Vitousek, 1987). Further experimental work is required to quantify the relationship between exchangeable NH_4^+ concentrations, nitrification rates, and denitrification efficiency in aquatic sediments.

ECOSYSTEM LEVEL CONSIDERATIONS

In addition to decreasing N recycled for primary production within a system, denitrification is a significant sink in many aquatic systems for external nitrogen inputs, decreasing the transport of nitrogen to downstream or offshore waters. In streams and rivers, estimates of the magnitude of denitrification as a sink for external N inputs range from 1% annually (10% in summer) over a 1 km stretch in two Danish lowland streams receiving large inputs of NO_3^- in groundwater (Christensen and Sørensen, 1988) to 75% in Duffin Creek, Ontario (Hill, 1981) (Table 2). Streams/rivers are a major transport pathway of N between aquatic

Table 2. N-removal via denitrification of external N inputs to streams and rivers.

Location	% N Loading Removed by Denitrification	Comments and Reference
Swifts Brook, Ontario	60%*	June-Oct, 2 km length, Kaushik and Robinson 1976
Duffin Creek, Ontario	75%	May-Oct (% of NO_3 input)
	50%	May-Oct (% of TN input)
	<5%	annual; Hill 1979, 1981, 1983
Nottawasaga River	13%	May-Oct (% of NO_3 input)
	<5%	annual; Hill 1983
Delaware River	20%	summer (% of DIN input); tidal fw portion; Seitzinger 1988
Potomac River	35%	summer (% of DIN input); tidal portion; Seitzinger 1987c
Purukohukohu Basin streams, New Zealand	20%	annual, 75 m length of stream channel; Cooper and Cooke 1984
Gelbaek and Rabis Baek Denmark	10%	summer (% of NO_3 input)
	1%	annual; 1 km stretch, Christensen and Sørensen 1988

*Denitrification not measured but calculated from N mass balance.

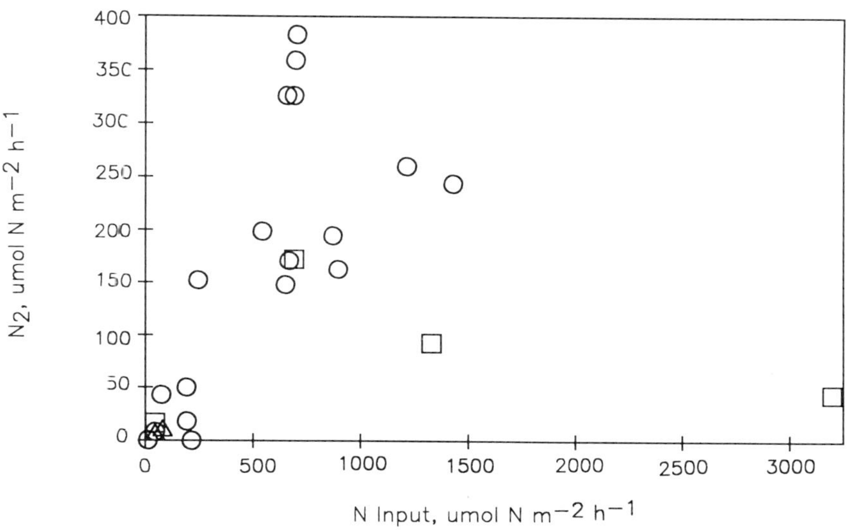

Figure 7. Denitrification vs external N loading rate to various lakes. Denitrification rates were determined by mass balance (○), nitrate uptake from overlying water only (□), or method used that measured denitrification of nitrate from overlying water and nitrification in sediments (△) (data from Tables 3 and 7, Seitzinger, 1988).

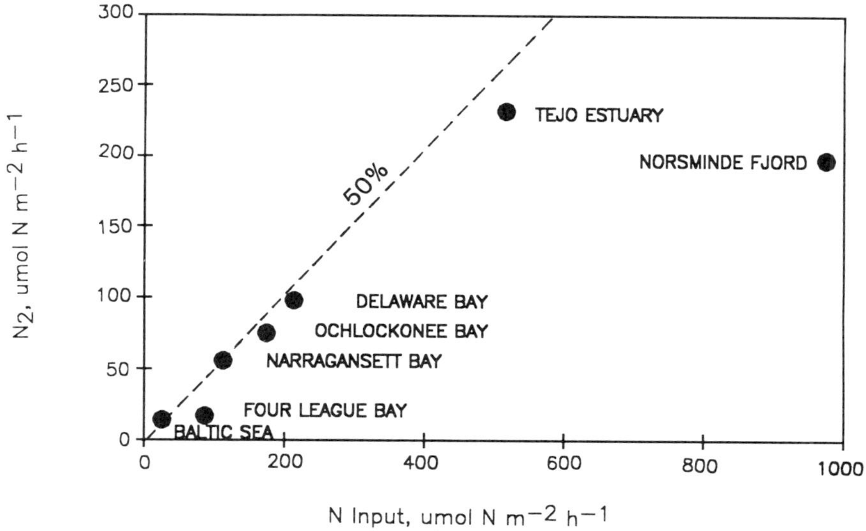

Figure 8. Denitrification rates vs external N loading rates in estuaries. Data for Norsminde Fjord (Jørgensen and Sørensen, 1988), Four League Bay (Smith et al., 1985), Baltic Sea (Shaffer and Ronner, 1984), other data summarized in Seitzinger, 1988 (figure modified from Seitzinger, 1988).

systems. Comprehensive studies are needed from the headwaters to the mouth of streams/rivers to assess the importance of denitrification within the various reaches as a sink for N. In lakes, the denitrification sink for nitrogen has been estimated to range from 1 to 62% (Fig. 7). Many of the estimates are based on mass balance N budgets for the lakes or on denitrification measurements that did not include nitrification sources within the sediments. The Lake Okeechobee N budget is one of the most comprehensive for denitrification; several approaches to measuring denitrification were used and were combined with ecosystem mass balance calculations (Messer and Brezonik, 1983). Denitrification was estimated to remove an amount of nitrogen ranging from 9 to 23% of external N inputs. In estuaries and nearshore coastal marine systems, denitrification often removes between 40 to 50% of external inputs of dissolved inorganic N (DIN) (Fig. 8), thus decreasing by about half the amount of terrestrially derived DIN transported to offshore waters (Seitzinger, 1988).

The similarity in the percent of the N loading that is lost through denitrification in estuaries is striking, considering the range of techniques used to measure denitrification, uncertainties in quantifying external N loading rates, extrapolating denitrification rates measured over small spatial scales to system-wide rates, etc. It is not clear what is controlling the linkage between external N loading and denitrification. However, it may be a function of the relationship between primary production and N loading. Denitrification rates appear to be related to the rate of benthic metabolism (Fig. 3) which is relatively well correlated with phytoplankton production plus allochthonous inputs (Nixon, 1981). Because the relationship between N loading rates and primary production rates in estuaries is quite variable (Boynton et al., 1982), it seems likely that there are other factors involved. Smith et al. (1989) developed a model of C-controlled N cycling in estuaries and argue that the rate of denitrification in a system is a function of the amount of net heterotrophy (net organic carbon oxidation) in the system.

Continental shelf sediments (50 Tg N/yr) are the major sedimentary denitrification sink in the global marine N budget (Christensen et al., 1987a). Approximately 50% of the total N inputs (particulate plus dissolved inputs from rivers, atmospheric deposition and N fixation) to the oceans is removed by denitrification in shelf sediments (Christensen et al., 1987a), with considerably less (~5 Tg N/yr; Hattori, 1983) removed in deep sea sediments. While denitrification is an important sink for N within estuaries, denitrification in estuarine and near shore coastal systems is only removing approximately 2.5 Tg N/yr, assuming 50% removal of the river inputs of DIN (Fig. 8) (5 Tg N/yr; Meybeck 1982) and thus does not appear to be a major sink in the global marine N budget.

SUMMARY

Denitrification is an active and ubiquitous process in freshwater and marine sediments. In the marine environment denitrification rates are lowest in open ocean sediments, with rates increasing by approximately an order of magnitude in continental shelf and again in near shore and estuarine areas. Some of the highest denitrification rates occur in eutrophic river sediments. Sediment nitrification is a major source of nitrate for denitrification in aquatic sediments; however, particularly in systems with a high concentration of nitrate in the water, diffusion of nitrate into the sediments is an important source. The coupling between nitrification and denitrification in deep sea sediments is not very efficient as nitrification exceeds denitrification by an order of magnitude. This contrasts with most other aquatic systems, where a large percentage of the nitrate produced in the sediments is denitrified. The seasonal pattern of denitrification in an ecosystem depends on a number of factors in addition to temperature, including the supply of nitrate from the sediments or water column and the supply of organic matter. In many freshwater systems, a higher percentage (75-100%) of organic nitrogen mineralized in the sediments is lost through coupled nitrification/denitrification than in near shore marine sediments (40-60%). Denitrification is a major sink in the overall N budget of most aquatic ecosystems.

REFERENCES

Aller, R.C., Yingst, J.Y., and Ullman, W.J., 1983, Comparative biogeochemistry of water in intertidal Onuphis (polychaeta) and Upogebia (crustacea) burrows: temporal patterns and causes, J. Mar. Res., 41:571-604.

Aller, R.C., et al., 1985. Early chemical diagensis, sediment-water solute exchange and storage of reactive organic matter near the mouth of the Changjiang, East China Sea, Cont. Shelf Res., 4:227-251.

Andersen, J.M., 1971, Nitrogen and phosphorus budgets and the role of sediments in six shallow Danish lakes, Arch. Hydrobiol., 74:528-550.

Andersen, J.M., 1977, Rates of denitrification of undisturbed sediment from six lakes as a function of nitrate concentration, oxygen and temperature, Arch. Hydrobiol., 80:147-159.

Andersen, T.K., Jensen, M.H., and Sørensen, J., 1984, Diurnal variation of nitrogen cycling in coastal marine sediments I. Denitrification, Mar. Biol., 83:171-176.

Bender, M.L., Fanning, K.A., Froelich, P.N., Heath, G.H., and Maynard, V., 1977, Interstitial nitrate profiles and oxidation of sedimentary organic matter in the eastern equatorial Atlantic, Science, 198:605-609.

Billen, G. 1978, A budget of nitrogen recycling in North Sea sediments off the Belgian coast, Est. Coast. Mar. Sci., 7:127-146.

Blackburn, T.H., and Henriksen, K., 1983, Nitrogen cycling in different types of sediments from Danish waters, Limnol. Oceanogr., 28:477-493.

Boynton, W.R., Kemp, W.M., and Keefe, C.W., 1982, A comparative analysis of nutrients and other factors influencing estuarine phytoplankton production, in: "Estuarine Comparisons," V.S. Kennedy, ed., Academic Press, NY.

Cattaneo, A., and Kalff. J., 1980, The relative contribution of aquatic macrophytes and their epiphytes to the production of macrophyte beds, Limnol. Oceanogr., 25:280-289.

Chan, Y.K., and Campbell, N.E.R., 1980, Denitrification in Lake 227 during summer stratification. Can. J. Fish. Aquat. Sci., 37(3):506-512.

Chen, R.L., Keeney, D.R., and McIntosh, T.H., 1983, The role of sediments in the nitrogen budget of lower Green Bay, Lake Michigan, J. Great Lakes Res., 9(1):23-31.

Chatarpaul, L., Robinson, J.B., and Kaushik, N.K., 1980, Effects of tubificid worms on denitrification and nitrification in stream sediment, Can. J. Fish. Aquat. Sci., 37:656-663.

Christensen, J.P., and Rowe, G.T., 1984, Nitrification oxygen consumption in northwest Atlantic deep-sea sediments, J. Mar. Res., 42:1099-1116.

Christensen, J.P., Murray, J.W., Devol, A.H., and Codispoti, L.A., 1987a, Denitrification in continental shelf sediments has major impact on the oceanic nitrogen budget, Global Biogeochem. Cycles, 1:97-116.

Christensen, J.P., Smethie, Jr., W.H., and Devol, A.H., 1987b, Benthic nutrient regeneration and denitrification on the Washington continental shelf, Deep-sea Res., 34:1027-1049.

Christensen, P.B., and Sørensen, J., 1986, Temporal variation of denitrification activity in plant-covered littoral sediment from Lake Hampen, Denmark, Appl. Environ. Microbiol., 51:1174-1179.

Christensen, P.B., and Sørensen, J., 1988, Denitrification in sediment of lowland streams: regional and seasonal variation in Gelbaek and Rabis Baek, Denmark, FEMS Microbiol. Ecol., 53:335-344.

Christensen, P.B., Nielsen, L.P., Revsbech, N.P., and Sørensen, J., 1989, Microzonation of denitrification activity in stream sediments as studied with a combined oxygen and nitrous oxide microsensor. Appl. Environ. Microbiol., 55(5):1234-1241.

Cooper, A.B., and Cooke, J.G., 1984, Nitrate loss and transformation in 2 vegetated headwater streams, N. Z. J. Mar. Freshwater Res., 18:441-450.

Crutzen, P.J., 1970, The influence of nitrogen oxides on the atmospheric ozone content, Q. J. R. Meteorol. Soc., 96:320- 325.

Den Hartog, C., and Polderman, P.J.G., 1975, Changes in seagrasses populations of the Dutch Waddenzee, Aquat. Bot. 1:141-147.

Duff, J.H., Triska. F.J., and Oremland, R.S., 1984, Denitrification associated with stream periphyton: Chamber estimates from undisrupted communities, J. Environ. Qual. 13(4):514-518.

Fendinger, N.J., and Adams, D.D., 1987, Nitrogen gas supersaturation in the recent sediments of Lake Erie and two polluted harbors, Wat. Res., 21:1371-1374.

Garber, J.H., 1982, "^{15}N-tracer and other laboratory studies of nitrogen remineralization in sediments and waters from Narragansett Bay, Rhode Island," Doctoral dissertation, Univ. of Rhode Island.

Gardner, W.S., Nalepa, T.F., and Malczyk, J.M., 1987, Nitrogen mineralization and denitrification in Lake Michigan sediments, Limnol. Oceanogr., 32:1226-1238.

Gerhart, D.Z., and Likens, G.E., 1975, Enrichment experiments for determining nutrient limitation: four methods compared, Limnol. Oceanogr., 20:649-653.

Goloway, F., and Bender, M., 1982, Diagenetic models of interstitial nitrate profiles in deep sea suboxic sediments, Limnol. Oceanogr., 27:624-638.

Goreau, T.H., et al., 1980, Production of NO_2^- and N_2O by nitrifying bacteria at reduced concentrations of oxygen, Appl. Environ. Microbiol., 40:526-532.

Grimm, N.B., and Fisher, S.G., 1986, Nitrogen limitation in a Sonoran Desert stream, J. N. Am. Benthol. Soc., 5(1):2-15.

Grundmanis, V., and Murray, J.W., 1977, Nitrification and denitrification in marine sediments from Puget Sound, Limnol. Oceanogr., 22(5):804-813.

Hahn, J., and Crutzen, P.J., 1982, The role of fixed nitrogen in atmosphere photochemistry, Phil. Trans. R. Soc. Lond. Ser. B., 296:521-541.

Haines, J.R., Atlas, R.M., Griffiths, R.P., and Morita, R.Y., 1981, Denitrification and nitrogen fixation in Alaskan continental shelf sediments, Appl. Environ. Microbiol., 41:412-421.

Hattori, A., 1983, Denitrification and dissimilatory nitrate reduction, in: "Nitrogen in the Marine Environment," E.J. Carpenter and D.G. Capone, eds., Academic Press, New York.

Henriksen, K., Hansen, J.I., and Blackburn, T.H., 1980, The influence of benthic infauna on exchange rates of inorganic nitrogen between sediment and water, Ophelia. Suppl., 1:249-256.

Henriksen, K., Rasmussen, M.B., and Jensen, A., 1983, Effect of bioturbation on microbial nitrogen transformations in the sediment and fluxes of ammonium and nitrate to the overlaying water, Ecol. Bull., Stockholm, 35:193-205.

Henriksen, K. and Kemp, W.M., 1988, Nitrification in estuarine and coastal marine sediments: methods, patterns and regulating factors, in: "Nitrogen Cycling in Coastal Marine Sediments" J. Sørensen, T.H. Blackburn, and T. Rosswall, eds.,

Hill, A.R., 1979, Denitrification in the nitrogen budget of a river ecosystem, Nature, 281:291-292.

Hill, A.R., 1981, Nitrate-nitrogen flux and utilization in a stream ecosystem during low summer flows, Can. Geogr., 25(3):225-239.

Hill, A.R., 1983, Nitrate-nitrogen mass balances for two Ontario rivers, in: "Dynamics of Lotic Ecosystems," T.D. Fontaine, and S.M. Bartell, eds., Ann Arbor Science, Michigan.

Iizumi, H., Hattori, A., and McRoy, C.P., 1980, Nitrate and nitrite in interstitial waters of eelgrass beds in relation to the rhizosphere, J. Exp. Mar. Biol. Ecol., 47:191-201.

Jenkins, M.C., and Kemp, W.M., 1984, The coupling of nitrification and denitrification in two estuarine sediments, Limnol. Oceanogr., 29:609-619.

Jensen, H.B., Jørgensen, K.S., and Sørensen, J., 1984, Diurnal variation of nitrogen cycling in coastal, marine sediments II. Nitrous oxide emission, Mar. Biol., 83:177-183.

Jensen, M.H., Andersen, T.K., Sørensen, J., 1988, Denitrification in coastal bay sediment: regional and seasonal variation in Aarhus Bight, Denmark, Mar. Ecol. Prog. Ser., 48:155-162.

Jørgensen, K.S., Jensen, H.B., and Sørensen, J., 1984, Nitrous oxide production from nitrification and denitrification in marine sediment at low oxygen concentrations, Can. J. Microbiol., 30:1073-1078.

Jørgensen, B.B., and Revsbech, N.P., 1985, Diffusive boundary layers and the oxygen uptake of sediments and detritus, Limnol. Oceanogr., 30(1):111-122.

Jørgensen, B.B., and Sørensen, J., 1985, Seasonal cycles of O_2, NO_3^-, and SO_4^{2-} reduction in estuarine sediments: the significance of a NO_3^- reduction maximum in spring, Mar. Ecol. Prog. Ser., 24:65-74.

Jørgensen, K.S., and Sørensen, J., 1988, Two annual maxima of nitrate reduction and denitrification in estuarine sediment, (Norsminde Fjord, Denmark), Mar. Ecol. Prog. Ser., 48:147-154.

Kaspar, H.F., 1982, Denitrification in marine sediment: measurement of capacity and estimate of in situ rate, Appl. Environ. Microbiol., 43(3):522-527.

Kaspar, H.F., 1983, Denitrification, nitrate reduction to ammonium and inorganic nitrogen pools in intertidal sediments, Mar. Biol., 74:133-139.

Kaspar, H.F., Asher, R.A., and Boyer, I.C., 1985a, Microbial nitrogen transformations in sediments and inorganic nitrogen fluxes across the sediment/water interface on the South Island West Coast, New Zealand, Estuar. Coast. Shelf Sci., 21:245-255.

Kaspar, H.F., Gillespie, P.A., Boyer. I.C., and MacKenzie, A.L., 1985b, Effects of mussel aquaculture on the nitrogen cycle and benthic communities in Kenepuru Sound, Marlborough Sounds, New Zealand, Mar. Biol., 85:127-136.

Kaushik, N.K., and Robinson, J.B., 1976, Preliminary observations on nitrogen transport during summer in a small spring-fed Ontario stream, Hydrobiol., 49(1):59-63.

Keeney, D.R., 1973, The nitrogen cycle in sediment-water systems, J. Environ. Qual., 2:15-29.

Keeney, D.R., Chen, R.L., and Graetz, D.A., 1971, Importance of denitrification and nitrate reduction in sediments to the nitrogen budgets of lakes, Nature, 233:66:67.

Kemp, W.M., and Murray, L., 1986, Oxygen release from roots of the submersed macrophyte Potamogeton perfoliatus L: regulating factors and ecological implications, Aquat. Bot., 271-283.

Knowles, R., 1982, Denitrification, Microbiol. Rev., 46(1):43-70.

Koike, I., Hattori, A., and Goering, J.J., 1978, Controlled ecosystem pollution experiment: effect of mercury on enclosed water columns. VI. Denitrification by marine bacteria, Mar. Sci. Comm., 4:1-12.

Koike, I., and Hattori, A., 1979, Estimates of denitrification in sediments of the Bering Sea shelf, Deep-sea Res., 26A:409-415.

Koike, I., and Sørensen, J., 1988, Nitrate reduction and denitrification in marine sediments, in: "Nitrogen Cycling in Coastal Marine Environments. Scope 33," T.H. Blackburn and J. Sørensen, eds., John Wiley & Sons.

Kuenen, J.G., and Robertson, L.A., 1987, Ecology of nitrification and denitrification, in: "The Nitrogen and Sulphur Cycles. Society for General Microbiology Symposium 42. Society for General Microbiology Limited", J.A. Cole and S. Ferguson, eds., Cambridge University Press.

Liao, C.F.-H., and Lean, D.R.S, 1978, Seasonal changes in nitrogen compartments of lakes under different loading conditions, J. Fish Res. Bd. Can., 35(8):1095-1101.

Likens, G.E., and Loucks, O.L., 1978, Analysis of five North American lake ecosystems III. Sources, loading and fate of nitrogen and phosphorus, Verh. Internat. Verein. Limnol., 20:568-573.

Madsen, P.P., 1979, Seasonal variation of denitrification rate in sediment determined by use of ^{15}N, Water Res., 13:461-465.

Matson, P.A., and Vitousek, P.M., 1987, Cross-system comparisons of soil nitrogen transformations and nitrous oxide flux in tropical forest ecosystems, Global Biogeochem. Cycles, 1(2):163-170.

Messer, J.J., and Brezonik, P.L., 1983, Comparison of denitrification rate estimation techniques in a large, shallow lake, Water Res., 17:631-640.

Messer, J.J., and Brezonik, P.L., 1984, Laboratory evaluation of kinetic parameters for lake sediment denitrification models, Ecol. Model., 21:277-286.

Meybeck, M., 1982, Carbon, nitrogen, and phosphorus transport by world rivers, Am. J. Sci., 282:401-450.

Nishio, T., Koike, I., and Hattori, A., 1982, Denitrification, nitrate reduction and oxygen consumption in coastal and estuarine sediments, Appl. Environ. Microbiol., 43:648- 653.

Nishio, T., Koike, I., and Hatorri, A., 1983, Estimates of denitrification and nitrification in coastal and estuarine sediments, Appl. Environ. Microbiol., 45:444-450.

Nixon, S.W., 1981, Remineralization and nutrient cycling in coastal marine ecosystems, in: "Estuaries and Nutrients," B.J. Neilson and L.E. Cronin, eds., Humana Press, Clifton, NJ.

Nixon, S.W., Pilson, M.E.Q., Oviatt, C.A., Donaghay, P., Sullivan, B., Seitzinger, S., Rudnick, D., and Frithsen, J., 1984, Eutrophication of a coastal marine ecosystem - an experimental study using the MERL microcosms, in: "Flows of Energy and Materials in Marine Ecosystems", M.J.R. Fasham, ed., Plenum Publishing Corp.

Oremland, R.S., Umberger, C., Culbertson, C.W., and Smith, R.L., 1984, Denitrification in San Francisco Bay intertidal sediments, Appl. Environ. Microbiol., 47(5):1106-1112.

Oren, A., and Blackburn, T.H., 1979, Estimation of sediment denitrification rates at in situ nitrate concentrations, Appl. Environ. Microbiol., 37:174-176.

Orth, R.J., and Moore, K.A., 1983, Submerged aquatic vegetation of the Chesapeake Bay: Past, present and future, Science, 222:51-53.

Owens, M., Garland, J.H.N., Hart, I.C., and Wood, G., 1972, Nutrient budgets in rivers, Symp. Zool. Soc. London, 29:21-40.

Reimers, C.E., Fisher, K.M., Merewether, R., Smith Jr., K.L., and Jahnke, R.A., 1986, Oxygen microprofiles measured in situ in deep ocean sediments, Nature, 320:741-744.

Revsbech, N.P., Sørensen, J., Blackburn, T.H., and Lomholt, J.P., 1980, Distribution of oxygen in marine sediments measured with microelectrodes, Limnol. Oceanogr., 25:403-411.

Revsbech, N.P., Nielson, L.P., Christensen, P.B., and Sørensen, J., 1988, Combined oxygen and nitrous oxide microsensor for denitrification studies, Appl. Environ. Microbiol., 54(9):2245-2249.

Robinson, J.B., Whiteley, H.R., Stammers, W., Kaushik, N.K., and Sain, P., 1979, The fate of nitrate in small streams and its management implications, in: "1979 Proc. 10th Annu. Ag. Waste Man. Conf.", Cornell Univ., Rochester, New York.

Rosenfeld, J.K., 1979, Ammonium adsorption in nearshore anoxic sediments, Limnol. Oceanogr., 24(2):356-364.

Ryther, J.H., and Dunstan, W.M., 1971, Nitrogen, phosphorus and eutrophication in the coastal marine environment, Science, 171:1008-1013.

Sand-Jensen, K., and Søndergaard, M., 1981, Phytoplankton epiphyte development and their shading effect on submerged macrophytes in lakes of different nutrient status, Int. Revue ges. Hydrobiol., 66:529-552.

Sand-Jensen, K., Prahl, C., and Stokholm, H., 1982, Oxygen release from roots of submerged aquatic macrophytes, Oikos, 38:349-354.

Seitzinger, S.P., Nixon, S.W., and Pilson, M.E.Q., 1984, Denitrification and nitrous oxide production in a coastal marine ecosystem, Limnol. Oceanogr., 29:73-83.

Seitzinger, S.P., and Nixon, S.W., 1985, Eutrophication and the rate of denitrification and N_2O production in coastal marine sediments, Limnol. Oceanogr., 30:1332-1339.

Seitzinger, S.P., 1987a, "The effect of pH on the release of phosphorus from Potomac River sediments," CBP/TRS 15. U.S. Environmental Protection Agency, Chesapeake Bay Program, Annapolis, MD.

Seitzinger, S.P., 1987b, Benthic nutrient cycling and oxygen consumption in the Delaware Estuary, in: "Ecology and restoration of the Delaware River basin,' S.K. Majumdar, E.W. Miller, and L.E. Sage, eds., Pennsylvania Academy of Science.

Seitzinger, S.P., 1987c, Nitrogen biogeochemistry in an unpolluted estuary: the importance of benthic denitrification, Mar. Ecol. Progr. Ser., 37:65-73.

Seitzinger, S.P., 1988, Denitrification in freshwater and coastal marine ecosystems: ecological and geochemical significance, Limnol. Oceanogr., 33(4, part 2):702-724.

Shaffer, G., and Ronner, U., 1984, Denitrification in the Baltic proper deep water, Deep-Sea Res., 31:197-220.

Slater, J.M. and D.G. Capone. 1987. Denitrification in aquifer soil and nearshore marine sediments influenced by ground-water nitrate. Appl. Environ. Microbiol. 53:1292-1297.

Smethie, Jr., W.M., 1987, Nutrient regeneration and denitrification in low oxygen fjords, Deep-Sea Res., 34:983-1006.

Smith, C.J., R.D. DeLaune and W.H. Patrick, Jr. 1985. Fate of riverine nitrate entering an estuary: I. Denitrification and nitrogen burial. Estuaries 8:15-21.

Smith, M.S. and K. Zimmerman. 1981. Nitrous oxide production by nondenitrifying soil nitrate reducers. Soil Sci. Soc. Am. J., 45:865-871.

Smith, S.V., Hollibaugh, J.T., Dollar, S.J., and Vink, S., 1989, Tomales Bay, California: a case for carbon-controlled nitrogen cycling, Limnol. Oceanogr. 34:37-52.

Sørensen, J., 1978, Denitrification rates in a marine sediment as measured by the acetylene inhibition technique, Appl. Environ. Microbiol., 36:139-143.

Sørensen, J., Jørgensen. B.B., and Revsbech, N.P., 1979, A comparison of oxygen, nitrate and sulfate respiration in coastal marine sediments, Mar. Ecol., 5:105-115.

Sørensen, J., Tiedje, V.M., and Firestone, R.B., 1980, Inhibition by sulfide of nitric and nitrous oxide reduction by denitrifying Pseudomonas fluorescens, Appl. Environ. Microbiol., 39:105-108.

Sørensen, J., Hydes, D.J., and Wilson, T.R.S., 1984, Denitrifi-cation in a deep-sea sediment core from the eastern equatorial Atlantic, Limnol. Oceanogr., 29(3):653-657.

Sweerts. J-P.R.A., and de Beer, D., 1989, Microelectrode measurements of nitrate gradients in the littoral and profundal sediments of a meso-eutrophic lake (Lake Vechten, The Netherlands), Appl. Environ. Microbiol., 55(3):754-757.

Taylor, B.F., 1983, Assays of microbial transformation, in: "Nitrogen in the Marine Environment," E.J. Carpenter and D.G. Capone, eds., Academic Press, New York.

Twilley, R.R., Kemp, W.M., Staver, K.W., Stevenson, J.C., and Boynton, W.R., 1985, Nutrient enrichment of estuarine submersed vascular plant communities. 1. Algal growth and effects on production of plants and associated communities, Mar. Ecol. Prog. Ser., 23:179-191.

Vanderborght, J. and Billen, G., 1975, Vertical distribution of nitrate concentration in interstitial water of marine sediments with nitrification and denitrification, Limnol. Oceanogr., 20:953-961.

Vanderborght, J., Wollast, R., and Billen, G., 1977, Kinetic models of diagenesis in disturbed sediments. Part 2. Nitrogen diagenesis, Limnol. Oceanogr., 22:794-803.

Vincent, W.F., Vincent, C.L., Downes, M.T., and Richerson, P.J., 1985, Nitrate cycling in Lake Titicaca (Peru-Bolivia): the effects of high-altitude and tropicality, Freshwater Biol., 15:31-42.

Wang, W.C., Yung, Y.L., Lacis, A.A., Mo, J., and Hansen, J.E., 1976, Greenhouse effects due to man-made perturbations of trace gases, Science, 194:685-690.

Wilson, T.R.S., 1978, Evidence for denitrification in aerobic pelagic sediments, Nature, 274:354-356.

DENITRIFICATION MODEL FOR MARINE SEDIMENT

T. Henry Blackburn

Department of Ecology and Genetics
University of Aarhus
DK 8000 Aarhus C
Denmark

INTRODUCTION

The purpose of this presentation is to describe the application of a simulation model in the investigation of some aspects of denitrification in marine sediments and to determine the factors conducive to high denitrification rates, preferably resulting in the denitrification of all NH_4^+, that was produced from organic nitrogen mineralization in the sediment. No evidence will be offered to show that the model can accurately describe the processes which occur in a sediment in situ. This proof will be addressed in future publications.

Reference should be made to recent review articles for a more complete survey of denitrification in marine sediments (Aller, 1988; Billen and Lancelot, 1988; Koike and Sørensen, 1988, Lancelot and Billen, 1985; Seitzinger, 1988; chapter by Seitzinger in this book).

MODEL

The model was constructed in STELLA (High Performance Systems, New Hampshire, USA), a software package for Apple Macintosh computers. It is a numerical model, whose main components are seen in Figure 1. The model is based on reactions and interactions in the nitrogen, carbon, oxygen and sulfur cycles in the upper layers of a marine sediment. Figure 1 shows only the uppermost sediment stratum, connected to the water and the underlying sediment by diffusional or mixing processes. There are four sediment layers, all constructed in a similar manner; the lowest layer does not have diffusional connections with

Denitrification in Soil and Sediment, Edited by N. P. Revsbech and J. Sørensen, Plenum Press, New York, 1990

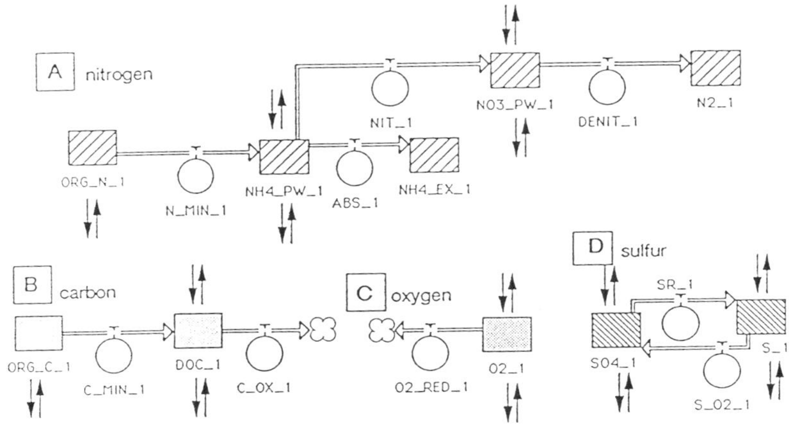

Fig. 1. An outline of the model showing the main components in the upper sediment layer. A: nitrogen cycle. B: carbon cycle. C: oxygen cycle. D: sulfur cycle. The double arrows represent diffusional or mixing fluxes.

the sediment below it, except for the flux of NH_4^+ being made equal to the flux between layers 3 and 4. Pools (concentrations) of the various components are depicted as boxes and the regulators of flows between the pools as circles. The pools in the nitrogen cycle (Figure 1A) are organic nitrogen (ORG_N_1), porewater NH_4^+ (NH4_PW_1), exchangeable NH_4^+ (NH4_EX_1), porewater NO_3^- (NO3_PW_1) and N_2 (N2_1). The flow regulators are the rate of organic nitrogen mineralization (N_MIN_1), the factor determining the ratio between free and exchangeable NH_4^+ (ABS_1), the rate of nitrification (NIT_1) and the rate of denitrification (DENIT_1). The terms used in the carbon (Fig. B), oxygen (Fig. 1C) and sulfur cycles (Fig. 1D) are self explanatory. All the initial concentrations and equations relating to the construction of the model are presented in the Appendix. Each simulation was run for 4 to 8 days, at which time steady state had been reached, except for the NH_4^+ pools, which were still increasing, particularly in the lower strata. Porosity was equal to unity. Oxygen concentration in the overlying water was set at 250 μM.

There was no input of organic matter. The organic nitrogen pool was made large and the rate constant for its mineralization was small. This resulted in an almost constant supply of NH_4^+ and dissolved organic carbon (DOC). The depth intervals were 2, 2, 4 and 32 mm for the sediment layers #1 to #4, respectively. The rates of nitrification, denitrification, DOC oxidation, sulfate reduction and sulfide oxidation were defined by first order rate constants (k), rather than by V_{max} and K_m terms. These were regulated by the concentration of oxygen, as described by the equations in the Appendix. DOC was not allowed to diffuse; all diffusion coefficients were set at 1.4×10^{-5} $cm^2\ s^{-1}$ (1.2 cm^2 day^{-1}). This would be the diffusion coefficient for O_2 in sediment with a porosity of 0.7. The diffusion was increased to 1.8 $cm^2\ day^{-1}$ in the water and top 2 mm of sediment. The sampling interval (dt) was 0.001 day. The simulations were run using Runge-Kutta 2.

The factors that were altered in the simulations were the depth of the overlying water, the concentrations of NH_4^+ and NO_3^- in this water, the C:N ratio of the organic matter, diffusion coefficients, and the k-values for organic nitrogen mineralization, nitrification and denitrification. The simulations are presented in 3 series: in the first, changes in the k-values for nitrification and denitrification were examined, followed by an investigation of alteration in nitrogen mineralization rates and finally by an examination of the effect of NH_4^+ and NO_3^- in the overlying water.

RATE CONSTANTS FOR NITRIFICATION AND DENITRIFICATION

The k-values for nitrification and denitrification were altered, while the following parameters were kept constant: NH_4^+ concentration in the overlying water = 0 µM, NO_3^- concentration in the overlying water = 0 µM, water depth = 20 m, C:N molar ratio in organic matter = 4, and nitrogen mineralization = 6.51 mmol m^{-2} day^{-1}. The results of these simulations are seen in Fig.2. When the k-value for denitrification was kept constant at 100 day^{-1}, increasing the k-value for nitrification resulted in decreasing NH_4^+ fluxes and in increasing NO_3^- fluxes and increasing rates of denitrification (Fig. 2A). These rates tended to level off at nitrification k-values greater than 200, but presumably with very high k-values all NH_4^+ could be oxidized to NO_3^-. Under these present conditions the maximum rate of denitrification (ca. 3 mmol m^{-2} day^{-1}) was slightly higher than the NO_3^- flux and was ca. 50% of the rate of the nitrogen mineralization rate. The denitrification rate was elevated further by increasing the k-value for denitrification, while maintaining k-nitrification at 250 (Fig. 2B). There was a corresponding decrease in the NO_3^- flux. The increase in denitrification rate achieved by this increase in k-denitrification from 100 to 250 was relatively modest (<10%).

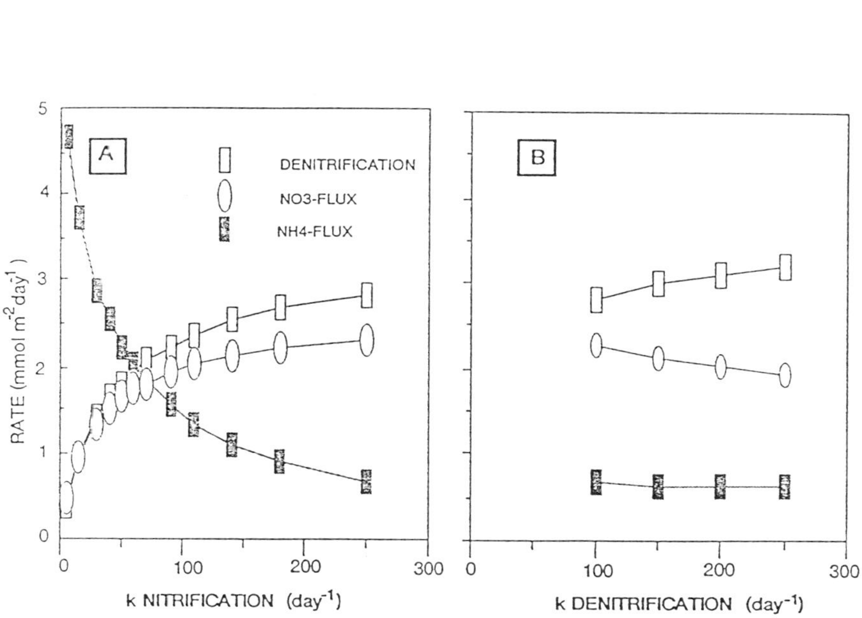

Fig. 2. The effect of changes in the rate constants for nitrification at k-denitrification = 100 (A) and denitrification at k-nitrification = 250 (B) on the rates of denitrification and fluxes of ammonium and nitrate.

The rate of organic carbon mineralization (ca. 25 mmol m^{-2} day^{-1}) in these simulations would correspond to a moderately dynamic sediment which might be expected to have values of 2 and 1 mmol m^{-2} day^{-1} for nitrification and denitrification rates respectively (Billen and Lancelot, 1988; Koike and Sørensen, 1988; Seitzinger, 1988). A nitrification rate of this magnitude would have been achieved in this simulation by a k-nitrification of <10. Values >100 were, therefore, selected in further simulations in order to obtain maximum nitrification rates. Similarly, almost maximum rates of denitrification (ca. 3 mmol m^{-2} day^{-1}) were achieved at a k-denitrification of 100 . It should be noted that denitrification did not greatly exceed 50% of the rate of nitrogen mineralization in these simulations.

The sediment depth profiles of NH_4^+, O_2, NO_3^-, nitrification and denitrification for the simulations, when k-nitrification was 5 or 250, are shown in Fig. 3. The high k-nitrification resulted in the consumption of much NH_4^+ and some O_2 (Fig. 3 A and B). The high k-nitrification resulted in higher NO_3^- concentrations, penetrating to a slightly greater depth in the sediment (Fig. 3B). This increased NO_3^- concentration resulted from a higher rate of nitrification (Fig. 3C), which did not penetrate much below 0.5 mm, limited by O_2 availability. The increased NO_3^- concentration and penetration produced a much higher rate of denitrification, partially in the anoxic sediment strata, but also in the partially oxidized zone; the highest rate was in the 2-4 mm stratum (Fig. 3D). Denitrification was allowed to proceed at half maximal rate in the presence of 20 μM O_2 (see appendix). The profiles provided no clues as to how the major portion of mineralized nitrogen might be denitrified. Clearly factors other than alteration of the rate constants for nitrification and denitrification must be important in determining the proportion of NH_4^+ that is denitrified.

RATE CONSTANTS FOR MINERALIZATION

The effect of alteration in the rate of organic nitrogen mineralization on the fluxes of NH_4^+ and NO_3^- and on the rates of nitrification and denitrification were examined. In these simulations, the parameters were those used previously, except that k-nitrification was set at 100 and k-denitrification at 200, while k-N-mineralization was altered, giving NH_4^+ production rates of 0.7 to 19.5 mmol m^{-2} day^{-1}. There was a linear increase in the products of N-mineralization (fluxes of NH_4^+ and NO_3^-, and denitrification) up to 15 mmol N m^{-2} day^{-1}. NO_3^- flux reached a maximum at mineralization rates <10 mmol m^{-2} day^{-1}; denitrification rates peaked at ca. 10 mmol m^{-2} day^{-1}(Fig. 4A). This is seen more clearly in Fig. 4B. The rate of nitrification coincided with the peak in NO_3^- flux. The displacement between the peaks of nitrification and denitrification is difficult to explain, but diffusional or other transport loss of NO_3^- from the sediment is the factor that determines the extent to which NH_4^+, via NO_3^-, is finally converted to N_2. In these simulations, 100% NH_4^+ could be oxidized to NO_3^- at low mineralization rates, but denitrification could not exceed 40%. At high mineralization rates, nitrification and denitrification accounted for only a very small proportion of the mineralized nitrogen (ca. 5%). The model predicts that denitrification rates reach a maximum of 1.5 mmol m^{-2} day^{-1} and that this rate decreases with increased organic loading (>40 mmol m^{-2} day^{-1}). Increases in the amount

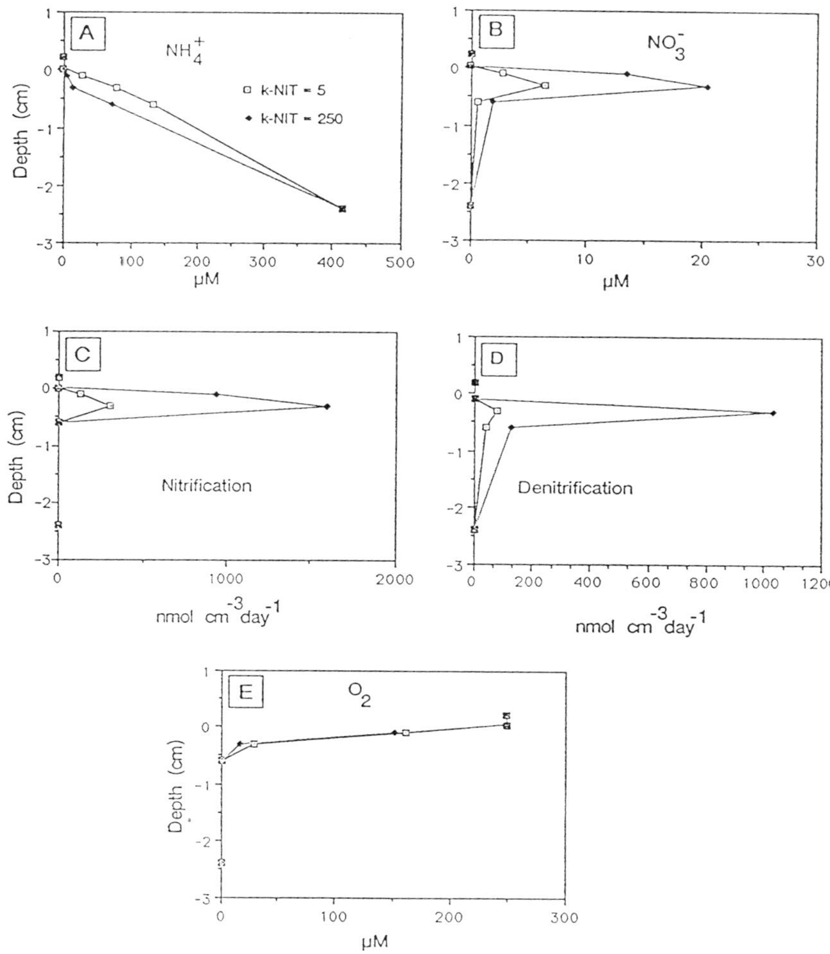

Fig. 3 Depth profiles of porewater concentrations of ammonium (A) and of nitrate (B), nitrification rate (C), denitrification rate (D) and porewater concentrations of oxygen (E). Results are shown for simulations run at k-denitrification = 100, and for k-nitrification = 5 or 250.

of organic material added to the sediment would be expected to result in decreases in the rate of nitrification and of denitrification.

These simulations were performed with the depth of the overlying water set at 20 m and this water was completely mixed. Over the 8 day simulation, there was no appreciable buildup in the concentration of NH_4^+ or NO_3^- in the water. Many of the reports (see Seitzinger 1988) in which high denitrification/ mineralization rates have been quoted, were performed in closed systems, having only a limited quantity of overlying water, which might have allowed modest increases in NH_4^+ or in NO_3^- concentrations. These higher concentrations in the overlying water could then have relaxed the gradients of these species within the sediment, thereby allowing higher concentrations, particularly of NO_3^-,

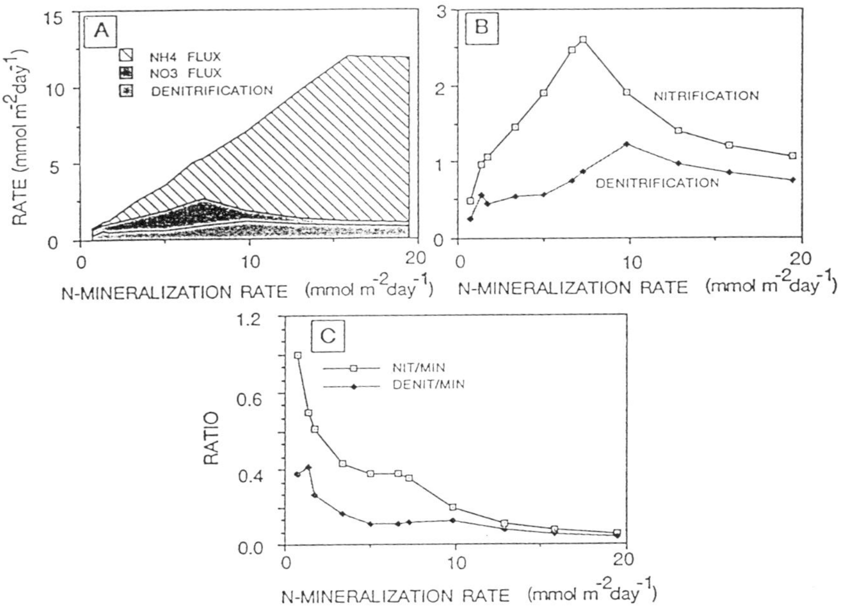

Fig. 4 The effect of increasing N-mineralization rate: (A) on denitrification rate, flux of ammonium and flux of nitrate: (B) on nitrification and denitrification rates: (C) on the proportion of the mineralized ammonium that was nitrified or finally denitrified.

to be operated on by the denitrifying bacteria. This effect was investigated in the next series of simulations.

EFFECT ON SEDIMENT OF NH_4^+ AND NO_3^- IN THE OVERLYING WATER

The parameters which were changed in these simulations are given in Table 1, as are the most significant results. All the simulations were run for 4 days and in all but one run, the k-denitrification was 200. The results are arranged in 3 groups: Group 1 runs #1-#7, Group 2 runs #8-#15, Group 3 runs #16-#25.

In Group 1 , the effect of variations in the first four parameters on the rate of denitrification was investigated. The overlying water height was only 5 cm. The concentrations of NH_4^+ and NO_3^- in this water changed during the 4 day simulation. In #1 and #2 a high mineralization rate was used at k-nitrification values of 100 and 300. The %denitrification (expressed as a proportion of the N-mineralization rate) was 21 and 39%, respectively. There was a build-up of NH_4^+ in the overlying water. The C:N ratio was set at 6, to allow for a high rate of NH_4^+ production,without excessive O_2 consumption. Reduction of this N-mineralization rate, at the same C:N ratio (#3 and #4), resulted in high rates of nitrification and of denitrification; denitrification was 81 and 88% of the mineralization rate, for k-nitrification 100 and 300, respectively. In the overlying water, NH_4^+ was 4 μM and NO_3^- was 42 μM at the end of the simulation. Increasing the rate of carbon oxidation (#5 and #6, C:N=12 thus doubling the carbon mineralization while keeping nitrogen mineralization constant) reduced the % denitrification to 75 and 81%, respectively for the k-nitrification of 100 and 300. NH_4^+ concentrations were higher (15 and 5 μM) and NO_3^- concentrations were lower (31 and 34 μM), reflecting the more reduced conditions in the sediment. Finally, in #7 the k-denitrification was increased to 400, other conditions being maintained as in #6. The result was a very high rate of denitrification, 85%, with NH_4^+ increasing to 5 μM and NO_3^- to 27 μM in the overlying water. These simulations indicated that the accumulation of NH_4^+ and NO_3^- in the overlying water was associated with very high denitrification rates, provided that the rate of N-mineralization (and C- mineralization) was not excessively high and that high potential rates of nitrification and denitrification were present.

In Group 2, the effect of changes in diffusion coefficient were examined, keeping C:N ratios and k-values for mineralization, nitrification and denitrification constant. The water height was 5 cm for #8-#11, and 20 m for #12-#15. Reducing the diffusion coefficient by half (#8) resulted in an accumulation of NH_4^+ in the water, associated with a zero nitrification rate and a low denitrification rate, which was dependent on NO_3^- released to the water in the early part of the simulation. At this low diffusion rate, steady state conditions were not attained. Increasing the diffusion coefficient x2 and x3 resulted in increased rates of nitrification, denitrification and %nitrogen denitrified (81, 91 and 106% for diffusion coefficients of 1.2, 2.4 and 3.6 cm^2 day^{-1}, respectively). These high rates were associated with an increase of NO_3^- in the water to ca. 30 μM.

A similar pattern was seen for the parallel simulations, in which the water depth was 20 m. There was, however, only a modest increase in NH_4^+ and NO_3^- in the water and nitrification and denitrification rates were not so high. Increasing diffusion coefficients (1.2, 2.4, and 3.6 cm^2 day^{-1}) resulted in lower % denitrification (19, 29 and 43%, respectively) in #13-#15, compared to #9-#11.. This increased nitrification and denitrification, due to increasing the diffusion coefficient, was not associated with the small increase in NH_4^+ and NO_3^- in the water. This was indicated by the fact that lowering of the concentration build-up in the water, by increasing the water depth to 1000 m, gave exactly the same response as for #15.It was clear, that in order to achieve a very high percentage of denitrification, the concentrations of NO_3^- had to be high in the water.

Increased diffusion led to some increase in nitrification and denitrification, irrespective of the increase of NO_3^- or NH_4^+ in the overlying water. However, in order to attain a high percentage for nitrification and denitrification it was necessary that there should be a high concentration of inorganic nitrogen in the overlying water.

In Group 3 simulations, the effect of increasing NO_3^- and NH_4^+ in the water was investigated (#16-#25). The initial concentrations were adjusted to the values indicated, all other factors were kept constant. The results of the simulations were grouped in ascending order of % denitrification achieved. The results were surprising as it seemed that increases in NH_4^- resulted in the same increase in % denitrification as did equivalent increases in NO_3^- concentration (25 and 26% for 5 μM; 34, 34 and 35% for 10 μM NH_4^+, 5 μM

Table 1. EFFECT OF CHANGING PARAMETERS ON DENITRIFICATION

#	C:N (mol)	N-min. rate	nit. (k)	denit. (k)	depth (m)	NH_4^+ µM	NO_3^- µM	Diffus. cm^2d^{-1}	nit. rate	denit. rate	denit./ N-min. (%)
1	6	9.80	100	200	0.05	124	5	1.2	2.06	2.10	21
2	6	9.80	300	200	0.05	77	9	1.2	3.75	3.80	39
3	6	3.38	100	200	0.05	12	44	1.2	2.82	2.73	81
4	6	3.38	300	200	0.05	4	42	1.2	3.00	2.96	88
5	12	3.38	100	200	0.05	15	31	1.2	2.50	2.55	75
6	12	3.38	300	200	0.05	5	34	1.2	2.60	2.75	81
7	12	3.38	300	400	0.05	5	27	1.2	2.68	2.87	85
8	12	3.38	300	200	0.05	33	2	0.6	0.00	0.95	28
9	12	3.38	300	200	0.05	5	34	1.2	2.60	2.75	81
10	12	3.38	300	200	0.05	4	25	2.4	3.18	3.09	91
11	12	3.38	300	200	0.05	2	31	3.6	3.91	3.58	106
12	12	3.38	300	200	20.00	0	0	0.6	0.88	0.72	21
13	12	3.38	300	200	20.00	0	0	1.2	1.96	0.65	19
14	12	3.38	300	200	20.00	0	0	2.4	2.29	0.99	29
15	12	3.38	300	200	20.00	0	0	3.6	3.10	1.47	43
16	12	3.38	300	200	20.00	0	0	1.2	1.96	0.65	19
17	12	3.38	300	200	20.00	5	0	1.2	2.66	0.86	25
18	12	3.38	300	200	20.00	0	5	1.2	1.96	0.89	26
19	12	3.38	300	200	20.00	10	0	1.2	3.29	1.14	34
20	12	3.38	300	200	20.00	5	5	1.2	2.66	1.16	34
21	12	3.38	300	200	20.00	0	10	1.2	2.03	1.20	35
22	12	3.38	300	200	20.00	15	0	1.2	3.92	1.36	40
23	12	3.38	300	200	20.00	0	15	1.2	2.03	1.44	43
24	12	3.38	300	200	20.00	10	10	1.2	3.29	1.66	49
25	12	3.38	300	200	20.00	15	15	1.2	3.92	2.16	64

All rates are in mmol m^{-2} day^{-1} . k-Values are day^{-1}

+ 5 µM NO_3^- and 10 µM NO_3^-). The % denitrification was increased from 19% for the zero concentration (#16) to 64% for 15 µM NH_4^+ + 15 µM NO_3^- (#25). When the initial concentrations were set at 5 µM NH_4^+ and 34 µM NO_3^-, the final concentrations in simulation #6, 71% denitrification was achieved, close to the 81% in #6. It was observed in these simulations that there was little flux of NO_3^- either into or out of the sediment (efflux 0.31 mmol m^{-2} day^{-1}) and no flux of NH_4^+. It is concluded that the increased nitrification and denitrification was due to changes in the gradients of NO_3^- and of NH_4^+ within the sediment. This is consistent with the fact that NH_4^+ in the water was as effective as NO_3^- in stimulating denitrification, provided that nitrification rates were high. In this context, when the diffusion coefficient in #17 was raised to 3.6 cm^2 day^{-1}, the amount of denitrification was raised to 54%, compared to 25% (#17) or 43% (#15) for the same diffusion coefficient, but with zero NH_4^+ in the water.

The aim of this presentation has been to look for conditions that would result in the highest possible convertion of NH_4^+ to N_2 in sediment. The k-denitrification values were high (up to 400 day^{-1}), but how do they compare with published values? Values of 0.024 and 1.2 day^{-1} were used in models of nitrogen cycling in sediments to give realistic simulations of denitrification (Aller, 1988; Billen and Lancelot, 1988). It would, therefore, seem to be certain that 200 day^{-1} was likely to have been over the maximum. This is also confirmed by the relationship between organic input and k-denitrification established by Lancelot and Billen (1985), in which the maximum k-denitrification was ca. 100 day^{-1}. In order to stretch denitrification potential to its limits, a simulation was run, a modification of #13, Table 1, in which k-nitrification and k-denitrification were raised to 3000 day^{-1}. This simulation resulted in an increase in denitrification rate from 0.65 to 1.11 mmol m^{-2} day^{-1} (from 19 to 43% NH_4^+ was denitrified). It is concluded that increases in denitrification potential alone cannot increase the proportion of NH_4^+ denitrified to 50%, under these conditions. Denitrification percentage cannot be increased significantly except by having NH_4^+ or NO_3^- in the overlying water.

SUMMARY

There are two main conclusions to be drawn from this investigation, both relating to coupled nitrification and denitrification in sediments. The first is that increasing organic mineralization rates do not result in parallel increases in denitrification rates. Beyond a certain point, due to oxygen consumption by the increased carbon availability, denitrification rates are reduced. This is not in total agreement with the predictions and observation made by Lancelot and Billen (1988), who found an increasing k-denitrification with increased organic loading. This would certainly be true for the situation where denitrification was dependent on a supply of NO_3^- from the overlying water. Secondly, <50% sediment NH_4^+ can be denitrified if NH_4^+ and NO_3^- are absent from the overlying water. The presence of quite low concentrations of NH_4^+ or NO_3^- in the overlying water can significantly increase denitrification and this effect is increased by factors that increase diffusion or mixing. The 100% denitrification of net NH_4^+ production in sediments, quoted by Seitzinger (1988), are possible with 30 μM NO_3^- in the overlying water and with x2 to x3-fold increases in diffusion coefficients.

REFERENCES

Aller, R. C., 1988, Benthic fauna and biogeochemical processes in marine sediments: the role of burrow structures, in: "Nitrogen cycling in coastal marine environments," T. H. Blackburn, and J. Sørensen, eds., John Wiley and Sons Ltd., Chichester.

Billen, G., and Lancelot, C., 1988, Modelling benthic nitrogen cycling in temperate coastal ecosystems, in: "Nitrogen cycling in coastal marine environments," T. H. Blackburn, and J. Sørensen, eds., John Wiley and Sons Ltd., Chichester.

Koike, I., and Sørensen, J., 1988, Nitrate reduction and denitrification in marine sediments, in: "Nitrogen cycling in coastal marine environments," T. H. Blackburn, and J. Sørensen, eds., John Wiley and Sons Ltd., Chichester.

Lancelot, C., and Billen, G., 1985, Carbon-Nitrogen relationships in nutrient metabolism of coastal marine ecosystems. Adv. Aquat. Microbiol., 3: 263.

Seitzinger, S. P., 1988, Denitrification in freshwater and coastal marine ecosystems: Ecological and geological significance. Limnol. Oceanogr. 33: 702.

APPENDIX

Concentrations in the sediment are expressed in μM
Integrated concentrations are in nmol cm^{-2}
Rates are in μM day^{-1}
Integrated rates are in nmol cm^{-2} day^{-1}
Rate constants are day^{-1}

```
DOC_1 = DOC_1 + dt * ( C_MIN_1 - C_OX_1 )
INIT(DOC_1) = 0
DOC_2 = DOC_2 + dt * ( C_MIN_2 - C_OX_2 )
INIT(DOC_2) = 0
DOC_3 = DOC_3 + dt * ( C_MIN_3 - C_OX_3 )
INIT(DOC_3) = 20
DOC_4 = DOC_4 + dt * ( C_MIN_4 - C_OX_4 )
INIT(DOC_4) = 20
N2_1 = N2_1 + dt * ( DENIT_1 )
INIT(N2_1) = 0
N2_2 = N2_2 + dt * ( DENIT_2 )
INIT(N2_2) = 0
N2_3 = N2_3 + dt * ( DENIT_3 )
INIT(N2_3) = 0
N2_4 = N2_4 + dt * ( DENIT_4 )
INIT(N2_4) = 0
NH4_EX_1 = NH4_EX_1 + dt * ( ABS_1 )
INIT(NH4_EX_1) = 0
NH4_EX_2 = NH4_EX_2 + dt * ( ABS_2 )
INIT(NH4_EX_2) = 0
NH4_EX_3 = NH4_EX_3 + dt * ( ABS_3 )
INIT(NH4_EX_3) = 20
NH4_EX_4 = NH4_EX_4 + dt * ( ABS_4 )
INIT(NH4_EX_4) = 40
NH4_PW_1 = NH4_PW_1 + dt * ( -NIT_1 - ABS_1 + N_MIN_1 + NH4_FL_2 - NH4_FL_1 )
INIT(NH4_PW_1) =6{nmol/cm3}
NH4_PW_2 = NH4_PW_2 + dt * ( N_MIN_2 - NIT_2 - ABS_2 - NH4_FL_2 + NH4_FL_3 )
INIT(NH4_PW_2) = 40
NH4_PW_3 = NH4_PW_3 + dt * ( N_MIN_3 + NH4_FL_4 - NH4_FL_3 - ABS_3 - NIT_3 )
INIT(NH4_PW_3) = 65
NH4_PW_4 = NH4_PW_4 + dt * ( N_MIN_4 - NH4_FL_4 - ABS_4 - NIT_4 + NH4_replace )
INIT(NH4_PW_4) = 230
NH4_W = NH4_W + dt * ( NH4_FL_1 )
INIT(NH4_W) =0{nmol/cm2}
NO3_PW_1 = NO3_PW_1 + dt * ( NIT_1 - NO3_FL_1 + NO3_FL_2 - DENIT_1 )
INIT(NO3_PW_1) = 0
NO3_PW_2 = NO3_PW_2 + dt * ( NIT_2 - NO3_FL_2 - DENIT_2 + NO3_FL_3 )
INIT(NO3_PW_2) = 0
NO3_PW_3 = NO3_PW_3 + dt * ( NIT_3 - DENIT_3 + NO3_FL_4 - NO3_FL_3 )
```

```
INIT(NO3_PW_3) = 0
NO3_PW_4 = NO3_PW_4 + dt * ( NIT_4 - DENIT_4 - NO3_FL_4 )
INIT(NO3_PW_4) = 0
NO3_W = NO3_W - dt * ( NO3_FL_1 )
INIT(NO3_W) = 0{nmol/cm2}
O2_1 = O2_1 + dt * ( O2_FL_1 - O2_FL_F2 - O2_RED_1 )
INIT(O2_1) = 0
O2_2 = O2_2 + dt * ( O2_FL_F2 - O2_FL_F3 - O2_RED_2 )
INIT(O2_2) = 0
O2_3 = O2_3 + dt * ( O2_FL_F3 - O2_FL_F4 - O2_RED_3 )
INIT(O2_3) = 0
O2_4 = O2_4 + dt * ( O2_FL_F4 - O2_RED_4 )
INIT(O2_4) = 0
ORG_C_1 = ORG_C_1 + dt * ( Corg_FL_2 - C_MIN_1 )
INIT(ORG_C_1) = ORG_N_1*CNratio
ORG_C_2 = ORG_C_2 + dt * ( Corg_FL_2 - Corg_FL_3 - C_MIN_2 )
INIT(ORG_C_2) = ORG_N_2*CNratio
ORG_C_3 = ORG_C_3 + dt * ( Corg_FL_3 - Corg_FL_4 - C_MIN_3 )
INIT(ORG_C_3) = ORG_N_3*CNratio
ORG_C_4 = ORG_C_4 + dt * ( Corg_FL_4 - C_MIN_4 )
INIT(ORG_C_4) = ORG_N_4*CNratio
ORG_N_1 = ORG_N_1 + dt * ( -N_MIN_1 - Norg_FL_2 )
INIT(ORG_N_1) =100000
ORG_N_2 = ORG_N_2 + dt * ( Norg_FL_2 - Norg_FL_3 - N_MIN_2 )
INIT(ORG_N_2) = 80000
ORG_N_3 = ORG_N_3 + dt * ( Norg_FL_3 - Norg_FL_4 - N_MIN_3 )
INIT(ORG_N_3) = 40000
ORG_N_4 = ORG_N_4 + dt * ( Norg_FL_4 - N_MIN_4 )
INIT(ORG_N_4) = 5000
S_1 = S_1 + dt * ( SR_1 - S_O2_1 + S_FL_2 - S_FL_1 )
INIT(S_1) = 0
S_2 = S_2 + dt * ( SR_2 - S_O2_2 + S_FL_3 - S_FL_2 )
INIT(S_2) = 0
S_3 = S_3 + dt * ( SR_3 - S_O2_3 + S_FL_4 - S_FL_3 )
INIT(S_3) = 0
S_4 = S_4 + dt * ( SR_4 - S_O2_4 - S_FL_4 )
INIT(S_4) = 0
S_W = S_W + dt * ( S_FL_1 )
INIT(S_W) =0{nmol/cm2}
ABS_1 = (NH4_PW_1-NH4_EX_1/ABS_FACT)*100
ABS_2 = (NH4_PW_2-NH4_EX_2/ABS_FACT)*100
ABS_3 = (NH4_PW_3-NH4_EX_3/ABS_FACT)*100
ABS_4 = (NH4_PW_4-NH4_EX_4/ABS_FACT)*100
ABS_FACT = ex_pw_ratio
aC_MIN_FR = aN_MIN_FR*a_CN_diff_hyd
aN_MIN_FR = .005
a_CN_diff_hyd = 1
a_COX_k = 30
```

```
a_DENIT_k = 200
a_DEPTH =20{m}
a_MIX_k = .003
a_NIT_k = 300
a_SOX_k = 800
a_SR_k = 10
CNratio = 12
CN_1 = C_MIN_1/N_MIN_1
CN_2 = C_MIN_2/N_MIN_2
CN_3 = C_MIN_3/N_MIN_3
CN_4 = C_MIN_4/N_MIN_4
CN_ORG_1 = ORG_C_1/ORG_N_1
CN_ORG_2 = ORG_C_2/ORG_N_2
CN_ORG_3 = ORG_C_3/ORG_N_3
CN_ORG_4 = ORG_C_4/ORG_N_4
CN_SUM = C_MIN_SUM/N_MIN_SUM
Corg_FL_2 = Corg_FL_F2
Corg_FL_3 = Corg_FL_F3
Corg_FL_4 = Corg_FL_F4
Corg_FL_F2 = (ORG_C_1-ORG_C_2)/INT1_2*a_MIX_k
Corg_FL_F3 = (ORG_C_2-ORG_C_3)/INT2_3*a_MIX_k
Corg_FL_F4 = (ORG_C_3-ORG_C_4)/INT3_4*a_MIX_k
C_FR = aC_MIN_FR
C_MIN_1 = C_MIN_F1
C_MIN_2 = C_MIN_F2
C_MIN_3 = C_MIN_F3
C_MIN_4 = C_MIN_F4
C_MIN_F1 = ORG_C_1*C_FR*DEP_1
C_MIN_F2 = ORG_C_2*C_FR*DEP_2
C_MIN_F3 = ORG_C_3*C_FR*DEP_3
C_MIN_F4 = ORG_C_4*C_FR*DEP_4
C_MIN_SUM = C_MIN_1+C_MIN_2+C_MIN_3+C_MIN_4
C_O2_F1 = DOC_1*O2_STIM_1*a_COX_k*DEP_1
C_O2_F2 = DOC_2*O2_STIM_2*a_COX_k*DEP_2
C_O2_F3 = DOC_3*O2_STIM_3*a_COX_k*DEP_3
C_O2_F4 = DOC_4*O2_STIM_4*a_COX_k*DEP_4
C_O2_SUM = C_O2_F1+C_O2_F2+C_O2_F3+C_O2_F4
C_OX_1 = C_OX_F1
C_OX_2 = C_OX_F2
C_OX_3 = C_OX_F3
C_OX_4 =C_OX_F4
C_OX_F1 = DENIT_F1+C_O2_F1+SR_F1
C_OX_F2 = DENIT_F2+C_O2_F2+SR_F2
C_OX_F3 = DENIT_F3+C_O2_F3+SR_F3
C_OX_F4= DENIT_F4+C_O2_F4+SR_F4
DENIT_1 = DENIT_F1
DENIT_2 = DENIT_F2
DENIT_3 = DENIT_F3
```

```
DENIT_4 = DENIT_F4
DENIT_F1 = O2_INH_1*a_DENIT_k*NO3_PW_1*DEP_1
DENIT_F2 = O2_INH_2*a_DENIT_k*NO3_PW_2*DEP_2
DENIT_F3 = O2_INH_3*a_DENIT_k*NO3_PW_3*DEP_3
DENIT_F4 = O2_INH_4*a_DENIT_k*NO3_PW_4*DEP_4
DENIT_SUM = DENIT_F1+DENIT_F2+DENIT_F3+DENIT_F4
DEP_1 = 0.2
DEP_2 = 0.2
DEP_3 = 0.4
DEP_4 = 3.2
DIFF = 1.2
ex_pw_ratio = 2
INT0_1 = (DEP_1)/2
INT1_2 = (DEP_1+DEP_2)/2
INT2_3 = (DEP_2+DEP_3)/2
INT3_4 = (DEP_3+DEP_4)/2
NH4_FL_1 = NH4_FL_F1
NH4_FL_2 = NH4_FL_F2
NH4_FL_3= NH4_FL_F3
NH4_FL_4 = NH4_FL_F4/3
NH4_FL_F1 = (NH4_PW_1-uM_NH4)/INT0_1*top_DIFF
NH4_FL_F2 = (NH4_PW_2-NH4_PW_1)/INT1_2*DIFF
NH4_FL_F3 = (NH4_PW_3-NH4_PW_2)/INT2_3*DIFF
NH4_FL_F4 = (NH4_PW_4-NH4_PW_3)/INT3_4*DIFF
NH4_replace = NH4_FL_4
NIT_1 =NIT_F1
NIT_2 = NIT_F2
NIT_3 = NIT_F3
NIT_4 = NIT_F4
NIT_F1 = O2_STIM_1*a_NIT_k*NH4_PW_1*DEP_1
NIT_F2 = O2_STIM_2*a_NIT_k*NH4_PW_2*DEP_2
NIT_F3 = O2_STIM_3*a_NIT_k*NH4_PW_3*DEP_3
NIT_F4 = O2_STIM_4*a_NIT_k*NH4_PW_4*DEP_4
NIT_SUM = NIT_F1+NIT_F2+NIT_F3+NIT_F4
NO3_FL_1 = NO3_FL_F1
NO3_FL_2 = NO3_FL_F2
NO3_FL_3 = NO3_FL_F3
NO3_FL_4 = NO3_FL_F4
NO3_FL_F1 = (NO3_PW_1-uM_NO3)/INT0_1*top_DIFF
NO3_FL_F2 = (NO3_PW_2-NO3_PW_1)/INT1_2*DIFF
NO3_FL_F3= (NO3_PW_3-NO3_PW_2)/INT2_3*DIFF
NO3_FL_F4 = (NO3_PW_4-NO3_PW_3)/INT3_4*DIFF
Norg_FL_2= Norg_FL_F2
Norg_FL_3 = Norg_FL_F3
Norg_FL_4 = Norg_FL_F4
Norg_FL_F2= (ORG_N_1-ORG_N_2)/INT1_2*a_MIX_k
Norg_FL_F3 = (ORG_N_2-ORG_N_3)/INT2_3*a_MIX_k
Norg_FL_F4 = (ORG_N_3-ORG_N_4)/INT3_4*a_MIX_k
```

```
N_FR = aN_MIN_FR
N_MIN_1 = N_MIN_F1
N_MIN_2 = N_MIN_F2
N_MIN_3 = N_MIN_F3
N_MIN_4 = N_MIN_F4
N_MIN_F1 =ORG_N_1*N_FR*DEP_1
N_MIN_F2 = N_FR*ORG_N_2*DEP_2
N_MIN_F3 = N_FR*ORG_N_3*DEP_3
N_MIN_F4 = N_FR*ORG_N_4*DEP_4
N_MIN_SUM = N_MIN_1+N_MIN_2+N_MIN_3+N_MIN_4
O2_FL_1 = (250-O2_1)/INT0_1*top_DIFF
O2_FL_F2 = (O2_1-O2_2)/INT1_2*DIFF
O2_FL_F3 = (O2_2-O2_3)/INT2_3*DIFF
O2_FL_F4 = (O2_3-O2_4)/INT3_4*DIFF
O2_RED_1 = O2_RED_F1
O2_RED_2 = O2_RED_F2
O2_RED_3 = O2_RED_F3
O2_RED_4 = O2_RED_F4
O2_RED_F1 = NIT_1*2+C_O2_F1+S_O2_1
O2_RED_F2 = NIT_2*2+C_O2_F2+S_O2_2
O2_RED_F3 = NIT_3*2+C_O2_F3+S_O2_3
O2_RED_F4 = NIT_4*2+C_O2_F4+S_O2_4
O2_RED_SUM = O2_RED_1+O2_RED_2+O2_RED_3+O2_RED_4
SR_1 = SR_F1
SR_2 = SR_F2
SR_3 = SR_F3
SR_4 = SR_F4
SR_F1 = DOC_1*O2_INH_1*a_SR_k
SR_F2 = a_SR_k*O2_INH_2*DOC_2
SR_F3 = DOC_3*O2_INH_3*a_SR_k
SR_F4 = DOC_4*O2_INH_4*a_SR_k
SR_SUM = SR_1+SR_2+SR_3+SR_4
S_FL_1 = S_FL_F1
S_FL_2 = S_FL_F2
S_FL_3 = S_FL_F3
S_FL_4 = S_FL_F4
S_FL_F1 = (S_1-uM_S)/INT0_1*top_DIFF
S_FL_F2 = (S_2-S_1)/INT1_2*DIFF
S_FL_F3 = (S_3-S_2)/INT2_3*DIFF
S_FL_F4 = (S_4-S_3)/INT3_4*DIFF
S_O2_1 = S_O2_F1
S_O2_2 = S_O2_F2
S_O2_3 = S_O2_F3
S_O2_4 = S_O2_F4
S_O2_F1 = O2_STIM_1*S_1*a_SOX_k*DEP_1
S_O2_F2 = O2_STIM_2*S_2*a_SOX_k*DEP_2
S_O2_F3 = O2_STIM_3*S_3*a_SOX_k*DEP_3
S_O2_F4 = O2_STIM_4*S_4*a_SOX_k*DEP_4
```

top_DIFF = DIFF*top_factor

top_factor = 1.5

uM_NH4 = NH4_W/(a_DEPTH*100)

uM_NO3 = NO3_W/(a_DEPTH*100)

uM_S = S_W/(a_DEPTH*100){uM..}

O2_INH_1 = graph(O2_1)(0.0,1.00),(30.00,0.100),(60.00, 0.0),(90.00, 0.0),(120.00, 0.0),(150.00, 0.0),(180.00, 0.0),(210.00, 0.0),(240.00, 0.0),(270.00, 0.0),(300.00, 0.0)

O2_INH_2 = graph(O2_2)(0.0,1.00),(30.0,0.100),(60.0, 0.0),(90.0, 0.0),(120, 0.0),(150, 0.0),(180, 0.0),(210, 0.0),(240, 0.0),(270, 0.0),(300, 0.0)

O2_INH_3 = graph(O2_3)(0.0,1.00),(30.0,0.100),(60.0, 0.0),(90.0, 0.0),(120, 0.0),(150, 0.0),(180, 0.0),(210, 0.0),(240, 0.0),(270, 0.0),(300, 0.0)

O2_INH_4 = graph(O2_4)(0.0,1.00),(30.0,0.100),(60.0, 0.0),(90.0, 0.0),(120, 0.0),(150, 0.0),(180, 0.0),(210, 0.0),(240, 0.0),(270, 0.0),(300, 0.0)

O2_STIM_1=graph(O2_1)(0.0,0.0),(30.00,0.800),(60.00,0.880),(90.00,0.900),(120.00,0.920),(150.00,0.940),(180.00,0.950),(210.00,0.960),(240.00,0.980),(270.00,0.990),(300.00,1.00)

O2_STIM_2=graph(O2_2)(0.0,0.0),(30.00,0.800),(60.00,0.880),(90.00,0.900),(120.00,0.920),(150.00,0.940),(180.00,0.950),(210.00,0.960),(240.00,0.980),(270.00,0.990),(300.00,1.00)

O2_STIM_3=graph(O2_3)(0.0,0.0),(30.00,0.800),(60.00,0.880),(90.00,0.900),(120.00,0.920),(150.00,0.940),(180.00,0.950),(210.00,0.960),(240.00,0.980),(270.00,0.990),(300.00,1.00)

O2_STIM_4=graph(O2_4)(0.0,0.0),(20.00,0.800),(40.00,0.880),(60.00,0.900),(80.00,0.920),(100.00,0.940),(120.00,0.950),(140.00,0.960),(160.00,0.980),(180.00,0.990),(200.00,1.00)

Contributors

(only senior authors listed)

Arah, Jonathan R. M., Edinburgh School of Agriculture, West Mains Road, Edinburgh EH9 3JG, U.K.

Blackburn, T. Henry, Department of Ecology and Genetics, University of Aarhus, Ny Munkegade, DK-8000 Århus C, Denmark

Christensen, Søren, Department of Population Biology, Copenhagen University, Universitetsparken 15, DK-2100 Copenhagen Ø, Denmark

Cole, Jeff A., Department of Biochemistry, University of Birmingham, P.O. Box 363, Birmingham B15 2TT, U.K.

Conrad, Ralf, Department of Biology, University of Konstanz, P.O. Box 5560, D-7750 Konstanz, F.R.G.

Coyne, Mark S., Department of Crop and Soil Sciences, Michigan State University, East Lansing, MI 48824, U.S.A.

Herbert, Rodney A., Department of Biological Sciences, University of Dundee, Dundee DD1 4HN, U.K.

Klemedtsson, Leif, Department of Microbiology, Swedish University of Agricultural Sciences, S- 750 07 Uppsala, Sweden

Knowles, Roger, Department of Microbiology, Macdonald College of McGill University, 21,111 Lakeshore Road, Ste. Anne de Bellevue, Quebec H9X 1C0, Canada

Koike, Isao, Ocean Research Institute, University of Tokyo, 1-15-1 Minamidai, Nakano-ku, Tokyo 164, Japan

Kroneck, Peter M. H., Department of Biology, University of Konstanz, P.O. Box 5560, D-7750 Konstanz, F.R.G

Myrold, David D., Department of Soil Science, Oregon State University, Strand Agricultural Hall 202, Corvallis, OR 97731-2213, U.S.A.

Parkin, Timothy B., Unites States Department of Agriculture-ARS, Environment Chemistry Laboratory, Beltsville Agricultural Research Center-West, Beltsville, MD 20705, U.S.A.

Revsbech, Niels Peter, Department of Ecology and Genetics, University of Aarhus, Ny Munkegade, DK-8000 Århus C, Denmark

Robertson, Lesley A., Kluyver Laboratory of Biotechnology, Delft University of Technology, Julianalaan 67a, 2628 BC Delft, The Netherlands

Seitzinger, Sybil P., Academy of Natural Sciences, Division of Environment Research, 19th St. and Parkway, Philadelphia, PA 19103, U.S.A.

Smith, Keith A., Edinburgh School of Agriculture, West Mains Road, Edinburgh EH9 2JG, U.K.

Sørensen, Jan, Department of Ecology and Molecular Biology, Section of Microbiology, Royal Veterinary and Agricultural University, Rolighedsvej 21, DK-1958 Frederiksberg C, Denmark

Yoshinari, Tadashi, Wadsworth Center, New York State Department of Health, State University of New York, Corning Tower, Albany, NY 12201-0509, U.S.A.

Zumft, Walter G., Department of Microbiology, University of Karlsruhe, Kaiserstrasse 12, D-7500 Karlsruhe 1, F.R.G.

INDEX